T0231762

CEREALS AND CEREAL-BASED FOODS

Functional Benefits and Technological
Advances for Nutrition and Healthcare

Innovations in Plant Science for Better Health: From Soil to Fork

CEREALS AND CEREAL-BASED FOODS

Functional Benefits and Technological
Advances for Nutrition and Healthcare

Edited by
Megh R. Goyal, PhD
Kamaljit Kaur, PhD
Jaspreet Kaur, PhD

APPLE
ACADEMIC
PRESS

First edition published 2022

Apple Academic Press Inc.
1265 Goldenrod Circle, NE,
Palm Bay, FL 32905 USA
4164 Lakeshore Road, Burlington,
ON, L7L 1A4 Canada

CRC Press
6000 Broken Sound Parkway NW,
Suite 300, Boca Raton, FL 33487-2742 USA
2 Park Square, Milton Park,
Abingdon, Oxon, OX14 4RN UK

© 2022 Apple Academic Press, Inc.

Apple Academic Press exclusively co-publishes with CRC Press, an imprint of Taylor & Francis Group, LLC

Library and Archives Canada Cataloguing in Publication

Title: Cereals and cereal-based foods : functional benefits and technological advances for nutrition and healthcare / edited by Megh R. Goyal, PhD, Kamaljit Kaur, PhD, Jaspreet Kaur, PhD.
Names: Goyal, Megh R., editor. | Kaur, Kamaljit (Professor of food science), editor. | Jaspreet Kaur, 1971- editor.
Series: Innovations in plant science for better health.
Description: First edition. | Series statement: Innovations in plant science for better health : from soil to fork | Includes bibliographical references and index.
Identifiers: Canadiana (print) 20200385887 | Canadiana (ebook) 20200386085 | ISBN 9781771889445 (hardcover) | ISBN 9781003081975 (ebook)
Subjects: LCSH: Grain. | LCSH: Cereal products. | LCSH: Functional foods.
Classification: LCC SB189 .C47 2021 | DDC 641.3/31—dc23

Library of Congress Cataloging-in-Publication Data

Names: Goyal, Megh R., editor. | Kaur, Kamaljit (Professor of food science), editor. | Jaspreet Kaur, 1971- editor.
Title: Cereals and cereal-based foods : functional benefits and technological advances for nutrition and healthcare / Megh R. Goyal, Kamaljit Kaur, Jaspreet Kaur.
Other titles: Innovations in plant science for better health.
Description: First edition. | Palm Bay, FL, USA : Apple Academic Press, 2021. | Series: Innovations in plant science for better health : from soil to fork | Includes bibliographical references and index. | Summary: "This book volume sheds light on the health benefits of selected cereal grains, processing technologies of cereals, specific roles of bioactive compounds of cereals in chronic disease prevention, and traditional and latest technologies to improve the functional benefits of cereal-based products. It presents a thorough review of the functional components of some lesser known or forgotten cereals and their role in maintaining good health. With advancements in cereal science and technology, new methods of processing have emerged that help to preserve or even enhance the health-benefitting properties of cereal grains. Further, plant breeding and biotechnology have contributed greatly in improving nutritional quality and functionality of these grains. This book provides comprehensive information on the simple as well as advanced methodologies for enhancing the properties of cereals that benefit human health. Some new approaches such as bio-fortification and extraction of bioactives from cereals are also included in the text. Cereals and Cereal-Based Foods: Functional Benefits and Technological Advances for Nutrition and Healthcare looks at the current scenario of the research in cereals and cereal-based products and covers the importance of pseudocereals, whole grains, and multigrain under four main parts: Functional Benefits of Cereal Grains discusses the health benefits of some selected grains that constituted staple food in the past but have lost their importance in the modern times. Technological Advances in Cereals covers new approaches in the processing of cereals that ensure maximum retention of health-benefitting components of cereal grains. Novel Strategies to Enhance Bioactives in Cereals integrates traditional as well as modern technologies with new approaches for improving functional benefits of cereal based foods. Role of Cereals in Disease Management elaborates on the specific role of cereal bioactives in prevention of some chronic diseases. This book will serve as a useful reference for teachers and students as well as those who are involved in research on development of functional products from cereal grains. It would also benefit technologists and industrialists engaged in health food manufacture"-- Provided by publisher.
Identifiers: LCCN 2020050417 (print) | LCCN 2020050418 (ebook) | ISBN 9781771889445 (hardcover) | ISBN 9781003081975 (ebook)
Subjects: LCSH: Grain. | Cereal products. | Cereals as food. | Functional foods.
Classification: LCC SB189 .C434 2021 (print) | LCC SB189 (ebook) | DDC 633.1--dc23
LC record available at https://lccn.loc.gov/2020050417
LC ebook record available at https://lccn.loc.gov/2020050418

ISBN: 978-1-77188-944-5 (hbk)
ISBN: 978-1-77463-788-3 (pbk)
ISBN: 978-1-00308-197-5 (ebk)

OTHER BOOKS ON PLANT SCIENCE FOR BETTER HEALTH BY APPLE ACADEMIC PRESS, INC.

Book Series: *Innovations in Plant Science for Better Health: From Soil to Fork*
Editor-in-Chief: Hafiz Ansar Rasul Suleria, PhD

Assessment of Medicinal Plants for Human Health: Phytochemistry, Disease Management, and Novel Applications
Editors: Megh R. Goyal, PhD, and Durgesh Nandini Chauhan, MPharm

Bioactive Compounds of Medicinal Plants: Properties and Potential for Human Health
Editors: Megh R. Goyal, PhD, and Ademola O. Ayeleso

Bioactive Compounds from Plant Origin: Extraction, Applications, and Potential Health Claims
Editors: Hafiz Ansar Rasul Suleria, PhD, and Colin Barrow, PhD

Health Benefits of Secondary Phytocompounds from Plant and Marine Sources
Editors: Hafiz Ansar Rasul Suleria, PhD, and Megh Goyal, PhD

Human Health Benefits of Plant Bioactive Compounds: Potentials and Prospects
Editors: Megh R. Goyal, PhD, and Hafiz Ansar Rasul Suleria, PhD

Phytochemicals and Medicinal Plants in Food Design: Strategies and Technologies for Improved Healthcare
Editors: Megh R. Goyal, PhD, Preeti Birwal, PhD, and Santosh K. Mishra, Ph

Phytochemicals from Medicinal Plants: Scope, Applications, and Potential Health Claims
Editors: Hafiz Ansar Rasul Suleria, PhD, Megh R. Goyal, PhD, and Masood Sadiq Butt, PhD

Plant- and Marine-Based Phytochemicals for Human Health: Attributes, Potential, and Use
Editors: Megh R. Goyal, PhD, and Durgesh Nandini Chauhan, MPharm

Plant-Based Functional Foods and Phytochemicals: From Traditional Knowledge to Present Innovation
Editors: Megh R. Goyal, PhD, Arijit Nath, PhD, and Hafiz Ansar Rasul Suleria, PhD

Plant Secondary Metabolites for Human Health: Extraction of Bioactive Compounds
Editors: Megh R. Goyal, PhD, P. P. Joy, PhD, and Hafiz Ansar Rasul Suleria, PhD

The Role of Phytoconstitutents in Healthcare: Biocompounds in Medicinal Plants
Editors: Megh R. Goyal, PhD, Hafiz Ansar Rasul Suleria, PhD, and Ramasamy Harikrishnan, PhD

The Therapeutic Properties of Medicinal Plants: Health-Rejuvenating Bioactive Compounds of Native Flora
Editors: Megh R. Goyal, PhD, PE, Hafiz Ansar Rasul Suleria, PhD, Ademola Olabode Ayeleso, PhD, T. Jesse Joel, and Sujogya Kumar Panda

ABOUT THE SENIOR EDITOR-IN-CHIEF

Megh R. Goyal, PhD

Retired Professor in Agricultural and Biomedical Engineering, University of Puerto Rico, Mayaguez Campus; Senior Acquisitions Editor, Biomedical Engineering and Agricultural Science, Apple Academic Press, Inc.

Megh R. Goyal, PhD, PE, is a Retired Professor in Agricultural and Biomedical Engineering from the General Engineering Department in the College of Engineering at the University of Puerto Rico–Mayaguez Campus; and Senior Acquisitions Editor and Senior Technical Editor-in-Chief in Agriculture and Biomedical Engineering for Apple Academic Press, Inc. He has worked as a Soil Conservation Inspector and as a Research Assistant at Haryana Agricultural University and Ohio State University.

During his professional career of 52 years, Dr. Goyal has received many prestigious awards and honors. He was the first agricultural engineer to receive the professional license in Agricultural Engineering in 1986 from the College of Engineers and Surveyors of Puerto Rico. In 2005, he was proclaimed as "Father of Irrigation Engineering in Puerto Rico for the Twentieth Century" by the American Society of Agricultural and Biological Engineers (ASABE), Puerto Rico Section, for his pioneering work on micro irrigation, evapotranspiration, agroclimatology, and soil and water engineering. The Water Technology Centre of Tamil Nadu Agricultural University in Coimbatore, India, recognized Dr. Goyal as one of the experts "who rendered meritorious service for the development of micro irrigation sector in India" by bestowing the Award of Outstanding Contribution in Micro Irrigation. This award was presented to Dr. Goyal during the inaugural session of the National Congress on "New Challenges and Advances in Sustainable Micro Irrigation" held at Tamil Nadu Agricultural University. Dr. Goyal received the Netafim Award for Advancements in Microirrigation: 2018 from the American Society of Agricultural Engineers at the ASABE International Meeting in August 2018.

A prolific author and editor, he has written more than 200 journal articles and textbooks and has edited over 80 books. He is the editor of three book series published by Apple Academic Press: Innovations in Agricultural & Biological Engineering, Innovations and Challenges in Micro Irrigation, and Research Advances in Sustainable Micro Irrigation. He is also instrumental in the development of the new book series Innovations in Plant Science for Better Health: From Soil to Fork.

Dr. Goyal received his BSc degree in engineering from Punjab Agricultural University, Ludhiana, India; his MSc and PhD degrees from Ohio State University, Columbus; and his Master of Divinity degree from Puerto Rico Evangelical Seminary, Hato Rey, Puerto Rico, USA.

ABOUT CO-EDITORS

Kamaljit Kaur
*Assistant Professor, Department of Food Science
and Technology, Punjab Agricultural University
Ludhiana 141001, India*

Kamaljit Kaur, PhD, is Assistant Professor in Department of Food Science and Technology, Punjab Agricultural University, Ludhiana, India. She has acquired proficiency in the field of food technology through her experience in food industries, teaching, and research at various levels for more than 12 years. She made a significant contribution in the field of research in food technology and has published more than 40 research papers, abstracts in conferences, book chapters, and practical manuals. She has attended advanced specialized training programs and has delivered invited and general lectures to scientific faculty, students, young entrepreneurs, and farmers. She has been awarded with best oral presenter and best poster presenter awards at several national and international conferences. Presently she is working on by-product utilization of baby corn in collaboration with the food industry; encapsulation of micronutrients and their stability studies in food; and cereals and grain processing.

Readers may contact her at: kamalbhella@pau.edu

Jaspreet Kaur
*Assistant Food Technologist Department of Food
Science and Technology Punjab Agricultural
University, Ludhiana 141001, India*

Jaspreet Kaur, PhD, is Assistant Food Technologist
in the Department of Food Science and Technology
at Punjab Agricultural University, Ludhiana, India.
Dr. Kaur has published more than 30 research
papers in reputed high-impact journals. She has been
actively involved in research projects. She has also worked as a research
associate at the Indian Council of Agricultural Research–Central Institute of
Post-Harvest Engineering and Technology. She has worked in several areas,
such as in product development, advanced quality analysis, development of
commercially viable technologies, convenience foods, composite foods, food
adulteration, etc. She is currently working on processing and value addition
of cereals and pulses. She has also been active in innovations in traditional
Indian dairy products and is guiding postgraduate students in this area.
She has won many awards and merit certificates during her study. She has
attended several training programs and has delivered lectures to wide array
of audiences, including scientific faculty, students, young entrepreneurs, and
farmers.

Readers may contact her at: jaspreet@pau.edu

CONTENTS

CONTRIBUTORS

Simran Kaur Arora
Assistant Professor, Department of Food Science and Technology, College of Agriculture,
Govind Ballabh Pant University of Agriculture and Technology, Pantnagar, Udham Singh Nagar
263145, Uttarakhand, India; Mobile: +91-8884965321; E-mail: sim_n@rediffmail.com

Diksha Bassi
M.Sc. Student, Department of Food Science and Technology, Punjab Agricultural University,
Ludhiana 141001, India; Mobile: +91-9229700009; E-mail: dikshabassi94@gmail.com

Suresh Bhise
Assistant Professor, College of Food Processing Technology & Bio-Energy, Anand Agricultural
University, Anand 388110, India; Mobile: +91-9872486692; E-mail: sureshbhise_cft@yahoo.co.in

Jyoti Bohra
PhD Research Scholar, Department of Foods and Nutrition, Punjab Agricultural University,
Ludhiana 141 004, Punjab, India. Mobile: +91-74668 38925; E-mail: jyoti.bohra@ gmail.com

Harinderjeet Kaur Bhullar
PhD Research Scholar, Department of Food Science and Technology, Punjab Agricultural University
Ludhiana, Punjab 141004, India; Mobile: +91-8059860824; E-mail: bhullar6266@gmail.com

Manreet Singh Bhullar
PhD Research Scholar, Department of Food Science and Human Nutrition, Iowa State University,
Ames 50011, IA, USA; Mobile +1-6159441715; E-mail: bhullar@iastate.edu

Megh R. Goyal
Retired Faculty in Agricultural and Biomedical Engineering, College of Engineering at the University
of Puerto Rico, Mayaguez Campus; and Senior Technical Editor-in-Chief in Agricultural and
Biomedical Engineering for Apple Academic Press Inc.; PO Box 86, Rincon, PR 006770086, USA;
E-mail: goyalmegh@gmail.com

Samneet Kashyap
Graduate Research Assistant, Department of Plant and Environmental Sciences, Clemson University,
Clemson 29631, SC-USA; Mobile: +1-843-617-8219; E-mail: samneek@clemson.edu

Amarjeet Kaur
Senior Milling Technologist, Department of Food Science and Technology, Punjab Agricultural
University Ludhiana 141004, Punjab, India; Mobile: +91-9888466677; E-mail: foodtechak@gmail.com

Gursharan Kaur
Assistant Professor, Department of Food Science and Technology, Khalsa College, Amritsar 143001,
India; Mobile: +91-9815075148; E-mail: doc_gursharan@yahoo.co.in

Jaspreet Kaur
Assistant Food Technologist, Department of Food Science and Technology, Punjab Agricultural
University (PAU), Ludhiana 141001, India; Mobile: +91-9915141584; E-mail: jaspreet@pau.edu

Kamaljit Kaur
Assistant Professor, Department of Food Science and Technology, Punjab Agricultural University,
Ludhiana 141001, India; Mobile: +91-9876152061; E-mail: kamalbhella@pau.edu

Prabhjot Kaur
MSc Student, Department of Food Science and Technology, Punjab Agricultural University,
Ludhiana 141 004, Punjab, India. Mobile: +91-7087804465; E-mail: jotprabh102@gmail.com

Ramandeep Kaur
PhD Research Scholar, Department of Food Science and Technology, Punjab Agricultural University,
Ludhiana 141004, Punjab, India; Mobile: +91-8054428270; E-mail: rsandhu047@gmail.com

Shubhpreet Kaur
Assistant Professor, Department of Food Science and Technology, Khalsa College, Amritsar 143002,
India; Mobile: +91-9815088906; E-mail: shubhpreet.k@yahoo.com

Ashwani Kumar
Assistant Professor, Department of Food Technology and Nutrition, Lovely Professional University,
Phagwara 144411, Punjab, India; Mobile: +91-8146941884; E-mail: ashwanichandel480@gmail.com

Akanksha Pahwa
Assistant Professor, Amity Institute of Food Technology, Amity University, Noida 201313, UP-India;
Mobile: +91-8130012326; E-mail: akankshapahwa91@gmail.com

Rasane Prasad
Assistant Professor, Department of Food Technology and Nutrition, Lovely Professional University,
Phagwara 144411, Punjab, India; Mobile: +91-7011421664; E-mail: rasane.18876@lpu.co.in

Ritu Priya
Assistant Professor, Department of Food Science and Technology, Khalsa College, Amritsar 143001,
India; Mobile: +91-9780716900; E-mail: ritupriya587@gmail.com

Reshu Rajput
PhD Research Scholar, Department of Food Science and Technology, Punjab Agricultural University
Ludhiana 141004, India; Mobile: +91-8319773069; E-mail: reshu.rajput21@gmail.com

Ravneet Sandhu
Graduate Research Assistant, Department of Horticultural Sciences, Gulf Coast Research and
Education Center, University of Florida, Wimauma 33598, FL - USA; Mobile: +1-319-671-4088;
E-mail: rsandhu@ufl.edu

Sandeep Singh
Assistant Professor, Department of Food Science & Technology, Khalsa College, Amritsar 143001,
India; Mobile: +91-9465672276; E-mail: sandeep_sing@yahoo.com

Sandeep Singh Sandhu
Scientist (Agronomy), School of Climate Change and Agricultural Meteorology, Punjab Agricultural
University, Ludhiana 141004, Punjab, India; Mobile: +91-8146300110; E-mail: ssandhu@pau.edu

Sarabjit Singh
M.Sc. Student, Department of Food Technology and Nutrition, Lovely Professional University,
Phagwara 144411, Punjab, India; Mobile: +91-9876203797; E-mail: sarabjitsinghm55@gmail.com

Tarvinder Pal Singh
Assistant Plant Breeder, Office of Director (Seeds), Punjab Agricultural University, Ludhiana 141001,
India; Mobile: +91-9872428072; E-mail: tpsingh@pau.edu

Mandeep Tayal
Graduate Research Assistant, Department of Biology, University of Texas Rio Grande Valley,
Edinburg 78539, TX-USA; Mobile: +1-9569002331; E-mail: mandeep.tayal01@utrgv.edu

Vidisha Tomer
Assistant Professor, Department of Food Technology and Nutrition, Lovely Professional University,
Phagwara 144411, Punjab, India; Mobile: +91-7011421664; E-mail: vidishatomer@gmail.com

ABBREVIATIONS

AACC	American Association of Cereal Chemists
AACCI	American Association of Cereal Chemists International
ACC	ammonium carboxymethyl cellulose
AD	Anno Domini
ALA	alpha linolenic acid
AO	antioxidant activity
ATP	adenosine triphosphate
AUC	area under the curve
BC_1F_8	back crossed filial generations
BHA	butylated hydroxyl toluene
BMI	body mass index
BHT	butylated hydroxyl toluene
CD	celiac disease
CGIAR	Consultative Group on International Agricultural Research
CHD	coronary heart disease
CoA	co-enzyme A
CVD	cardio vascular disease
d.b.	dry basis
DASH	dietary approach to stop hypertension
DBP	diastolic blood pressure
DF	dietary fiber
DHA	docosa-hexaenoic acid
DNA	deoxyribonucleic acid
DPPH	2,2-diphenyl-1-picrylhydrazyl
Dwb	dry weight basis
EAA	essential amino acids
FAO	Food and Agriculture Organization
FDA	Food and Drug Administration
G×E	gene by environment
GAE	gallic acid equivalent
GCC	gluten containing cereals
GFSC	gluten free staple cereals
GGE	glycemic glucose equivalent
GI	glycemic index

GL	glycemic load
GNA	galanthus nivalis agglutinin
GRAS	generally regarded as safe
HAW	high amylose wheat
HDL	high-density lipoprotein
HDL-c	high-density lipoprotein-cholesterol
HDPE	high-density polyethylene
IAUC	incremental area under the curve
IBD	inflammatory bowel diseases
IBS	irritable bowel syndrome
IDA	iron deficiency anemia
IDF	insoluble dietary fiber
IgE	immunoglobulin E
IRS	insulin resistant syndrome
KI	potassium iodide
KIO_3	potassium iodate
KNO_3	potassium nitrate
LDL	low-density lipoprotein
LDPE	low-density polyethylene
MAP	modified atmospheric packaging
NAFLD	nonalcoholic fatty liver disease
NASH	nonalcoholic steatohepatitis
NCGS	nonceliac gluten sensitivity
NW-Europe	Northwest Europe
P_2O_5	phosphorus pentoxide
PCC	potassium carboxymethyl cellulose
PCOS	poly cystic ovarian syndrome
PDI	pellet durability index
PUFA	poly unsaturated fatty acids
PEM	protein-energy malnourished
RDA	recommended dietary allowance
RDS	rapidly digestible starch
RGI	relative glycemic impact
RS	resistant starch
RTE	ready-to-eat
SBP	systolic blood pressure
SCC	sodium carboxymethyl cellulose
SCFA	short-chain fatty acid
SD LDL-c	small dense low density lipoprotein cholesterol

SDF	soluble dietary fiber
SDS	slowly digestible starch
TTIs	time–temperature indicators
T2DM	type 2 diabetes
TAG	triacyl glycerides
TPC	total phenol content
tTG	transglutaminase
VCEAC	vitamin C equivalent antioxidant activity
w.b.	wet basis
WDEIA	wheat-dependent, exercise-induced anaphylaxis
WG	whole grain

PREFACE

The book is divided into four main parts: Part I, Functional Benefits of Cereal Grains, discusses the health benefits of some selected grains that constituted staple food in the past but have lost their importance in modern times. Besides providing knowledge of the bioactive components and health benefits of these, this part also contains valuable information on the innovative products that have been reported in recent literature. Part II of the book, Technological Advances in Processing of Cereals for Healthcare, covers new approaches in processing of cereals that ensure maximum retention of health benefitting components of cereal grains. Part III, Novel Strategies to Enhance Bioactive Compounds in Cereals, integrates traditional as well as modern technologies with new approaches in improving functional benefits of cereal based foods. Part IV of the book, Role of Cereals in Disease Management, elaborates on the specific role of cereal bioactives in prevention of some chronic diseases.

The primary theme of this book is the exploration of cereals and their products for their health benefits. The book presents a thorough review of the functional components of lesser known or forgotten cereals and their role in maintaining good health. Since time immemorial, humans have been consuming cereals as their staple diet and have realized the role of these in their well-being. Ancient man domesticated a large number of cereals in different parts of the world and knew a great deal about their role in health management. However, with time, some of this knowledge was forgotten, and cultivation of such grains was given up for the more product friendly grains such as wheat and rice, especially for their refined flours. Our emphasis is on some of these underutilized cereals and their bioactive components. These components have either curative or preventive health benefits. Considering the rising incidences of chronic disorders such as diabetes, heart diseases, and cancers, the importance of such plant foods has increased manifold. These have now gained status of "super foods."

With advancements in cereal science and technology, new methods of processing have emerged that help to preserve or even enhance the health benefitting properties of cereal grains. Further, plant breeding and biotechnology have contributed greatly in improving nutritional quality and functionality of these grains. This book contains comprehensive information on the simple as well as advanced methodologies for enhancing the health-benefitting properties of cereals. Some new approaches such as biofortification and extraction of bioactives from cereals are also included in the text.

This book will serve as a useful reference for teachers and students as well as those who are involved in research on development of functional products from cereal grains. It would also benefit technologists and industrialists engaged in health food manufacture.

The editors are highly thankful to the contributing authors for their hard work and cooperation.

Although every care has been taken to provide the latest and most comprehensive information on the topic, any suggestions on improvement are always welcome.

The contributions by the cooperating authors to this book volume have been most valuable in the compilation. Their names are mentioned in each chapter and in the list of contributors. We appreciate you all for having patience with our editorial skills. This book would not have been written without the valuable cooperation of these investigators, many of whom are renowned scientists who have worked in the field of food science and biochemistry and food engineering throughout their professional career.

We would like to thank editorial and production staff and Ashish Kumar, Publisher and President at Apple Academic Press, Inc., for making every effort to publish this book when all are concerned with health issues.

We express our admiration to our families and colleagues for understanding and collaboration during the preparation of this book volume.

—Editors

PART I
Functional Benefits of Cereal Grains

CHAPTER 1

BARLEY-BASED FUNCTIONAL FOODS

PRABHJOT KAUR, JASPREET KAUR, KAMALJIT KAUR, and
JYOTI BOHRA

ABSTRACT

Barley (*Hordeum vulgare*) is nutritious, and it is rich in carbohydrates, good quality proteins, fats, vitamins, and minerals. It is particularly a good source of soluble fiber, that is, β-glucan, which on regular consumption has several positive effects on human health. Barley is used to prepare several products, such as, barley bread, cookies, and beverages. It can be fermented and malted. Barley malt is used in beer production. Barley grain has lost its importance in recent years due to consumer preference for other cereals, such as wheat and rice. Recently, however, barley is fast emerging as a base material for health foods due to increased awareness of the nutritional and health benefits of this grain.

1.1 INTRODUCTION

The cereal grains are important food commodities for consumption and income generation. These are main energy sources for majority of the world population. Barley (*Hordeum vulgare*) is a historic cereal grain, which is mainly cultivated for animal feed. After wheat, maize, and rice crops, barley is the world's fourth largest grain crop in terms of quantity produced. As human food, it is also consumed as meal and is used for brewing and malting for manufacture of beer. In regions of extreme climate like Ethiopia, Morocco, and Himalayas, barley is principal source of food. It is also used for bread making [36].

Archeological evidence suggests that barley was domesticated around 8000 BC in the Himalayan region, Morocco, and Ethiopia. Barley is hardier than wheat and can inherently adapt itself admirably well under limited inputs, marginal lands, and dry climates. It is more suited to cooler and drier climates. Its growing period is about 5 months in the plains and lasts for about 6–7 months in high hills. It is an important cereal in the world with an annual production of around 148 million metric tons, ranking next to maize, wheat, and rice. Russia is the leading producer of barley with a production of 20.6 million metric tons. It is grown widely in several European countries and Canada. In India, it is an important winter season (*Rabi*) cereal crop grown in Punjab, Rajasthan, Madhya Pradesh, eastern Uttar Pradesh, and Bihar. It occupies an area of 47 million ha. Annual production of barley in India was 1.75 million metric tons in 2017 [22].

This chapter focuses on the nutritional and bioactive components of barley and its health benefits. The chapter also provides an insight into the traditional and new functional food products from barley.

1.2 TYPES OF BARLEY

Barley can be classified on the basis of number of rows of grains on the head [9]:

- Barley with two rows of grains on the head is called two-row barley.
- Barley with four rows is called four-row barley.
- Barley with six rows is called six-row barley.

Six-row type is the most cultivated barley and two-row barley is the wild variety. The protein content of six-row barley is higher than two-row barley. This makes six-row barley more suitable for animal feeding. Malt barley has lower protein. Another way of classification of barley is by describing its beards (awns) covering the kernels. Barley can also be classified based on hull as hulless (naked) or hulled; based on usage as feed or malt type; based on seed color as white, yellow, blue, and colorless; and based on height. Following is the list of some varieties of barley:

- (Normal) hooded.
- Awnless or awnletted in central rows and lateral rows.
- Elevated hooded.
- Elevated hoods in central row, and awnless in lateral rows.
- Long awned.
- Long awned in central row, awnletted or awnless in lateral rows.
- Short awned.
- Short awned in central row, awnletted or awnless in lateral rows.
- Subjacent hooded.

1.3 NUTRITIONAL COMPOSITION OF BARLEY

Barley has gained increased attention among researchers and consumers due to its nutritional composition. As for other cereals, the major components of barley grain include carbohydrates (e.g., starch and fiber), high-quality protein, lipids, and minerals (Table 1.1), which vary with the type of barley. For instance, the hull-less varieties of barley have higher content of protein, starch, and β-glucan compared to hulled barley. Barley is a good source of soluble fiber and β-glucan [14].

TABLE 1.1 Chemical Composition (Per 100 g Edible Portion) of Barley Grain

Component	Unit	Value	Component	Unit	Value
Moisture	g	9.77	**Water-soluble vitamins**		
Protein	g	10.94	Thiamine	mg	0.36
Ash	g	1.06	Riboflavin	mg	0.18
Total fat	g	1.30	Niacin	mg	2.84
Carbohydrate	g	61.29	Pantothenic acid	mg	0.14
Energy	kJ	1321	Biotin	μg	2.38
Dietary fiber			Total folates	μg	31.58
Total	g	15.64	**Carotenoids**		
Insoluble	g	9.98	Lutein	μg	5.39
Soluble	g	5.66	Zeaxanthin	μg	1.90
Minerals			Total carotenoids	μg	69.87
Magnesium	mg	48.97	**Minerals**		
Manganese	mg	1.24	Selenium	μg	18.61
Phosphorus	mg	178	Sodium	mg	7.56
Potassium	mg	268	Zinc	mg	1.50

Source: Indian food composition tables, 2017 [40].

Barley grain has a high content of carbohydrates with an average of 61.29 g carbohydrates per 100 g of edible portion of grain (Table 1.1). It has 60%–64% of starch, 1%–2% sucrose, and 1% of sugars in different forms. The starch in barley is easily digestible and is a source of energy. Out of total starch, the amylose content is 20%–30%. In high-amylose barley, amylose content can be as high as 45% [55]. Glacier barley has nearly 42% of amylose content. In barley, carbohydrates occur both in soluble and insoluble forms. It contains higher amount of insoluble fiber. Among cereals and millet groups, barley has the highest content of total fiber. It contains 8%–10% of insoluble fiber and 1%–1.5% of water-soluble polysaccharides [41]. Milled barley contains about 25.12% total dietary fiber and 6.72% of β-glucan, which is higher than the dehulled barley [39]. β-Glucan is a soluble fiber and its content in barley is highest among cereal grains. The amount of protein and water-soluble polysaccharides in barley is proportional to the content of β-glucan in grains of barley [31].

The protein content of barley is comparable to other cereals, such as wheat. The whole grain contains 8%–20% protein. The protein quality is good with amino acids, such as, lysine (an essential amino acid), which is deficient in wheat [19]. Barley proteins may be classified as storage and nonstorage types. Storage proteins include prolamins (hordeins) and globulins (as per Osborne protein classification) [50]. In barley, range of protein content is nearly the same as present in pearl millet and wheat, but higher than ragi, rice, and maize. The total content of globulins and albumins protein is about 3.5%, whereas glutelins and hordeins content is 3%–4% each [41].

Barley contains 2%–3% lipids, which are mostly concentrated in the germ. The major fatty acids in barley are [13] palmitic (16:0), linoleic (18:2), and oleic (18:1).

Barley is a good source of niacin (2.84 mg/100 g), thiamine (36 mg/100 g), and riboflavin (0.18 mg/100 g). The amount of these in barley is even more than milk. Compared to other cereals and millets, it contains moderate amount of biotin (2.38 µg/100 g) but a lower amount of panthothenic acid (0.14 mg/100 g). Barley is a poor source of vitamins C and D (Table 1.1).

On average, barley contains 2% of minerals, which include magnesium, manganese, potassium, phosphorus, selenium, zinc, and sodium (Table 1.1).

1.4 BIOACTIVE COMPONENTS OF BARLEY

1.4.1 β-GLUCAN

Dietary fiber can be classified into two groups according to their water solubility: insoluble and soluble. The β-glucan, a soluble fiber in cereals, is a branched polysaccharide having glycosidic bond at β-(1-3 and 1-4) of glucopyranose units. Barley contains the highest level of β-glucan (2%–10%) among other cereal grains. β-Glucan is present in endosperm, which encloses matrix protein, starch, and lipids reserves of grain. The cell walls of endosperm cells are high in β-glucan. Endosperm cell walls in barley contain approximately 75% β-glucans, 20% protein, and arabinoxylans [24, 25, 30].

β-Glucan influences the functional and nutritional properties of food. Soluble dietary fibers are known for their hypocholestrolemic effects. β-Glucans in barley grains have important role in controlling diabetes.

1.4.2 PHENOLIC ACIDS

The major phenolic acids in barley include benzoic acids and cinnamic acids that are characterized by the presence of unsaturated carboxylic group. These are mostly in the bound form with components of cell wall [31, 53]. The phenolic components may be present in the bound, conjugated, and free forms and the total content of phenolic acids ranges from 604 to 1346 µg/g [2]. These acids are concentrated in outer part of barley kernel. Free phenolic acids include vanillic acid, syringic acid, and ferulic acid that act as natural antioxidants and have antiproliferative properties. These are considered as important functional components of barley grain.

1.4.3 FLAVONOIDS

Flavonoids in barley include anthocyanins, flavanols, and proanthocyanidins. Cyanidin 3-glucosode has been reported as the most common anthocyanin in purple barley [1]. The content of flavonoids depends upon color of the grains. Normally, blue and purple varieties possess higher content of flavonoids compared to others. In barley, flavonoids content has been reported to range from 62.0 to 300.8 µg/g [38]. Anthocyanins are water-soluble compounds and are present in the pericarp of grains. Flavonoids are important in preventing damage from UV radiation. They also have bioactivities to prevent cancer and heart diseases.

1.4.4 OTHER PHYTOCHEMICALS

Apart from the above discussed compounds in this section, barley contains lignans, tocols, phytosterols, and folates that play important physiological roles. Lignans are polyphenolic compounds having antioxidant, antibacterial, and antiviral effects apart from other health benefits. Lignans are similar in structure to estradiol [44, 45]. Barley is also known to have antioxidant compounds (such as tocopherols and tocotrienols), contents of which are highest in barley among cereals. These are mainly located in the germ portion. Phytosterols are present in free and bound forms in barley grains. Folates are present in high concentration in outer layers of barley grain. These compounds are essential in many metabolic activities of our body.

1.5 BARLEY AND HUMAN HEALTH

Barley is being used for preparation of functional foods due to high content of bioactive compounds. Being rich source of dietary fiber, barley is antidiabetic [42, 48], can treat constipation [21], and helps in smooth movement of bowel [47]. According to the research study, blood sugar can be regulated by eating wholegrain barley. The grains possess nutraceutical properties and cause lowering of postprandial blood glucose level [57, 59]. Adequate intake of fiber helps in improving glucose metabolism in type-2 diabetes patients and helps in lowering plasma lipid concentration [17]. The viscosity of the fiber is directly proportional to the lowering of postprandial glycemic index [35, 61]. Due to increase in viscosity, there is a decrease in the rate of absorption of food in the intestine. This balances the insulin response and postprandial glucose [60, 62].

For achieving health benefits with barley, Food and Drug Administration (FDA) has advocated a daily intake of 3 g of β-glucan to significantly reduce risk of coronary heart diseases. β-Glucan is also known to reduce incidence of hyperinsulinemia and hyperglycemia [15] and risk factors for degenerative diseases, such as cardiovascular diseases (CVDs) [37], hyperlipidemia [63], obesity, cancer [51], and hypertension [6, 26, 43, 54]. Soluble β-glucan helps to lower the cholesterol level in the blood [11, 16], which is further beneficial in the management and prevention of various CVDs [34]. In the colon region, colonic microorganisms can ferment some portion of soluble fiber β-glucan [12, 58], which forms short chain fatty acids in large intestine.

1.6 PROCESSING OF BARLEY GRAIN

Barley is processed for three main purposes: feed, malt, and food. It has been popular as animal feed. Nearly 70% of the harvested barley is used for animal feed. It is a rich source of good quality carbohydrates and proteins and an excellent source of nutrition for animals [4]. For feeding the animals, threshed grain is subjected to cracking or grinding operations. However, malt and food purpose require special postharvest handling procedure. The major postharvest operations of barley processing are given in Figure 1.1.

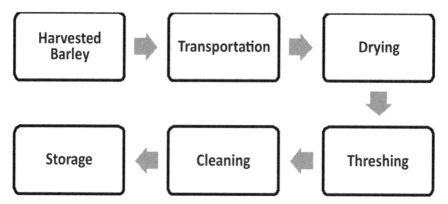

FIGURE 1.1 Major postharvest operations of barley processing.

Hulled barley needs to be dehulled before further processing. Subsequently, it is pearled, flaked, blocked, polished or ground to flour, grits, or flakes before being used directly for product development. Barley grains are subjected to a rotating abrasive surface. As a result of abrasion, barley grain is dehulled, polished, and pearled. By steam conditioning and hot rolling of pearled barley, barley flakes are made, which are used in soups or in breakfast cereals. Barley flour is prepared from pearled barley by hammer milling or roller milling. Whole barley grain can also be milled in flour. It is used for baby foods, leavened, and unleavened breads. Flour is also a by-product of cutting, polishing, and pearling processes. Barley grits are prepared by cutting the pearled grain into portions, which are further graded by size and are polished.

Malting of barley is a popular practice for preparation of barley malt. Hulled barley is preferred for malting. Malt is made by soaking and drying of barley kernels. It is a controlled germination process (Figure 1.2). Barley malt is used for beer production. It is also used in syrups and extracts for enhancing color, flavor, or sweetness and in baked products, cereals, confectionary, and beverages.

1.7 FUNCTIONAL FOODS FROM BARLEY

The functional foods contain bioactive compounds to provide health benefit beyond their nutritional value. Barley, being rich in dietary fiber, has excellent potential for formulation into functional foods.

Barley is consumed in most parts the world in the form of breads, beverages, and as an ingredient in various cuisines. It is also used in beverages, baked products, porridge and malt, and in a variety of food products (e.g., crackers, breads, cookies, extruded snacks, fruit-filled cereal bars, pastas etc.). Development of health foods from barley has become an area of interest among research workers and health experts. Therefore, variety of health promoting products from barley have been reported in the recent past.

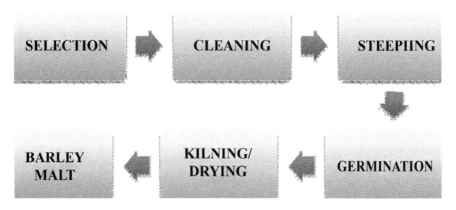

FIGURE 1.2   Preparation of barley malt.

1.7.1 BARLEY BREAD

Barley bread is made from barley flour mainly for improving the nutritional and functional characteristics of bread. It is also a low-cost alternative to wheat. It provides fiber, vitamins, minerals, and selenium (trace mineral). An Egyptian product called as baladi bread was developed by supplementing wheat flour with whole barley flour and gelatinized corn. Inclusion of barley and corn significantly improved the nutritional and rheological parameters of bread dough. It was found that barley flour may replace wheat flour by 30% for preparation of bread of acceptable quality [33]. In another study, β-glucan extracted from barley by hot-water method was used as a supplement in bread to significantly improve the sensory characteristics of bread [3].

1.7.2 BARLEY PORRIDGE

Porridge is a popular traditional food in several countries. Barley porridge is a breakfast cereal, which is highly nutritious and is made from raw, broken coarse barley grains. It is rich in fiber that helps in weight loss. Recent developments in barley porridge have been the development of methods to prepare ready-to-eat canned porridge. For this purpose, the water uptake and volume changes in barley grain during hydrothermal processing of canned porridge was numerically modeled [56].

1.7.3   FUNCTIONAL BEVERAGES

β-Glucan extracted from barley has been used in preparation of various types of beverages. The functional beverages made from barley can play an important role in health promotion and the prevention of degenerative diseases [49]. Recently, a functional beverage was developed following fermentation of barley flour [20]. It was found that the beverage viscosity was improved significantly with the increasing of β-glucans (in the recipe). Also the highest viscosity was obtained with 1% of β-glucans compared to the lowest viscosity without β-glucans.

Beverages with β-glucan are rich in soluble fibers and provide smooth mouthfeel. β-Glucan gum can also be used as a thickening agent in beverages; thus replacing traditional beverage thickeners, such as alginates, pectin, xanthan, and carboxymethyl cellulose [29].

Barley tea is an infusion of roasted hulled barley grain, which is a staple beverage in Korea, Japan, and China. In Korea, it is known as *Boricha* and in Japan it is known as *Mugicha.* It is cleansing, cooling, and caffeine-free detoxifying drink. It has a nutty, aromatic, and warm flavor with slight bitter undertone. Barley tea provides anticoagulant, antioxidant, antidiabetic, and antibacterial properties. It is rich in fiber to ease constipation and healthy bowel movements. It aids in sleep and promotes male fertility.

"Boza" is traditional fermented beverage, which is prepared from wheat, millet, corn, rice, barley, oats, and other flours. It is light yellow in color, having a sweet or sour taste, and is highly viscous. The drink is commonly consumed in Balkan Peninsula, Bulgaria, and Turkey [18].

Barley water is a traditional drink that is prepared by boiling grain in water and by straining out. In the category of healthy beverages, barley water has high ranking because barley is a rich source of vitamins and soluble and insoluble fibers to further help in decreasing risk of CVDs and diabetes.

Instant barley drink is prepared from roasted cereal grains of barley and is generally consumed in Norway, Ireland, United Kingdom, Denmark, Malta, and Hong Kong. Generally, this drink is caffeine-free with pleasant taste. Instant barley drinks are available in granular and powder forms. They are rich source of fiber and low in calories. Natural ingredients can be added to enhance their properties and to make them tastier.

Barley soup is made from barley flour. Milk and honey are added to the dried barley powder. Barley soup originated in Arabian Peninsula because of healing effects for fevers. Generally, pearled barley grain is used for making the soup.

1.7.4   FERMENTED AND GERMINATED BARLEY

In recent years, many research works have been conducted to develop fermented and germinated barley especially with probiotic microorganism like *Lactobacillus acidophilus*. Bacterial enzymes can increase the bioavailability of amino acids via proteins hydrolysis.

Further, acids and alcohols are produced during fermentation to decrease pH. This inhibits the growth of common pathogenic microbes resulting in improved shelf-life of fermented foods [52]. The barley seeds are germinated and then are fermented with 5% probiotic curd to have significantly higher amount of lysine and a better amount of thiamine and niacin. On the other hand, there is a reduction in fiber content. β-Galactosidase enzyme released during germination of barley fractionally attacks galactomannan and yields galactose. The polysaccharide and mucilage components are broken down and are utilized by the growing sprouts [7].

1.7.5   OTHER FOODS

Complementary foods are given to 6-months to 2-year-old children in addition to mothers' milk. These foods should be high in nutrition and have good digestibility. Malting enhances nutritional value of a grain. Malted and extruded barley was blended with malted and extruded pearl millet, whey protein concentrate, skimmed milk powder, sugar, and refined vegetable oil to prepare weaning mix by response surface methodology [10]. Malted barley was used to prepare a complementary blend along with maize and peas. The complementary mix was nutritionally rich and had high sensory acceptability [23].

Barley kernels are husked, then flattened and rolled for making barley flakes. Barley flakes are used as breakfast cereal and look like rolled oats.

Biscuits are popular convenience foods and can serve as excellent vehicle for conveying bioactive components to humans [27]. Barley flour has been used to prepare cookies that can serve as functional food. The acceptability of barley biscuits can be increased by incorporating natural and artificial flavors. However, natural colorants and flavorants, such as turmeric and cocoa, not only increase the acceptability of cookies but also improve the nutritional and health benefits [5]. Barley flour can also be used as a component of healthy composite blend for biscuit manufacture. Biscuits prepared with wheat, barley, and buckwheat flours had a significantly higher calcium iron and zinc content compared to wheat flour biscuits [32].

In the pasta preparation, barley flour plays a leading role. The β-glucan component of barley forms a viscous mass that significantly improves pasta texture [28]. Barley can be used to prepare functional pasta as it is an excellent source of β-glucans.

Barley can be used to prepare highly nutritious snacks via extrusion process. The research has shown that hulled varieties of barley are more suited to expansion through extrusion and produce good quality extruded snacks [8]. Extrusion of barley along with sorghum and horse gram flours resulted in extrudates with high phenolic contents and antioxidant activities and improved hydration characteristics [46].

Barley casserole can be prepared by combining barley with celery, onion, carrots, mushrooms, and green pepper. Broth is added to the mixture that is boiled and then is baked for 45 min. Barley sausage, known as *pölsa* in Sweden, is made from pearled grain.

β-glucan has a property of thickening agent. Moreover, it is also known to be a stabilizing and emulsifying agent. These properties make it suitable for making ice creams, sauces, and salad dressings. To make the soup or stew healthier and more flavorful, barley is added. Cooked barley is added to broth with a variety of vegetables for making risotto.

1.8 SUMMARY

Barley is a rich source of carbohydrates, fiber, proteins, vitamins, and minerals. It is also the source of protective bioactive compounds, such as dietary fiber, proteins, and phytosterols, which play important role in prevention of chronic ailments, such as CVDs, diabetes, and cancers. Bran and germ parts of the whole grain should be retained in the processed products as they also exert same health effects. Research is needed to increase the palatability of barley wholegrain products to attract consumers to include it in their daily diets.

KEYWORDS

- barley
- cereal grain
- health food
- nutrition
- β-glucan

REFERENCES

1. Abdel-Aal, E. M.; Young, J. C.; Rabalski, I. Anthocyanin Composition in Black, Blue, Pink, Purple, and Red Cereal Grains. *Journal of Agricultural and Food Chemistry*, **2006**, *54*, 4696–4704.

2. Abdel-Aal, E. M.; Choo, T. M; Dhillon, S.; Rabalski, I. Free and Bound Phenolic Acids and Total Phenolics in Black, Blue, and Yellow Barley and their Contribution to Free Radical Scavenging Capacity. *Cereal Chemistry*, **2012**, *89* (4), 198–204.

3. Ahmad, A.; Anjum, F. M.; Zahoor, T.; Chatha, Z. A.; Nawaz, H. Effect of Barley β-Glucan on Sensory Characteristics of Bread. *Pakistan Journal of Agricultural Science*, **2008**, *45* (1), 88–94.

4. Akar, T.; Avki, M.; Dusunceli, F. Barley: Post Harvest Operations; **2004**; http://www.fao.org/inpho/content/compend/text/ch31.htm; Accessed on August 13, 2019.

5. Amer A. A.; El-Beltagi, H. S.; Ali, R. F. M. The Effects of Wheat Flour and Barley Flour on the Quality and Properties of Biscuits Colored with Synthetic and Natural Colorants. *Notulae Scientica Biologicae*, **2019**, *11* (1), 30–38.

6. Anderson, J. W. Dietary Fiber and Human Health. *Horticultural Sciences*, **1990**, *25*, 1488–1495.

7. Arora, S.; Jood, S.; Khetarpaul, N. Effect of Germination and Probiotic Fermentation on Nutrient Composition of Barley-based Food Mixtures. *Food Chemistry*, **2010**, *119*, 779–784.

8. Baidoo, E. A.; Murphy, K.; Ganjyal, G. M. Hulled Varieties of Barley Showed Better Expansion Characteristics Compared to Hull-less Varieties during Twin-screw Extrusion. *Cereal Chemistry*, **2019**, 1–14.

9. *Baik*, B. K.; *Ullrich*, S. E. *Barley* for Food: Characteristics, Improvement and Renewed Interest. *Journal of Cereal Science*, **2008**, *48*, 233–242.

10. Balasubramanian, S.; Kaur, J.; Singh, D. Optimization of Weaning Mix based on Malted and Extruded Pearl Millet and Barley. *Journal of Food Science and Technology*, **2014**, *51* (4), 682–690.

11. Behall, K. M.; Scholfield, D. J.; Hallfrisch, J. Diets Containing Barley Significantly Reduce Lipids in Mildly Hypercholesterolemic Men and Women. *American Journal of Clinical Nutrition*, **2004**, *80*, 1185–1193.

12. Bell, S.; Goldman, V. M.; Bistrian, B. R.; Arnold, A. H.; Ostroff, G.; Forse, R. Effect of β-Glucan from Oats and Yeast on Serum Lipids. *Critical Reviews in Food Science and Nutrition*, **1999**, *39*, 189–202.

13. Bhatty, R. S. Physiochemical and Functional (Bread Making) Properties of Hull-Less Barley Fractions. *Cereal Chemistry*, **1986**, *63*, 31–35.

14. Bhatty, R. S. The Potential of Hull-Less Barley. *Cereal Chemistry*, **1999**, *76*, 589–599.

15. Brennan, C. S.; Tudorica, C. M. The Role of Carbohydrates and Non-starch Polysaccharides in the Regulation of Postprandial Glucose and Insulin Responses in Cereal Foods. *Journal of Nutraceuticals, Functional and Medical Foods*, **2003**, *4*, 49–55.

16. Cavallero, A.; Empilli, S.; Brighenti, F.; Stanco, A. M. High (1→3,1→4)-β-Glucan Barley Fractions in Bread Making and Their Effects on Human Glycemic Response. *Journal of Cereal Science,* **2002**, *36*, 59–66.

17. Chandalia, M. Beneficial Effects of High Dietary Fiber Intake in Patients with Type 2 Diabetes Mellitus. *England Journal of Medicine*, **2000**, *342*, 1392–1398.

18. Chonova, V. M.; Karadzhov, G.; Chochkov, R. Rheological Properties of Fermented Beverage from Barley Flour. *Ukrainian Food Journal*, **2013**, *2*, 320–326.

19. Chung, O. K.; Pomeranz, Y. Amino Acids in Cereal Proteins and Protein Fractions. Chapter 5; In: *Digestibility, and Amino Acid Availability in Cereals and Oilseeds*; Finley, J. W., and Hopkins, D. T. (Eds.); AACC, Saint Paul, MN; **1985**; pages 169–232.

20. Din, A. *Development of Beverage from Barley.* Ph.D. Thesis; National Institute of Food Science and Technology, New Delhi: India; **2009**; pages 219.

21. Dohnalek, M. H. The Role of Fiber in Clinical Nutrition. In: *Dietary Fiber: Bioactive Carbohydrates for Food and Feed*; Van der Kamp, J. W., Asp, N. G., Miller, J. J., and Schaafsma, G. (Eds.); Wageningen Academic Publishers, Wageningen, The Netherlands; **2004**; pages 271–294.

22. FAO. *FAOSTAT: Crops*; **2018**; http://www.fao.org/faostat/en/#data/QC/metadata; Assessed on June 12, 2019.

23. Fikiru, O.; Bultosa, G.; Forsido, S. F.; Temesgen, M. Nutritional Quality and Sensory Acceptability of Complementary Food Blended from Maize (*Zea mays*), Roasted Pea (*Pisum sativum*), and Malted Barley (*Hordeum vulgare*). *Food Science and Nutrition,* **2017**, *5* (2), 173–181.

24. Fincher, G. B. Morphology and Chemical Composition of Barley Endosperm Cell Walls. *Journal of the Institute of Brewing*, **1975**, *81,* 116–122.

25. Fincher, G. B.; Stone, B. A. Cell Walls and Their Components in Cereal Grain Technology. In: *Advances in Cereal Science and Technology*; Pomeranz, Y. (Ed.); American Association of Cereal Chemists, Saint Paul, MN; **1986**; volume *8*; pages 207–295.

26. Foster, P.; Miller, K. J. B. International Tables of Glycemic Index. *American Journal of Clinical Nutrition*, **1994**, *59*, 66–69.

27. Frost, D. J.; Adhikari, K; Lewis, D. S. Effect of Barley Flour on the Physical and Sensory Characteristics of Chocolate Chip Cookies. *Journal of Food Science and Technology*, **2011**, *48,* 569–576.

28. Giacco, R.; Vitale, M.; Riccardi, G. Pasta: Role in Diet. In: *The Encyclopedia of Food and Health*; Caballero, B., Finglas, P., and Toldrá, F. (Eds.); Academic Press, Oxford, UK; **2016**; volume *4*; 242–245.

29. Giese, J. H. Hitting the Spot: Beverages and Beverage Technology. *Food Technology,* **1992**, *46*, 70–80.

30. Henry, R. J. Pentosans and (1-3), (1-4)-β-Glucan Concentrations in Endosperm and Whole Grain of Wheat, Barley, Oats and Rye. *Journal of Cereal Science*, **1987**, *6*, 253–258.

31. Holtekjolen, A. K.; Uhlen, A. K.; Brathen, E.; Sahlstrom, S.; Khnutesen, S. H. Contents of Starch and Non-starch Polysaccharides in Barley Varieties of Different Origin. *Food Chemistry*, **2006**, *94*, 348–358.

32. Hussain, A; Kaul, R.; Bhat, A. Development of Healthy Multigrain Biscuits from Buckwheat-Barley Composite Flour. *Asian Journal of Dairy and Food Research*, **2018**, *37* (2), 120–125.

33. Hussein, A.; Hussein, M. S.; Kamil, M. M.; Hegazy, N. A.; Abo El-Nor, S. A. H. Effect of Wheat Flour Supplemented with Barley and/or Corn Flour on Balady Bread Quality. *Polish Journal of Food Nutrition Sciences*, **2013**, *63* (1), 11–18.

34. Jenkins, D. J. A.; Axelsen, M.; Kendall, C. W. C.; Augustin, L. S. A.; Vuksan, V.; Smith, U. Dietary Fiber, Carbohydrates and the Insulin Resistant Diseases. *British Journal of Nutrition*, **2000**, *83*, 157–163.

35. Jenkins, D. J. A.; Wolever, T. M. S.; Leeds, A. R. Dietary Fibers, Fiber Analogues and Glucose Tolerance: Importance of Viscosity. *British Medical Journal*, **1978**, *1*, 1392–1394.

36. Kent, N. L.; Evers, A. D. *Kent's Technology of Cereals*. 4th edition; Elsevier, Oxford, UK; **1994**; pages 319.

37. Keogh, G. F.; Cooper; G. J. S.; Mulvey, T. B. Randomized Controlled Crossover Study of the Effect of a Highly β-Glucan Enriched Barley on Cardiovascular Disease Risk Factors in Mildly Hypercholesterolemic Men. *American Journal Clinical Nutrition*, **2003**, *78*, 711–718.

38. Kim, M. J.; Hyun, J. N.; Kim, J. A.; Park, J. C. Relationship Between Phenolic Compounds, Anthocyanins Content and Antioxidant Activity in Colored Barley Germplasm. *Journal of Agricultural Food Chemistry*, **2007**, *55*, 4802–4809.

39. Kiryluk, J.; Kawka, A.; Gasiorowski, H.; Chalcarz, A. Milling of Barley to Obtain β-Glucan Enriched Products. *Molecular Nutrition and Food Research*, **2000**, *44*, 238–241.

40. Longvah, T.; Ananthan, R.; Bhaskarachary, K.; Venkaiah, K. *Indian Food Composition Tables*. National Institute of Nutrition, Indian Council of Medical Research, Hyderabad, Telangana, India; **2017**; pages 219.

41. MacGregor, A. W.; Fincher, G. B. *Carbohydrates of the Barley Grain*. Chapter 3; In: *Barley: Chemistry and Technology*; MacGregor, A.W. and Bhatty, R. S. (Eds.); AACC, Saint Paul, MN; 1993; pages 73–130.

42. Maier, S. M.; Turner, N. D.; Lupton, J. R. Serum Lipids in Hypercholesterolemic Men and Women Consuming Oat Bran and Amaranth Products. *Cereal Chemistry*, **2000**, *77*, 297–302.

43. McIntosh, G. H., Whyte, J.; Whyte, M. R.; Nestle, P. J. Barley, and Wheat Foods: Influence on Plasma Cholesterol Concentrations in Hypercholesterolemic Men. *American Journal of Clinical Nutrition*, **1991**, *53*, 1205–1209.

44. Prasad, K., Jadhav, A. Prevention and Treatment of Atherosclerosis with Flaxseed-Derived Compound Secoisolariciresinol Diglucoside. *Current Pharmaceutical Design*, **2016**, *22*, 214–220.

45. Rhee, Y. Flaxseed Secoisolariciresinol Diglucoside and Enterolactone Down-Regulated Epigenetic Modification Associated Gene Expression in Murine Adipocytes. *Journal of Functional Foods*, **2016**, *23*, 523–531.

46. Rudra, S. G.; Jakhar, N.; Nishad, J.; Saini, N.; Sen, S. Extrusion Conditions and Antioxidant Properties of Sorghum, Barley and Horse Gram Based Snack. *International Journal of Plant Research*, **2015**, *28* (2), 171–182.

47. Sanjoaquin, M. A.; Appleby, P. N.; Spencer, E. A.; Key, T. J. Nutrition and Lifestyle in Relation to Bowel Movement Frequency: A Cross-Sectional Study of 20,630 Men and Women in EPIC-Oxford. *Public Health and Nutrition*, **2004**, 7, 77–83.

48. Schulze, M. B.; Liu, S.; Rimm, E. B.; Manson, J. E. Glycemic Index, Glycemic Load, and Dietary Fiber Intake and Incidence of Type 2 Diabetes in Younger and Middle-Aged Women. *American Journal of Clinical Nutrition*, **2004**, *80,* 348–356.

49. Shahidi, F. Functional Foods: Their Role in Health Promotion and Disease Prevention. *Journal of Food Science,* **2004**, *69,* 146–149.

50. Shewry, P. R. *Barley Seed Proteins*. Chapter 4; In: *Barley: Chemistry and Technology*; MacGregor, A. W. and Bhatty, R. S. (Eds.); AACC, Saint Paul, MN; **1993**; pages 131–197.

51. Sier, C. F.; Gelderman, K. A.; Prins, F. A.; Gorter, A. Beta-Glucan Enhanced Killing of Renal Cell Carcinoma Micro-Metastases by Monoclonal Antibody G250 Directed Complement Activation. *International Journal of Cancer*, **2004**, *109* (6), 900–908.

52. Sindhu, S. C.; Khetarpaul, N. Development, Acceptability and Nutritional Evaluation of an Indigenous Food Blend Fermented with Probiotic Organisms. *Nutrition and Food Science*, **2005**, *35*, 20–27.

53. Tang, Y.; Li, X.; Zhang, B.; Chen, P. X.; Liu, R; Tsao, R. Characterization of Phenolics, Betanins and Antioxidant Activities in Seeds of Three *Chenopodium quinoa* Wild Genotypes. *Food Chemistry*, **2015**, *166,* 380–388.

54. Tappy, L.; Gugolz, E.; Wursch, P. Effects of Breakfast Cereals Containing Various Amounts of Beta-Glucan Fibers on Plasma Glucose and Insulin Responses in NIDDM Subjects. *Diabetes Care*, **1996**, *19*, 831–834.

55. Tsochatzis, E.D.; Bladenopoulos, K.; Papageorgiou, M. Determination of Tocopherol and Tocotrienol Content of Greek Barley Varieties under Conventional and Organic Cultivation Techniques using Validated Reverse Phase High Performance Liquid Chromatography Method. *Journal of the Science of Food and Agriculture*, **2012**, *92*, 1732–1739.

56. Wang, L.; Wang, M.; Guo, M'; Ye, X.; Ding, T.; Liu, D. Numerical Simulation of Water Absorption and Swelling in Dehulled Barley Grains During Canned Porridge Cooking. *Processes*, **2018**, *6*, 230–235.

57. Wood, P. J. Physicochemical Characteristics and Physiological Properties of Oat $(1{\to}3)$ $(1{\to}4)$-β-D Glucan. In: *Oat Bran*; Wood, P.J. (Ed.); AOAC, Saint Paul, MN; **1993**; pages 83–112.

58. Wood, P. J.; Beer, M. U. Functional Oat Products. In: *Functional Foods: Biochemical and Processing Aspects*; Mazza G. (Ed.); Technomic Publishing, Lancaster, PA; **1998**; pages 1–37.

59. Wood, P. J.; Braaten, J. T.; Fraser, W. S.; Riedel, D.; Poste, L. Comparisons of the Viscous Properties of Oat Gum and Guar Gum and the Effects of these and Oat Bran on Glycemic Index. *Journal of Agricultural and Food Chemistry*, **1990**, *38*, 753–757.

60. Wood, P. J.; Beer, M. U.; Butler, G. Evaluation of Role of Concentration and Molecular Weight of Oat β-Glucan in Determining Effect of Viscosity on Plasma on Plasma Glucose and Insulin following an Oral Glucose Load. *British Journal of Nutrition*, **2000**, *84*, 19–23.

61. Wood, P. J.; Braaten, F. W.; Scott, F. W.; Riedel, K. D. Effect of Dose and Modification of Viscous Properties of Oat Gum on Plasma Glucose and Insulin Following an Oral Glucose Load. *British Journal of Nutrition*, **1994**, *72*, 731–743.

62. Wursch, P.; Sunyer, P. The Role of Viscous Soluble Fiber in the Metabolic Control of Diabetes: Review with Special Emphasis on Cereals Rich in Beta-Glucan. *Diabetes Care*, **1997**, *20*, 1774–1780.

63. Yang, J. L.; Kim, Y. H.; Lee, H. S.; Lee, M. S.; Moon, Y. K. Barley Beta-Glucan Lowers Serum Cholesterol Based on the Up-Regulation of Cholesterol 7 Alpha-Hydroxylase Activity and mRNA Abundance in Cholesterol Fed Rats. *Journal of Nutritional Science and Vitaminology*, **2003**, *49*, 381–387.

CHAPTER 2

PEARL MILLET: HEALTH FOODS FOR THE FUTURE

JASPREET KAUR, KAMALJIT KAUR, AMARJEET KAUR, and
SANDEEP SINGH SANDHU

ABSTRACT

Pearl millet is a rich source of carbohydrates, good quality protein, fats, minerals, and vitamins. Due to its phytochemical constituents, it is a useful functional ingredient in other foods. It has been a staple food for people of Africa and Asia. However, its production and consumption have fallen over time due to a preference for wheat and rice due to lack of convenience, off-flavor, color, and antinutritional factors. With the use of traditional and modern technologies, these shortcomings can be overcome. Pearl millet has a tremendous potential as nutri-cereal. In a scenario of climate change and the need for diversification, pearl millet can ensure food security and suitably fit as a health food for the future.

2.1 INTRODUCTION

Pearl millet [*Pennisetum glaucum* (L.) R.Br.] is known by several names in India such as *bajra* in Hindi, *bajri* in Gujrati and Marathi, *kambu* in Tamil, *sajjalu* in Telugu, and *kambam* in Malyalam. It is a diploid ($2n = 2x = 14$), C4, cereal crop. Domestication of pearl millet took place in the sub-Saharan region of Africa, about 4000–5000 years ago [38, 40]. Major growing areas of pearl millet include West Africa and the Indian subcontinent. The annual global production of pearl millet is more than 50% of the total millets produced. After maize, rice, wheat, barley, and sorghum, pearl millet is the sixth most important cereal crop in the world [22].

It has maximum drought tolerance compared to other domesticated cereals [25]. It is well-adapted to grow under conditions of low rainfall (200–600 mm), low soil fertility, and high temperature, where maize and wheat would fail to grow. Due to its adaptability to grow under harsh conditions, it is a major staple crop for over 90 million people in West Africa and the Thar desert in India [26, 70]. This accounts for nearly 29 million ha out of which nearly 15 million is in Africa and 11 million is in Asia [54]. Globally, developing countries contribute to more than 95% of pearl millet production. India is the largest producer of pearl millet in the world, having an annual production of 8.3 million tons [22] on an area of 9.8 million ha.

This chapter includes discussions on the nutritional value of pearl millet; cultural practices that can affect the nutritional quality of pearl millet grains; antinutritional factors; activities of phytocomponents with health benefits; processing of cereal grain for food; traditional food processing techniques; pearl millet-based traditional food products; and future potential of pearl millet for new products.

2.2   MORPHOLOGY OF PEARL MILLET GRAIN

The color of pearl millet grain may be white, pale yellow, brown, grey, slate blue, or purple. The grains are ovoid and 3–4 mm in length. These are bigger in size than other millets. The average grain weight is 8 g per 1000 grains, though it may vary from 2.5 to 14 g. The pearl millet kernel is about one-third the size of sorghum mainly because of higher germ to endosperm proportion in sorghum. Pearl millet grain consists of about 8% pericarp, 17% germ, and 75% endosperm [60]. The pericarp is covered by a thin waxy layer that protects the pericarp from weathering. Below the pericarp is the seed coat a single (one-cell thick) aleurone layer.

2.3   NUTRITIONAL PROFILE OF PEARL MILLET

The nutritional value of pearl millet is better than maize, wheat, or rice and even better than other millets. The nutritional profile of pearl millet and other important cereals is given in Table 2.1.

Due to high-fat content, its consumption leads to satiety. Pearl millet has a well-balanced protein. Its fiber content is higher than the most cereal grains. It has a high content of protein and superior amino acid balance [35]. It is a rich source of minerals (such as iron and calcium) and vitamins. As compared to sorghum, the concentration of threonine is high but leucine content is lower though adequate. As compared to other cereals, pearl millet has higher tryptophan content. However, it is deficient in lysine [15].

TABLE 2.1   Composition of Pearl Millet [Per 100 g of Edible Portion (12% w.b.)]

Nutrient	Units	Value	Nutrient	Units	Value
Ash	g	2.2	Fe	mg	11.0
Ca	Mg	42	Niacin	mg	2.8
Carbohydrate	g	67.0	Protein[a]	g	11.8
Crude fiber	g	2.3	Riboflavin	mg	0.21
Energy	Kcal	363	Thiamin	mg	0.38
Fat	g	4.8			

Source: [30, 69].

[a]Protein = N × 6.25.

2.3.1   ENERGY

The energy provided by 100 g of pearl millet ranges from 1646 to 1691 kJ (d.b.) [66] due to high content of fats and carbohydrates. Diets composed of pearl millet can thus be low-cost alternative to fulfill the energy requirements of under-nourished population, especially in the developing countries.

2.3.2   CARBOHYDRATES

The whole grain of pearl millet has plenty of carbohydrates that mainly include starch, dietary fiber, and soluble sugars. Starch is the major component of endosperm of pearl millet grain and consists of glucose, such as amylose and amylopectin. Total starch content is nearly 71.6% (d.b.) [66]. Pearl millet has a smaller endosperm compared to other cereals. It has a proportionally large germ. Dietary fiber content in pearl millet grain is approximately 8% (d.b.). This is again lower than that in other cereal grains. The content of soluble sugars (such as sucrose and raffinose) is low in the pearl millet grain [60].

2.3.3   PROTEIN

The content of protein in the pearl millet grain varies from 8.6% to 19.4%, with an average of 14.5% on dry basis [60], which is higher than most other cereals because pearl millet has larger germ in proportion to the endosperm. Hence, it is low in water-soluble and alcohol-soluble protein fractions (such as prolamins) compared to other cereals [11, 66]. Pearl millet protein has a balanced amino acid composition. However, the lysine content is higher than most other cereal grains. This increases the susceptibility to nonenzymatic browning, especially Maillard reaction when heated with sugars. Compared to maize, pearl millet has better essential amino acid profile. The content of lysine and methionine is 40% higher while that of threonine is 30% higher in pearl millet compared to maize [10].

2.3.4   ENZYMES

Pearl millet grain contains several enzymes, such as lipase [23]. Pericarp, aleurone layer, and germ contain most enzymes in the grain [36]. Lipase catalyzes the hydrolysis of fat to release free fatty acids. In contrast to other cereal grains, lipoxygenase enzyme is not found in pearl millet. The oxidation of unsaturated fatty acids is catalyzed by lipoxygenase.

2.3.5   LIPIDS

Lipids are hydrophobic compounds that may be classified as simple lipids (mainly free fatty acids), acyl lipids, and neutral lipids [8]. Lipids may also be present in bound form. The lipid content of a pearl millet may vary from 1.5% to 6.8% (d.b.) [66]. This is nearly 5% higher than most other cereal grains except maize. Lipids are mainly concentrated in the germ [1]. Lipids in the free form vary from 5.6% to 6.1% while that in the bound form comprise of 0.6%–0.9%. The grain of pearl millet also contains sterol-containing glycolipids and other phospholipids. Lipids are mainly present as triglycerides in pearl millet grain [45]. Vitamins A and E are fat-soluble vitamins present in pearl millet [60].

Linoleic and oleic are major proportion of fatty acids in pearl millet (Figure 2.1). As in other lipids, about 88% of the total lipids in the pearl millet grain are found in the germ portion [57]. Remaining 12% is distributed equally in the pericarp and endosperm [1]. Since lipids are present as triglycerides and polyunsaturated fatty acids, therefore these may cause deterioration in the quality of flour and products from pearl millet [13].

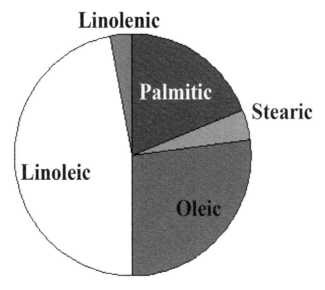

FIGURE 2.1   Relative proportion of major fatty acids in pearl millet [57].

2.4   CULTURAL PRACTICES AFFECTING NUTRITIONAL VALUE OF PEARL MILLET

Various cultural practices can affect the nutritional quality of pearl millet grains. Most affected quality parameter is protein content. Change in protein content may further affect other parameters.

A significant increase in protein content of pearl millet grains was observed with application of nitrogen compared to the control (no nitrogen application) [31, 52, 58]. Proper nutrition of pearl millet crop through application of 30 kg of P_2O_5 and 20 kg of zinc was able to improve its nitrogen content by 22.8% compared to that obtained without application of these nutrients [63]. Positive effect of zinc application on nitrogen content may be because zinc acts as a catalyst and/or co-enzyme in many physiological processes that might get activated due to proper nutrition. Increased nitrogen content leads to increase in protein content. The increase in protein content due to proper nutrition has also been reported in chickpea [48] and in fenugreek [32]. It has been reported that pre-sowing of seed hardening with KNO_3 significantly increased the grain nitrogen content of pearl millet [33].

Integrated nutrient management also had significant influence on protein content of pearl millet [49]. Application of 75% recommended fertilizer dose (RFD) + poultry manure @ 2 t ha^{-1} + Azospirillum @ 5 kg ha^{-1} + phosphate solubilizing bacteria (PSB) @ 5 kg ha^{-1} resulted in significantly higher protein content than that obtained with 75% RFD + farmyard manure @ 5 t ha^{-1} + Azospirillum @ 5 kg ha^{-1} + PSB @ 5 kg ha^{-1}. Similar results have also been reported by other investigators [39]. It is also reported that the integrated application of 30 kg of N + 20 kg of P_2O_5 + farmyard manure @ 6 t ha^{-1} gave significantly higher protein content of pearl millet compared to the control [47]. It was also found that pearl millet seed inoculation with *Azotobacter* significantly increased its grain protein content compared to noninoculated control [67].

Planting the crop on ridges enhanced the crude protein content in pearl millet grain to 12.90% compared to 12.06% in the broadcast method of sowing [61]. Similar results were also reported by others [18]. In another study, it was found that modification in the surface configuration as ridge and furrow and using plastic mulch significantly enhanced grain nitrogen content in pearl millet compared to the flat and dust mulch [33].

It was found that pearl millet cultivars differ in their grain protein content. Pioneer 86M86 cultivar had significantly higher protein content compared to Kaveri Super Boss and ICTP-8203 cultivars [52].

2.5 ANTINUTRITIONAL FACTORS

Although pearl millet is rich in carbohydrates, proteins, fats, vitamins, and minerals, yet its use is limited due to several antioxidant compounds that it possesses. The antinutritional factors in pearl millet are phytic acid, polyphenols, and fiber associated. These seriously affect bioavailability of nutrients and cause poor digestibility of proteins and carbohydrates [66].

2.5.1 PHYTIC ACID

Phytic acid is present as phytate, that is, myoinositol-1,2,3,4,5,6 hexakisphosphate. It is a site for storage of nearly 75% of total phosphorus of the kernel. Due to its chelating character, phytic acid binds cations, such as iron, zinc, magnesium, and calcium. The content of phytic acid in pearl millet varies with genetic factors and geographical location. One study reported a range of

179–306 mg/100 g of phytic acid. Phytic acid acts as an antinutritional factor that lowers the bioavailability of major and trace minerals, such as calcium, phosphorus, iron, zinc, and so on.

2.5.2 POLYPHENOLS

In addition to phytic acid, polyphenols act as antinutrients that limit the bioavailability of other nutrients in food. These may also lead to certain disease condition, such as goiter. C-glycosyl flavones (glucosyl vitexin, glucosyl orientin, and vitexin) are category of polyphenols that are considered goitrogenic. In certain areas of Asia and Africa, it has been observed that endemic goiter was more prevalent in people consuming pearl millet-based foods. Such condition occurs along with iron deficiency [24, 73]. Preliminary treatments, such as dry heating, blanching, malting, and acid treatment can affect the in-vitro iron availability. All treatments reduced total iron content but increased the total in-vitro iron bioavailability. Dry heating that comprised holding the sample in a hot air oven at 100 ± 2 °C for 2 h, had the maximum in-vitro iron availability among all treatments [9].

2.6 PHYTOCHEMICAL COMPONENTS VERSUS HEALTH BENEFITS

Pearl millet is an important staple food for a sizeable proportion of rural population especially in African and Asian countries. It is a source of vital nutrients, such as proteins with balanced amino acid composition, minerals, and vitamins as discussed earlier in this chapter. It may be regarded as a nutri-cereal. However, in addition to regular nutrients, it also contains phytocomponents with physiological benefits.

Regular consumption of foods prepared from whole millet grains is associated with protection against chronic and degenerative diseases, such as diabetes, cancers, heart ailments, Parkinson's disease, and others [12]. Such properties of food are referred to as functional properties and are result of consumption of plant-based phytochemicals. These include components of dietary fiber, resistant starch, lignans, phenolic components (such as flavonoids, tannins, and phytic acid). These are present in pearl millet but some of them have been regarded as antinutrients. Consumption of pearl millet is beneficial in specific health conditions as is discussed below.

2.6.1 DIABETES

Diabetes, a metabolic disorder that causes high blood sugar, requires diet changes so as to control postprandial blood glucose levels. Foods with low glycemic index (GI) need to be included in the daily diet to prevent glucose rise after meals. Blood sugar level can be maintained at safe levels and for a longer time with low GI foods [12]. Studies have suggested that pearl millet has low GI of 55 and can be used to manage onset and control of diabetes [37]. A high amylase activity and low fructose and glucose levels compared to wheat further indicate the use of pearl millet [46].

2.6.2 CANCER

Whole grains of pearl millet are rich in phenolic compounds (such as flavonoids) that have high antioxidant activity and are located in pericarp and testa [29]. These suppress development of tumors. Studies indicate prevention of breast cancers in premenopausal women. Pearl millet flour contains nearly 750 mg of phenolic components per 100 g [62]. Major phenolic components of pearl millet include hydroxybenzoic acids and flavonoids. Traditional methods favor flour development from whole grains with high content of phenolics.

2.6.3 CARDIOVASCULAR DISEASES

Phytic acid is known to help reduce cholesterol and maintain safe levels in the blood. Pearl millet is a good source of phytic acid. It is also a good source of methionine, niacin, and minerals (such as iron, zinc, and potassium). These nutrients play important roles in lowering high blood pressure and reduce risk of heart attack. Studies have demonstrated that whole pearl millet inhibited oxidation of low-density lipoprotein (LDL) cholesterol and liposome, due to presence of phenols [12].

2.6.4 GLUTEN INTOLERANCE

Celiac disease is an autoimmune disease and refers to intolerance to gliadin (a prolamine found in gluten). Unlike wheat, rye, and barley, pearl millet is a gluten-free cereal. Hence, foods developed from pearl millet are safe and recommended for consumption by celiac. Even after being cooked, pearl millet retains alkaline properties and is safe for wheat allergics.

2.6.5 OTHER PHYSIOLOGICAL CONDITIONS

Whole pearl millet flour is a rich source of insoluble fiber. Benefits of insoluble fiber include longer intestinal transit time and lesser the secretion of bile juices. It also lowers risk of stone formation in the gall bladder. High fiber also promotes satiety as the food remains longer in the intestine. This helps to control hunger and promotes weight loss.

2.7 GENERAL USES OF PEARL MILLET

Pearl millet has been used for food, feed, and forages [4] all over the world and especially in the African and Asian countries [41]. Studies have also been conducted on the use of pearl millet as feed and fodder for animals. It may also be used as a feed ingredient for broiler chickens [20]. The research on feed processing parameters has indicated that pearl millet has an increased grinding rate and has lower requirement of energy for grinding compared with maize. A reduction in pearl millet grain particle size caused improved pellet durability index percentage. There was no significant difference in the pelleting performance of pearl millet-based diets compared with soy-corn diet.

2.8 PROCESSING OF PEARL MILLET GRAIN

Processing of a cereal grain for food involves the partial separation and/ or modification of the three main parts of the cereal grain: the germ, the starch-containing endosperm, and the protective outer covering (pericarp). Processing is mainly done to ensure food safety, digestibility, shelf-life, nutrition, convenience, satiety, affordability, and to make it available in a wide variety of forms.

2.8.1 OBJECTIVES OF PROCESSING OF PEARL MILLET

1. The utilization of pearl millet for food is limited since it contains several antinutritional factors (e.g., phytic acid, tannins, etc.), poor

digestibility of proteins and carbohydrates, and low palatability. Various processing technologies, such as decortications, milling, soaking, and others reduce the antinutritive factors in millets and improve the nutritional value.

2. Pearl millet is rich in fats and also contains lipase enzyme. When whole grains are milled, the resultant flour becomes rancid rapidly. Rancidity reduces acceptability of flour and its products. Processing helps to inactivate the enzymes and enhances the storability of pearl millet. After processing, the following developments are delayed:

   • *Enzyme action*: Natural enzymes (such as lipases) and proteolytic enzymes in pearl millet grain can break down proteins, fats, and carbohydrates resulting in spoilage.

   • *Microbial action*: Raw and processed food products are susceptible to bacterial and fungal attack, which causes food to rot and formation of mold. These microbes can multiply under favorable conditions and food can become toxic.

   • *Oxidation*: Rancidity may be caused by exposure of food to air leading to formation of peroxides. In the presence of air, fats and proteins react with oxygen and become rancid, imparting an unpleasant taste to the food. This, too, needs to be controlled.

3. Processing of pearl millet tends to significantly enhance the flavor, taste, and odor. Certain processing treatments (such as roasting) result in formation and release of certain flavor compounds to enhance the acceptability of food products.

4. Convenience: Unlike wheat and rice, processing of pearl millet is time-consuming and people are unaware of methods to use to it for preparing their food at home. Availability of processed pearl millet products will save time spent in preparation and cooking. This will particularly benefit families, where adults are working away from home. Consumers benefit from ready-to-cook and ready-to-eat meals that need less time for heating and preparation.

5. There is huge demand for processed products from pearl millet. Pearl millet is a gluten-free cereal that can serve to provide vital nutrition for people with gluten intolerance.

6. Processing of pearl millet can enhance availability of nutrients as it would unlock the nutrients in the grain. Blending with legumes and fortification may be other strategies to add new nutrients.

7. Processing can help to fit pearl millet into an extremely modern varied diet of the modern man. Processed pearl millet products would

ease marketing and distribution tasks and would provide popularity to this highly underutilized crop.

2.9 POSTHARVEST PRACTICES

Harvesting of the crop is done when it starts to mature. The general indications of maturity of pearl millet crop include the yellowish color of leaves with dried appearance. Grains turn hard. In semiarid tropics of Africa and Asia, millets are grown for human food. These are processed using traditional methods that have several disadvantages as being laborious, monotonous, and are carried out by hand. These processing techniques enable preparation of foods that are acceptable to the local population in terms of taste and convenience. The traditional techniques comprise decortication (which consists of pounding and winnowing or occasionally sifting), roasting, flaking, fermentation, malting, and grinding. In addition to being laborious, these techniques result in a product of poor quality. There is a great scope in improvement of these methods. This would lead to better products. People would readily adopt this crop for food if sufficient good quality of pearl millet flour is made available. It would also increase its demand further.

It has been observed that pearl millet is not processed on a large scale as industrial methods of processing pearl millets are not well-developed. This is quite unlike cereals, for example, wheat and rice that have become more popular. There is a huge potential for processing of millets into food products. In several countries, attempts have been made to improve the existing processing techniques. In several African countries, custom milling was introduced some time back. This had a significant impact on the industrial processing of pearl millet. In Nigeria, about 80% of millets are custom milled into whole flour. This way, over 2.5 million tons of millets are processed each year [42]. The main postharvest operations for pearl millet include threshing, cleaning and winnowing, grading, dehulling or pearling, and milling (Figure 2.2).

The major primary processing operations of pearl millet are dehulling and milling. These operations involve reduction of pearl millet grain to kernel and flour. The flour thus obtained can be used for preparation of several popular food products, such as porridges and breads (leavened and unleavened). However, pearl millet flour has poor keeping quality as it becomes rancid on exposure to moisture and air [13]. Traditionally, pearl millet is manually pounded, generally using pestle and mortar.

FIGURE 2.2 Postharvest operations in pearl millet.

In several African countries, some traditional products prepared from pearl millet are quite popular. These include foods like gruels, porridges, couscous, doughs, flatbreads and nonalcoholic and alcoholic beverages like beer. Pearl millet is also suitable as an ingredient of poultry diets [17].

Recent surveys point out the changing pattern of pearl millet consumption in India. During 1972–1973, nearly 99% of the pearl millet produced in India was used as food and only 1% was diverted for nonfood use, such as feed. Strikingly in 2016–2017, only 27% was used as food and remaining 73% was used for feed purpose [55].

2.10 TRADITIONAL FOOD PROCESSING TECHNIQUES

Traditional food processing techniques have been commonly used in the developing countries, especially in the rural areas. As discussed previously, processing improves nutritional quality, increases digestibility and bioavailability of food nutrients, and helps in reduction or elimination of antinutrients. Such traditional methods include decortication, milling, soaking, cooking, germination, fermentation, malting, popping, and so on.

2.10.1 SOAKING

Grains are commonly soaked before cooking. This is a popular household practice that effectively reduces time of cooking and significantly lowers antinutritional components, such as phytic acid and tannins, and improves mineral bioavailability. Soaking can also be combined with other operations, such as dehulling and cooking. This significantly reduces the amount of antinutrients like phytic acid and phenolics. It also increases in vitro protein digestibility and mineral bioavailability of iron and zinc.

2.10.2 GERMINATION

Germination refers to sprouting of grain after dormancy. It activates enzymes that catalyze breakdown of complex molecules into simpler forms that have more bioavailability. Antinutritional factors (e.g., phytic acid, amylase inhibitors, and polyphenols) are effectively reduced as a result of germination [62]. Previous studies have shown that germination of pearl millet caused a reduction in content of phytates by 28%–46% and nearly 33% in content of tannins. It also increased the in vitro digestibility of protein (14%–26%) and starch (86%–112%) in pearl millet. There as significant increase in the in vitro extractability and bioaccessibility of minerals (such as calcium, iron, and zinc) in pearl millet due to germination [34]. In addition, contents of protein fractions, sugars, and soluble dietary fiber and vitamins (such as thiamine, niacin, lysine) showed augmented levels after germination [5].

2.10.3 FERMENTATION

Fermentation has been a popular food processing practice in many parts of the world and contributes to daily diet of significant proportion of the world population [3, 16]. It is a popular practice for producing a variety of products with improved flavor, texture, and nutritional properties. *Daegu* is a traditional fermented food product of pearl millet in Burkina Faso. Like germination, fermentation causes decrease in contents of antinutrients and increase in availability of protein with improved digestibility. Fermentation resulted in complete elimination of phytic acid and amylase inhibitors in pearl millet [21, 62]. However, tannin contents are more resistant to such treatments. Fermentation also caused increase in pearl millet protein digestibility.

Fermentation not only reduces the antinutrients but also increases the functional constituents in foods. Fermentation of pearl millet grains using *Aspergillus oryzae* resulted in increased phenolic contents and antioxidant activity of pearl millet along with increase in DNA damage protection that is indicative of prevention of chronic diseases, such as cancers [59]. Natural and starter fermentation of pearl millet increases its antioxidative activity [14].

2.10.4   MALTING

Since pearl millet is rich in carbohydrates, therefore it can be used to replace barley for malt production as a low-cost ingredient in foods and beer production. It has been observed that malting results in the increase in beta-amylase activity and free alpha-amino nitrogen. This increase is higher in pearl millet in comparison to sorghum. Several studies have been aimed at improvement of quality of beer produced from pearl millet. Several factors such as moisture and time during germination can affect quality of malt prepared from pearl millet. Moisture is a critical factor for quality malt production [51]. Pearl millet malt can also be used as a malt extender. In fact, it is more effective than sorghum malt and is more economically beneficial in developing countries [50]. Germination time of 3–5 days and temperature of 25–30 °C have been found suitable for obtaining good quality malt from pearl millet. This leads to high diastatic power, α- and β-amylase activity, good free α-amino nitrogen along with lower losses during malting. Malting and fermentation also affect the physicochemical and microstructural properties of pearl millet flour and its products, such as biscuits [2]. These include oil and water absorption properties, swelling capacity and bulk density, increase in crystallinity, and changed microstructure. Hence, malting and fermentation lead to improved product quality in addition to improved nutrition. Malting improves several properties of millet, making it a useful ingredient in health foods. These include physicochemical, nutritional, and functional properties. It is also effective in removing mousy odor of damp pearl millet. This improves the overall quality of malt as an ingredient nutritious, value-added convenience foods for the masses.

2.10.5   POPPING OR PUFFING

Popping is a traditional technique of heat processing food grains by use of sand or salt. It is a kind of high-temperature short-time method, during which the starch present in the endosperm gelatinizes. It also unlocks flavor and aroma compounds making the grain a pleasant food. Popping of pearl millet is also a low-cost and simple technique to prepare convenience foods for consumption by the masses.

2.11 TRADITIONAL FOOD PRODUCTS FROM PEARL MILLET

Pearl millet has been consumed as a traditional food grain by people of Africa and Asia. It has been used as porridges in thick or thin consistency. These are in the fermented or unfermented form. In several areas, flatbreads are also prepared from pearl millet. Again, these may be fermented or unfermented. Other traditional ways of consuming pearl millet are steaming, boiling, and others. In India, pearl millet has been used as *roti*, *idli*, and several other traditional products.

Other products like *bhakar, chakli, dashmi, kharvadi, khichadi, kurdaya, nagdive, papadi, shankarpali, shev, thalipeeth, and usal* have also been prepared from pearl millet [19].

2.12 POTENTIAL OF PEARL MILLET FOR NEW FOOD PRODUCTS

With rising awareness of role of diet in ensuring good health, people are increasingly moving toward whole grains, such as pearl millet. The functional foods have emerged important players in the world markets [56]. Pearl millet can be used alone or combined with other foods to produce functional foods.

2.12.1 *EXTRUDED PRODUCTS*

Pasta has been developed from composite flour comprising of pearl millet, barley, concentrate whey protein, and carboxymethylcellulose [72]. The developed pasta has augmented levels of iron, calcium, phosphorus, and beta glucan. Recent studies have focused on using pearl millet for developing extruded products, such as pasta, snacks, weaning mixes, and others. These include development of pasta by replacing wheat semolina with pearl millet and carrot pomace. Although solid's loss and firmness were increased, yet acceptable pasta could be prepared by partial replacement with pearl millet [28]. Composite flour containing finger millet and pearl millet along with wheat were also used to develop pasta. Carboxymethylcellulose was used as a stabilizer. The product had better nutritional characteristics, phenolic content, and antioxidant activity. Carboxymethylcellulose caused better interaction of starch with protein as revealed by study on microstructure of pasta [27].

Traditional technologies like germination, malting, and others can be combined with new technologies like extrusion processing to create innovative products. These have been used to prepare weaning foods. Weaning mix was developed from extrudates of plain and malted flours of pearl millet along with barley. The mix was optimized using response surface methodology [6].

2.12.2   BAKED PRODUCTS

Pearl millet is now being used to develop acceptable and innovative baked products. Biscuits have been prepared using orange peels and pearl millet for production of functional biscuits. The product had high content of bioactive ingredients (e.g., dietary fiber, phenols, flavonoids) with augmented antioxidant properties [43]. Nutritionally superior and organoleptically acceptable wafers have been developed [64]. The wafers were prepared using pearl millet and barnyard millets along with wheat flour.

Pearl millet varieties have been evaluated for bread-making performance [68]. It was found that acceptable bread could be prepared from composite flour containing 20% proportion of pearl millet flours. In another study, whole pearl millet flour and wheat flour (50% each) were used to prepare bread by incorporating dextran produced in-situ using *Weissella confusa*. The dextran incorporation improved the dough rheological properties and loaf volume and reduced crumb firmness and staling rate. In addition, the bread had higher phenolic contents, antioxidant activity, protein digestibility, and lower GI [71].

2.12.3   BEVERAGE POWDERS

Beverage powders are a category of instant foods made either by drying gelatinized paste or by extrusion processing. Combination of these two techniques has been used to prepare highly acceptable beverages with an increase in iron, calcium, energy values, and most of the amino acids. The processing treatment also increased the starch digestibility [44].

2.12.4 STARCH

Starch constitutes the major component of pearl millet (62.8%–70.5%). Pearl millet starch can be extracted by adjustment in pH and centrifugation [53]. This may be used in place of corn starch. Native pearl millet starch was isolated and was further modified by using hydrothermal, acidic, or enzymatic treatments [65]. It has been reported that modification of pearl millet starches by the above-mentioned three methods can lead to improved pasting properties, freeze-thaw stability, solubility, and water binding capacities [7]. Thus, starch isolated from pearl millet can be a useful ingredient for the food industry.

2.13 SUMMARY

Apart from good quality protein, fats, and carbohydrates, pearl millet contains several phytochemicals that prevent occurrence of chronic diseases. It is a hardy crop that can withstand climate change. It is apt for nutritional and health security of the masses in developed countries that have huge demands for gluten-free foods and beverages as the number of celiac patients is increasing day-by-day. Globally there are large numbers of people who are intolerant to wheat barley, or rye. There is a scope of improvement in the processing of pearl millet. Hence, food scientists and technologists need to look forward to develop acceptable food products from pearl millet to make it the food of the future.

KEYWORDS

- **antinutritional factors**
- **health food**
- **nutrition**
- **pearl millet**
- **phytochemicals**

REFERENCES

1. Abdelrahman, A.; Hoseney, R. C.; Varriano-Marston, E. The Proportions and Chemical Compositions of Hand-Dissected Anatomical Parts of Pearl Millet. *Journal of Cereal Science*, **1984**, *2*, 127–133.

2. Adebiyi, J. A.; Obadina, A. O.; Mulaba-Bafubiandi, A. F.; Adebo, O. A.; Kayitesi E. Effect of Fermentation and Malting on the Microstructure and Selected Physicochemical Properties of Pearl Millet (*Pennisetum glaucum*) Flour and Biscuit. *Journal of Cereal Science*, **2016**, *70,* 132–139.

3. Amadou, I.; Gounga, M. E.; Le, G. W. Millets: Nutritional Composition, some Health Benefits and Processing: A Review. *Emirates Journal of Food and Agriculture,* **2013**, *25*, 501–508.

4. Arora, P.; Sehgal, S.; Kawatra, A. Content and HCl-Extractability of Minerals as Affected by Acid Treatment of Pearl Millet. *Food Chemistry*, **2003**, *80* (1), 141–144.

5. Arora, S; Jood, S; Khetarpaul, N. Effect of Germination and Probiotic Fermentation on Nutrient Profile of Pearl Millet-Based Food Blends. *British Food Journal*, **2011**, *113* (4), 470–481.

6. Balasubramanian, S.; Kaur, J.; Singh, D. Optimization of Weaning Mix based on Malted and Extruded Pearl Millet and Barley. *Journal of Food Science and Technology,* **2014**, *51* (4), 682–690.

7. Balasubramanian, S; Sharma, R; Kaur, J; Bhardwaj, N. Characterization of Modified Pearl Millet (*Pennisetum typhoides*) Starch. *Journal of Food Science and Technology,* **2014**, *51* (2), 294–300.

8. Belitz, H. D.; Grosch, W.; Schieberle, P. *Food Chemistry*. 4th ed.; Springer-Verlag, Berlin Heidelberg, Germany; **2009**; p. 319.

9. Bhati, D; Bhatnagar, V; Acharya, V. Effect of Pre-milling Processing Techniques on Pearl Millet Grains with Special Reference to *In-Vitro* Iron Availability. *Asian Journal of Dairy and Food Research*, **2016**, *35* (1), 76–80.

10. Burton, G. W; Wallace, A. T.; Rachile, K. O. Chemical Composition and Nutritive value of pearl millet. *Crop Science*, **1972**, *12* (2), 187–188.

11. Chandna, M.; Matta, K. N. Characteristics of Pearl Millet Protein Fractions. *Journal of Biological Chemistry*, **1990**, *29*, 3395–3399.

12. Chandrasekara, A.; Shahidi, F. Anti-proliferative Potential and DNA Scission Inhibitory Activity of Phenolics from Whole Millet Grains. *Journal of Functional Foods*, **2011**, *3*, 159–170.

13. Chaudhary, P.; Kapoor, A. C. *Changes in the Nutritional Value of Pearl Millet Flour During Storage. Journal of the Science of Food and Agriculture*, **1984**, *35* (11), 1219–1224.

14. Chinenye, O. E.; Ayodeji, O. A.; Baba, A. J. Effect of Fermentation (Natural and Starter) on the Physicochemical, Anti-nutritional and Proximate Composition of Pearl Millet Used for Flour Production. *American Journal of Bioscience and Bioengineering.* **2017**, *5* (1), 12–16.

15. Chung, O. K.; Pomeranz, V. Amino Acids in Cereal Protein Fractions. In: *Digestibility and Amino Acid Availability in Cereals and Oil Seeds*; Finely, J. W., Hopkins, D. T. (Eds.); American Associations of Cereal Chemists, St. Paul, MN; **1985**; pp. 65–707.

16. Coulibaly, A.; Kouakou, B.; Chen, J. Extruded Adult Breakfast Based on Millet and Soybean: Nutritional and Functional Qualities, Source of Low Glycemic Food. *Journal of Nutrition and Food Science,* **2012**, *2*, 1–9.

17. Davis, A. J.; Dale, N.; Ferreira, F. J. Pearl millet as an Alternative Feed Ingredient in Broiler Diets. *Journal of Applied Poultry Research*, **2003**, *12,* 137–144.

18. Deshmukh, S. P.; Patel, J. G.; Patel, A. M. Ensuing Economic Gains from Summer Pearl Millet (*Pennisetum glaucum* L.) Due to Different Dates of Sowing and Land Configuration. *African Journal of Agricultural Research*, **2013**, *8* (48), 6337–6343.

19. Deshmukh, D. S.; Pawar, B. R.; Yeware, P. P.; Landge, V. U. Consumer's Preference for Pearl Millet Products. *Agriculture Update*, **2010**, *5* (1/2), 122–124.

20. Dozier, W. A., Hanna, W.; Behnke, K. Grinding and Pelleting Responses of Pearl Millet-Based Diets. *Journal of Applied Poultry Research*, **2005**, *14* (2), 269–274.

21. Elyas, S. H. A.; El Tinay, A. H.; Yousif, N. E.; Elsheikh, E. A. Effect of Natural Fermentation on Nutritive Value and *In-Vitro* Protein Digestibility of Pearl Millet. *Food Chemistry* **2002**, *78* (1), 75–79.

22. FAOSTAT. *Statistical database.* Food and Agricultural Organization of the United Nations; 2019*;* http://www.fao.org/faostat; Accessed on June 22, 2019.

23. Galliard, T. Rancidity in Cereal Products. In: *Rancidity in Foods*; Allen, J. C., Hamilton, R. J. (Eds.); Aspen Publishers: Gaithersburg, MD; **1999**; pp. 140–156.

24. GoncËalves; C. F. L.; de Freitas, M. L.; Ferreira, A. C. F. Flavonoids: Thyroid Iodide Uptake and Thyroid Cancer—A Review. *International Journal of Molecular Science*, **2017**, *18,* 1247–1252.

25. Govindaraj, M.; Shanmugasundaram, P.; Sumathi, P.; Muthiah, A. R. Simple, Rapid and Cost-effective Screening Method for Drought Resistant Breeding in Pearl Millet. *Electronic Journal of Plant Breeding*, **2010**, *1,* 590–599.

26. Gulia, S. K.; Wilson, J.; Carter, J.; Singh, B. P. *Progress in Grain Pearl Millet Research and Market Development.* In: *Issues in New Crops and New Uses*; Janick J. and Whipkey, A. (Eds.); ASHS Press, Alexandria, VA; **2007**; pp. 196–203.

27. Gull, A.; Prasad, K.; Kumar, P. Nutritional, Antioxidant, Microstructural and Pasting Properties of Functional Pasta. *Journal of the Saudi Society of Agricultural Sciences*, **2018**, *17*, 147–153.

28. Gull, A.; Prasad, K; Kumar, P. Effect of Millet Flours and Carrot Pomace on Cooking Qualities, Color and Texture of Developed Pasta. *LWT—Food Science and Technology*, **2015**, *63*, 470–474.

29. Huang, M. T.; Ferraro, T. Phenolic Compounds in Food and Cancer Prevention. In: *Phenolic Compounds in Food and Their Effects on Health, II*; Philadelphia, PA: ACS Symposium Series; **1992**; pp. 8–34.

30. Hulse, J. H.; Laing, E. M.; Pearson, O. E. *Sorghum and the Millets: Their Composition and Nutritive Value.* Academic Press, London; **1980**.

31. Jadhav, H, P.; Khafi, H. R.; Raj, A. D. Effect of Nitrogen and Vermicompost on Protein Content and Nutrients Uptake in Pearl Millet (*Pennisetum glaucum* (l.) R. Br. Emend stuntz). *Agricultural Science Digest*, **2011**, *31* (4), 319–321.

32. Jakhar, R. K., Yadav, B. L.; Choudhary, M. R. Irrigation Water Quality and Zinc on Growth and Yield of Fenugreek (*Trigonella foenumgraecum* L.). *Journal of Spices and Aromatic Crops*, **2013**, *22,* 170–173.

33. Kanwar, S.; Gupta, V.; Rathore, P. S.; Singh, S. P. Effect of Soil Moisture Conservation Practices and Seed Hardening on Growth, Yield, Nutrient Content, Uptake and Quality of Pearl Millet (*Pennisetum glaucum* (L.) R. Br.). *Journal of Pharmacognosy and Phytochemistry*, **2017**, *6* (4), 110–114.

34. Krishnan, R.; Meera, M. S. Pearl Millet Minerals: Effect of Processing on Bioaccessibility. *Journal of Food Science and Technology*, **2018**, *55* (9), 3362–3372.

35. Labetoulle, L. *Nutritional Survey of Rural Communities in the North Central Regions of Namibia*. Research Report to French Mission for Co-operation and DEES—North Central Division (MAWRD), CRIAA SA-DC, Windhoek; **2000**; p. 102.

36. Lai, C.; Varriano-Marston, E. Lipid Content and Fatty Acid Composition of Free and Bound Lipids in Pearl Millets. *Cereal Chemistry*, **1980**, *57*, 271–274.

37. Mani, U. V.; Prabhu, B. M.; Damle, S. S.; Mani, I. Glycemic Index of Some Commonly Consumed Foods in Western India. *Asia Pacific Journal of Clinical Nutrition*, **1993**, *2*, 111–114.

38. Manning, K.; Pelling, R.; Higham, T; Schwenniger, J. L.; Fuller, D. Q. 4500-Year Old Domesticated Pearl Millet (*Pennisetum glaucum*) from the Tilemsi Valley, Mali: New Insights into an Alternative Cereal Domestication Pathway. *Journal of Archaeological Science,* **2011**, *38*, 312–322.

39. Meena, R.; Gautam, R. C. Effect of Integrated Nutrient Management on Productivity, Nutrient Uptake and Moisture use Functions of Pearl Millet. *Indian Journal of Agronomy*, **2005**, *50* (4), 305–307.

40. Munson, P. J. Archaeological Data on the Origins of Cultivation in the Southwestern Sahara and Its Implications for West Africa. In: *The Origins of African Plant domestication*; Harlan, J. R., DeWet, J. M. J., and Stemler, A. B. L. (Eds.); Mouton Press, Hague, The Netherlands; **1975**; pp. 187–210.

41. Nambiar, V. S.; Dhaduk, J. J.; Neha, S.; Tosha, S.; Rujuta, D. Potential Functional Implications of Pearl Millet (*Pennisetum glaucum*) in Health and Disease. *Journal of Applied Pharmaceutical Science*, **2011**, *1* (10), 62–67.

42. Ngoddy, P.O. Sorghum Milling in Nigeria—A Review of Industrial Practice, Research and Innovations. Unpublished Paper Presented at the Symposium on the Current Status and Potential of Industrial Uses of Sorghum in Nigeria, Kano, Nigeria; 4–6 December of **1989**; p. 6.

43. Obafaye, R. O.; Omoba, O. S. Orange Peel Flour: A Potential Source of Antioxidant and Dietary Fiber in Pearl Millet Biscuit. *Journal of Food Biochemistry*, **2018**, *42*, 12523–12528.

44. Obilana, A. O.; Odhav, B.; Jideani, V. A. Nutritional, Biochemical and Sensory Properties of Instant Beverage Powder Made from Two Different Varieties of Pearl Millet. *Food and Nutrition Research*, **2018**, *62*, Online article: 10.29219/fnr.v62.1524.

45. Osagie, A. U.; Kates, M. Lipid Composition of Millet (*Pennisetum americanum*) Seeds. *Lipids*, **1984**, *19*, 958–965.

46. Oshodi, A. A.; Ogungbenle, H. N.; Oladimeji, M, O. Chemical Composition, Nutritionally Valuable Minerals and Functional Properties of Benni-Seed, Pearl Millet and Quinoa Flours. *International Journal of Food Science and Nutrition*, **1999**, *50*, 325–331.

47. Parihar, C. M.; Rana, K. S.; Parihar, M. D. Crop Productivity, Quality and Nutrient Uptake of Pearl Millet (*Pennisetum glaucum*) and Indian Mustard (*Brassica juncea*) Cropping System as Influenced by Land Configuration and Direct and Residual Effect of Nutrient Management. *Indian Journal of Agricultural Sciences*, **2009**, *79* (11), 927–930.

48. Pathak, G. C.; Gupta, B.; Pandey, N. Improving Reproductive Efficiency of Chickpea by Foliar Application of Zinc. *Brazilian Journal of Plant Physiology*, **2012**, *24* (3), 173–180.

49. Patil, P.; Nagamani, C.; Reddy, A. P. K.; Umamahesh, V. Effect of Integrated Nutrient Management on Yield Attributes, Yield and Quality of Pearl Millet (*Pennisetum glaucum*

(L.) R. br.emend. stuntz). *International Journal of Chemical Studies*, **2018**, *6* (4), 1098–1101.

50. Pelembe, L. A. M.; Dewar, J.; Taylor, J. R. N. Effect of Malting Conditions on Pearl Millet Malt Quality. *Journal of the Institute of Brewing*, **2002**, 108, 7–12.

51. Pelembe, L. A. M.; Dewar, J.; Taylor, J. R. N. Effect of Germination Moisture and Time on Pearl Millet Malt Quality with Respect to Its Opaque and Lager Beer Brewing Potential. *Journal of the Institute of Brewing*, **2004**, *110* (4), 320–325.

52. Prasad, S. K.; Samota, A.; Singh, M. K.; Verma, S. K. Cultivars and Nitrogen Levels Influence on Yield Attributes, Yield and Protein Content of Pearl Millet under Semi-arid Condition of Vindhyan Region. *The Ecoscan*, **2014**, *6,* 47–50.

53. Qian, J.; Rayas-Duarte, P.; Grant, L. Partial Characterization of Buckwheat (*Fagopyrum esculentum*) Starch. *Cereal Chemistry*, **1998**, *75,* 365–373.

54. Rathore, S.; Singh, K.; Kumar, V. Millet Grain Processing, Utilization and Its Role in Health Promotion: A Review. *International Journal of Nutrition and Food Sciences*, **2016**, *5* (5), 318–329.

55. Reddy, A. R.; Raju, S. S.; Sureas, A.; Kumar, P. Analysis of Pearl Millet Market Structure and Value Chain in India. *Journal of Agribusiness in Developing and Emerging Economies* **2017**, *8* (2), 1–19.

56. Roberfroid, M. B. Prebiotics and Probiotics: Are They Functional Foods? *American Journal of Clinical Nutrition*, **2000**, *71* (6), 1682–1687.

57. Rooney, L. W. Sorghum and Pearl Millet Lipids. *Cereal Chemistry*, **1978**, *55*, 584–590.

58. Sakarvadia, H. L.; Golakiya, B. A.; Parmar, K. B.; Polara, K. B.; Jetpara, P. I. Effect of Nitrogen and Potassium on Yield, Yield Attributes and Quality of Summer Pearl Millet. *Asian Journal of Soil Science*, **2012**, *7* (2), 292–295.

59. Salar, R. K.; Purewal, S. S. Improvement of DNA Damage Protection and Antioxidant Activity of Biotransformed Pearl Millet (*Pennisetum glaucum)* Cultivar PUSA-415 Using *Aspergillus oryzae* MTCC 3107. *Biocatalysis and Agricultural Biotechnology*, **2016**, *8*, 221–227.

60. Serna-Saldivar; Rooney L.W. Structure and Chemistry of Sorghum and Millets. In: *Sorghum and Millets: Chemistry and Technology*; Dendy, D. A. V. (Ed.); American Association of Cereal Chemists, St. Paul, MN; **1995**; pp. 69–124.

61. Sharma, B.; Kumari, R.; Kumari, P.; Meena, S. K.; Singh, R. M. Evaluation of Pearl Millet (*Pennisetum glaucum* L.) Performance Under Different Planting Methods at Vindhyan Region of India. *Advances in Bioresearch*, **2018**, *9* (3), 123–128.

62. Sharma, A.; Kapoor, A.C. Levels of Anti-nutritional Factors in Pearl Millet as Affected by Processing Treatments and Various Types of Fermentation. *Plant Foods for Human Nutrition*, **1996**, *49*, 241–252.

63. Singh, L.; Sharma, P. K.; Kumar, V.; Rai, A. Nutrient Content, Uptake and Quality of Pearl Millet Influenced by Phosphorus and Zinc Fertilization (*Pennisetum galaucum* L.) under Rainfed Condition. *International Journal of Chemical Studies*, **2017**, *5* (6), 1290–1294.

64. Sruthi, V.; Waghray, K.; Rathod A. N. Development of Wafers Incorporated with Pearl Millet Flour and Barnyard Millet Flour. *International Journal of Scientific Research in Science and Technology*, **2018**, *4* (5), 44–51.

65. Suma, P.; Urooj, A. Isolation and Characterization of Starch from Pearl Millet (*Pennisetum typhoidium*) Flour. *International Journal of Food Properties*, **2014**, *18,* 2675–2687.

66. Taylor, J. R. N. Millet: Pearl. In: *Encyclopedia of Grain Science*; Wrigley, C. (Ed.); Elsevier: Melbourne, Australia; **2004**; pp. 253–261.
67. Togas, R.; Yadav, L. R.; Choudhary, S. L.; Shisuvinahalli, G. V. Effect of *Azotobacter* on Growth, Yield and Quality of Pearl Millet. *Journal of Pharmacognosy and Phytochemistry*, **2017**, *6* (4), 889–891.
68. Tortoe, C.; Akonor, P. T.; Hagan, L.; Kanton, R. A. L. Assessing the Suitability of Flours from Five Pearl Millet (*Pennisetum americanum*) Varieties for Bread Production. *International Food Research Journal*, **2019**, *26* (1), 329–336.
69. United States National Research Council/National Academy of Sciences. *United States Canadian Tables of Feed Composition*. Third revision; National Academy Press, Washington, DC; **1982**; p. 218.
70. USAID. Pearl Millet. In: *Agricultural Adaptation to Climate Change in the Sahel: A Review of Fifteen Crops Cultivated in the Sahel*; African and Latin American Resilience to Climate Change (ARCC), Washington, D.C.; **2014**; pp. 8–11.
71. Wang, Y.; Compaoré-Sérémé, D.; Sawadogo-Lingani, H. Influence of Dextran Synthesized *In Situ* on the Rheological, Technological and Nutritional Properties of Whole Grain Pearl Millet Bread. *Food Chemistry*, **2019**, *285*, 221–230.
72. Yadav, D. N.; Balasubramanian, S.; Kaur, J.; Anand, T.; Singh, A. K. Non-wheat Pasta based on Pearl Millet Flour Containing Barley and Whey Protein Concentrate. *Journal of Food Science and Technology*, **2014**, *51* (10), 2592–2599.
73. Zimmermann, M. B. The Influence of Iron Status on Iodine Utilization and Thyroid Function. *Annual Reviews of Nutrition*, **2006**, *26*, 367–369.

CHAPTER 3

OATS: BIOCHEMICAL, HEALTH BENEFITS, AND FUTURE ASPECTS

SANDEEP SINGH, SHUBHPREET KAUR, and GURSHARAN KAUR

ABSTRACT

Utilization of oats (*Avena sativa* L.) was initially for feeding livestock animals, but now it is an important cereal as a food for human consumption because of high contents of proteins, carbohydrates, lipids, dietary fiber, phytochemicals, and phenolic compounds that can reduce risks of cardiovascular diseases, type 2 diabetes, obesity, celiac disease, and so on. Oats and oat-based food products, such as breakfast cereals, breads, cookies, and infant foods, are getting increased importance to utilize oats and their components to formulate healthy food products. This chapter presents an overview on biochemical components of oats with their health benefits and use of oats and value-added products as food and nonfood purposes.

3.1 INTRODUCTION

Avena sativa L. (oat) is an important cereal grain commonly used as a source of nutrition in the human diet, for industrial purpose and as livestock feed. It is the most common variety of oats and known by other names, such as white oats, common oats, covered oats, and hulled oats. It is a major variety among other cultivated species of oats in the world, such as:

- *Avena byzantina* C. Koch (red Oat),
- *Avena sativa* var. *nuda* (naked oat),
- *Avena strigosa* (bristle oat), and
- *Avena abyssinica* Hochst (Ethiopian oat).

The popular species of oats are diploid, tetraploid, or hexaploid [20]. *A. sativa* var. *nuda*, referred to as naked or hull-less variety, has a good grain quality, yet it is prone to mechanical damage during processing due to loosely attached hull to the groat [20].

World production of oat grain currently ranks sixth owing to production of popular cereals like rice, corn, wheat, barley, and sorghum. Utilization of oats for food purpose is nearly 10% of the total oats produced in the world and the consumption is lower than other common cereals, such as wheat, rice, corn, or barley. The average annual production has lowered down to almost half from 1961 (~49.5 million tons) to 2017 (25.9 million tons) [15]. This could be attributed to the increase in production of crops like wheat, rice, or corn that yield popular products and the simultaneous decline in the utilization of oats for animal feed. With the rise in consumption of oats for human foods, they are being grown all around the globe in Europe, North America, Asia, and Australia. Major countries in the world that contributed to the production of oats in the year 2017 were the Russian Federation, Canada, Australia, Poland, China, and Finland [15]. In comparison to other cereals, oats more suitably grow in the cool and moist climate. However, these are susceptible to heat and water stress at various stages of crop development [27].

Oats are popular among other cereals for providing various nutrients and bioactive compounds and are consumed whole or partially processed. They provide ample amounts of carbohydrates, proteins, lipids, fiber, minerals, vitamins, and several phytochemicals. Oats have higher proportion of proteins particularly globulins, which in turn provide a proper proportion of the required essential amino acids. The proteins lack gluten components and hence make it safer food for the individuals suffering from celiac disease [43]. β-Glucan is a soluble fiber in oats that provides several health benefits, such as lowering of blood glucose and serum cholesterol, provision of satiety, and so on.

As a feed, *Avena nuda* or naked oats are more suitable for feeding poultry birds, pigs, and racehorses. Besides their utility as an animal feed, they can be used as food products. For human food purpose, oats are generally consumed as breakfast cereals in various forms, such as rolled oats, oatmeal, oat flour, or, oat bran. These oat products can be further processed for manufacturing ready-to-eat cereals, beverages, breads, and infant foods.

This chapter presents an overview on biochemical components of oats with their health benefits and use of oats and value-added products as food and non-food purposes.

3.2 OAT CEREAL: ORIGIN AND HISTORY

Oats are regarded as one of the oldest cultivated grasses, which originated primarily in the Mediterranean and Middle East countries. Significant traces of oat grains were found in a cave of the southeastern region of Italy, even before the domestication of plants [21]. History of oat grains has been found linked with archeological relics obtained as old as 4000 years back and later around 100–200 years back in Egypt [33]. Utilization of oats as human food occurred quite later than wheat, rice, or barley. Rye and oats were introduced unintentionally as weed contaminants when the popularity of wheat and barley grains increased in the regions of Europe and Eastern areas of Asia [45].

A number of earlier records from the Roman period revealed oats being cultivated as an important cereal. The historians contended that people consumed oats in the form of oatmeal or porridge, whereas some Greek and Roman authors cited the plant as forage for livestock, weed, or medicinal herb in the literature dating back around ~AD 23–79 [53]. The early names for oats were probably given by the Greeks, however. Roman Empire probably used the term *Avena* long back for oats before a *genus* was assigned to it in 1700–1750. A thorough study further conducted in the 1950s helped to recognize and categorize several species and subspecies of oats.

The cool and humid climatic conditions of Northwest Europe favored the cultivation of oats about 6500–2500 years ago. Cultivation of oats as a crop probably flourished due to higher yield from oats compared to other crops cultivated under similar climatic and soil conditions. Since the climatic conditions for the growth of oat grains were highly favorable, they could easily grow in areas, where other crops like barley or wheat showed poor productivity. Further improvements in breeding practices and some crossbreeding led to the development of modern types of oats and turned it into a major cereal crop for food consumption. Northwestern Europe favored the cultivation of white oats and utilized them for making oatmeal, bread and beer, while red oats remained to be utilized as livestock feed in the Mediterranean regions.

Cultivation of white oats extended from Western Europe to other regions of Europe in the later years of Bronze Age. Red oats, on the other hand, gained popularity in the Mediterranean and Latin countries. Oats were first brought into the United States about 500 years ago [8], wherein white oat was more suitable for spring season and red oat for the autumn season. Presently, the popular oats in the United States are the ones developed by cross-breeding both the white and red oats [20].

Majority of the oats grown lately are hexaploid, such as white oats or red oats. The wild species like *A. abyssinica* is tetraploid, whereas *A. nuda* is diploid. *A. nuda*, the naked oats, is cultivated for its higher nutritional values owing to a higher proportion of essential amino acids than the other common cereals. Consequently, naked oats are beneficial for manufacturing specialty products that may fetch higher price as nutritional supplement foods. The naked oats are, however, poor in terms of their utility owing to their lower yield and unprotected grain makes them prone to mechanical damage and mold attack during harvesting [44].

The Spaniards probably passed on red oats into the western regions in late 1500 or around AD 1600 and were sown initially on an island called Cuttyhunk, near the coasts of Massachusetts. Oats could not gain early popularity owing to the presence of corn as a higher-yielding crop. However, by the 1880, oatmeal began to flourish as a breakfast food due to the increased demand among European immigrants and compelling the local mills for the production of breakfast cereals.

3.3   STRUCTURAL AND BIOCHEMICAL CHARACTERISTICS OF OATS

The physical structure of oat caryopsis or "groat" is quite comparable to that of wheat owing to the presence of a crease that extends toward the whole length of the groat and several fine hair-like protuberances known as "trichomes." The grains of oats are however elongated and leaner in width than that of wheat. The grain develops from flowers or florets that are enclosed in a pair of bracts or leaves, called lemma and palea. Among the covered or hulled oats like white or red oats, the groat is firmly enclosed in a covering called hull that develops from a pair of bracts or leaves, known as lemma and palea. These coverings, however, are loose and get removed during cleaning or threshing in the varieties of naked or hull-less oats.

The hull on the outside of the groat comprises of cellulose, hemicellulose, and lignin, and accounts for 25%–30% of the whole kernel [48]. Inner to the hull, oat groat consists of bran (10%–12%), germ (3%–4%) and endosperm (60%–65%) as major structural components [20], similar to those of other cereal grains. The variation in the composition of oats may be attributed to the differences in varieties and environmental factors.

Bran comprises of several cellular layers, such as pericarp, seed coat or testa, hyaline or nucellus layer, and aleurone cells. The lipase enzyme is present almost entirely in the pericarp layers of the grain. Aleurone cells form the innermost layer of the bran and completely surround the starchy endosperm as well as the majority of the germ or embryo portion. Aleurone layer is a rich source of several minerals and vitamins along with phytic acid and antioxidants [18]. This layer is morphologically a part of the endosperm and performs an important function during germination of grain. Aleurone along with scutellum produces several enzymes during germination, which in turn helps in the breakdown and transportation of components from the endosperm toward the germ portion. The starchy endosperm forms the largest portion of the grain that serves as a major reservoir of starch, proteins, lipids, and β-glucan. The endosperm consists of several cells that are packed with small-sized or aggregated granules of starch in a protein matrix. The germ or the embryo principally consists of embryonic axis and scutellum. Embryonic axis further consists of undeveloped leaves (plumule) and root system (radicle) that are fully capable of giving rise to a new plant during germination. Oat germ has a rich quantity of proteins, which further contain an antinutritional factor called phytin. Phytin binds most of the Phosphorous content along with smaller amounts of calcium and potassium.

3.3.1 STARCH

Like all other cereals, the endosperm of an oat groat contains starch as a major constituent that accounts for 50%–60% of the groat. Oats, however, are known for having a lower quantity of carbohydrates in comparison to other cereals. The starch content of oats may vary according to the protein content, the type of variety, or the growing conditions. Starch is composed primarily of amylose and amylopectin.

Amylopectin is the larger and highly branched fraction of starch, consisting of glucopyranose units ($\sim 2 \times 10^3$ to 2×10^5) that are linked together by alpha-1,4- and alpha-1,6-glycosidic bonds. Amylose is present in lower proportion (20%–34%) [26], and it essentially consists of long and straight chains of glucopyranose units, linked only by alpha-1,4-glycosidic bonds. In addition to these two fractions, oat starch granules also contain minute proportions of noncarbohydrate components, such as proteins (0.4%–0.9%), lipids (0.7%–2.5%), and phosphorous (0.15%–0.19%) [34]. In general, oat starch shows higher proportion of bound lipids (0.7%–2.5%) compared to starches obtained from wheat, rice, or corn (~0.5%–1.0%) [26].

Owing to the presence of amylose-lipid complexes, oat starch shows two endothermic transitions during differential scanning calorimeter analysis. The first endotherm at a lower temperature indicates the melting of starch crystallites, whereas the second endotherm at higher temperature relates to the melting of amylose-lipid complexes. Similar transitions can be observed in starches from other sources having amylose-lipid complexes [40].

The packing arrangement of amylose and amylopectin inside the starch granules is directly related to the crystallinity of starch granules. Similar to other cereal starches, oat starch also exhibits typical A-type pattern, in which double helices comprising the crystallites are densely packed, with low water content. Tuber starches show B-type pattern, in which the crystallites are not densely packed, whereas C-type is a blend of both A- and B-type patterns and observed in starches obtained from pulses.

The starch morphology of oat grains is quite different as the starch granules show weak birefringence. The granules are ovoid, polyhedral or irregularly shaped, and generally present in clusters or as aggregates. While in other cereal grains, like wheat and barley, solid and discrete starch granules are observed, which are optically clear bodies and fall into discrete size distributions. Oat starch granules have a diameter ranging from 1.9 to 2.4 µm [23], which is similar to rice but quite smaller than the starches of wheat, rye, barley, or corn [39].

3.3.2   PROTEINS

The uniqueness of oat groat lies in its high protein and lipid content in comparison to other cereals. The protein content of oat groat lies between 15% and 20%, depending on the variety and growing conditions. Oat proteins exhibit higher biological value due to higher proportion of lysine, which otherwise is found deficient in wheat or corn. Distribution of proteins in the oat groat is variable and greater part of the total proteins is found in the bran (~49%) and the endosperm portion (~45%). Most of the storage proteins in endosperm and aleurone are situated inside the protein bodies, which forms the protein matrix.

Based on their solubility, oat proteins have been categorized into four main fractions: albumins (~10%–20%), globulins (~50%–75%), prolamins (~5%–15%), and glutelins (~20%–25%) [30]. The percent proportions of different fractions may, however, vary depending upon the technique and method of extraction used. Similar to rice, oat proteins also show high proportion of globulins, whereas prolamins or avenins constitute the smallest fraction of the total proteins in oats. The amino acid composition of globulin as a major storage protein is more suitable for the nutrition of animals or humans due to a higher proportion of aspartic acid, arginine, histidine and lysine [13]. Compared to other fractions, the alcohol-soluble prolamin is characterized by a higher amount of glutamic acid as well as proline and a lower content of lysine. Higher quantity of globulin in oat proteins also contributes to the higher proportions of lysine, which otherwise is present in lower amounts in other cereals. Hence, the nutritional quality of oats is of better significance than the other cereals.

3.3.3   LIPIDS AND ENZYMES

Majority of the lipids of oats are present in the endosperm unlike the lipids present in the germ (corn) or bran (rice). The endosperm of oat grain perhaps contains the highest level of unsaturated lipids as compared to other cereal grains [18]. In most of the other cereals, lipids are primarily present in the germ portion. Among all other cereal grains, oat groats exhibit maximum amount of lipids, which may vary from 4% to 10%, depending on the variety [51]. The quantity and the type of lipids contribute to various nutritional, textural, organoleptic, and functional properties. Hence, the types of fatty acids present in the lipids play an important role in various characteristics of oats. Also, during germination, the lipids present in the grain contribute to the energy provision and nutrition for the growing embryo.

Among the various fatty acids, oat lipids consist high proportion of unsaturated fatty acids, such as oleic and linoleic acid, which are essential fatty acid for human nutrition. Palmitic acid is a saturated fatty acid; and along with oleic acid, linoleic acid, they contribute as three major fatty acids (90%–95%) in oat lipids. Besides these, oat lipid composition also consists of stearic acid and myristic acid as saturated fatty acids as well as linolenic acid and eicosenoic acid as unsaturated fatty acids. The fatty acid composition of oat lipids shows a balanced ratio of saturated fatty acids to polyunsaturated fatty acids due to the presence of higher oleic acid and linoleic acid content. Besides the major lipids, primarily as neutral triglycerides (50%–85%), oat lipids also consist of phospholipids as lecithins (~2%–5% of the total lipids) and minute amounts of sterols [46].

Even in its native or dormant state, oats exhibit the high-level activity of lipase enzyme. Owing to high activity of this enzyme, high lipid content in the oat groat poses hurdles during processing, such as development of free fatty acids or undesirable flavors due to oxidation. Presence of palmitic acid as a saturated fatty acid enhances the stability of oat lipids against oxidation. Oat lipids may remain stable in sound groats when stored under low moisture conditions. During processing, disruption of bran layers results in the release of lipase enzyme, which in turn produces free fatty acids. Milled or processed oat products tend to have a short shelf life due to the production and subsequent oxidation of fatty acids. The free fatty acids once produced further cause the development of bitter taste and incidence of rancidity. Hence to prevent the occurrence of these unwanted changes in oats and oat-based products, the lipase and other enzymes, such as lipoxygenase or lipoperoxidase, must be inactivated or denatured. Oat stabilization can be achieved through inactivation of these enzymes by heating directly or using steam at 90–100 °C with the moisture content at 12% or above [51].

3.3.4 β-GLUCAN

β-Glucan is a polysaccharide present in oats that plays a unique role as soluble or dietary fiber in human nutrition. It is a polymer having a straight chain of β-D-glucopyranosyl residues, which are linked together by (1→3) and (1→4) bonds in proportions of about 30%–70%, respectively [25]. Majority of the β-glucan is present in the subaleurone and aleurone cell walls of the endosperm, and its content may vary from 2% to 8%, depending on the variety and the environmental factors [11]. Within the endosperm, the cell walls form the seat of most of the β-glucan (~85%), and toward the outer regions of endosperm, that is, the subaleurone cell walls have the highest β-glucan concentration.

Total dietary fiber present in oats can be classified as soluble and insoluble type. β-Glucans, which are water soluble, have lower molecular weight and degree of polymerization that contributes for the major part of the dietary fiber (~55%), whereas the residual fiber contributing as insoluble (~45%) in oats may have a degree of polymerization >100.

Majority of β-glucans are obtained from oats or barley for their commercial production. They are commonly used in food formulations to enhance the functional and physicochemical properties. The physicochemical properties and structure of β-glucan can be different, depending on their source, molecular mass, degree of polymerization, solubility, and method of extraction. In a patented process [4], an enzymatic treatment of oat bran results in a white, flavorless powder with high β-glucan content. Such β-glucan rich sources can be used as a fat substitute to provide a smooth texture similar to the texture obtained using a fat emulsion. Similar products can be formulated using β-glucans for developing low-calorie food products that otherwise may contain rich proportion of fats.

3.3.5 PHENOLIC COMPOUNDS AS ANTIOXIDANTS

The majority of the phytochemicals present in oats are phenolic compounds, such as phenolic acids, flavonoids, avenanthramides, tocopherols, lignans, and others. Phenolics include phenolic acids and flavonoids that are characterized by the presence of one or more hydroxyl (–OH) groups attached to a single or more benzene rings. Phenolic acids are the derivatives of benzoic acid or cinnamic acid that are basically composed of a carboxylic or acid group attached to a phenolic ring. Ferulic acid, though in the bound form, is the most prevalent phenolic acid found in oats and oat products. Ferulic acid proportion in oat products may range from ~230 to ~270 mg/kg, which accounts for 75% of the total phenolic acids content. Apart from ferulic acid, oats also contain other phenolic acids, such as caffeic acid, protocatechuic acid, chlorogenic acid, coumaric acid, hydroxybenzoic acid, syringic acid, sinapic acid, and vanillic acid. These acids are usually found attached through ester linkages to different components of cell-walls, such as lignin, cellulose, or proteins.

Oats also contain flavonoids (~1.16 μmol/g of grain) and exhibit highest free flavonoids content compared to rice, corn, or wheat, which is ~0.45 micromoles of catechin equivalent per gram of grain [1].

Avenanthramides are alkaloid compounds exclusively found in oats and comprise of amide derivative of hydroxycinnamic acids and anthranilic acid. Among different forms of avenanthramides, the most prominent ones are avenanthramide-A, avenanthramide-B, and avenanthramide-C that exhibit strong antioxidant properties. The amounts of avenanthramide –A, –B, and –C in oats may range from 21 to 62 mg/kg, depending on the type of variety and environmental conditions during growth [12].

Lignans are present in pseudocereals, cereals, legumes, fruits, and vegetables. They are richly found in flax seeds. The predominant lignans found in oats are syringaresinol, pinoresinol, and lariciresinol. In oats, the total quantity of lignans may vary from ~800 to 2550 μg/100 g [41], which is higher than wheat, rice, or barley. Tocopherols and tocotrienols together contribute to form vitamin E that is known for its potent antioxidant properties. Tocopherols and tocotrienols exist in various forms (alpha–, beta–, gamma–, and delta–); however, their alpha-forms have the widest distribution and highest antioxidant activity. In oats, majority of vitamin E content was due to its α-tocotrienol content (~56 mg/kg), whereas the rest was contributed by α-tocopherol (~15 mg/kg) [29]. The α-tocopherol $(C_{29}H_{50}O_2)$ is a derivative of 6-hyroxychroman ring, to which a phytol radical is attached at C-2 (Figure 3.1).

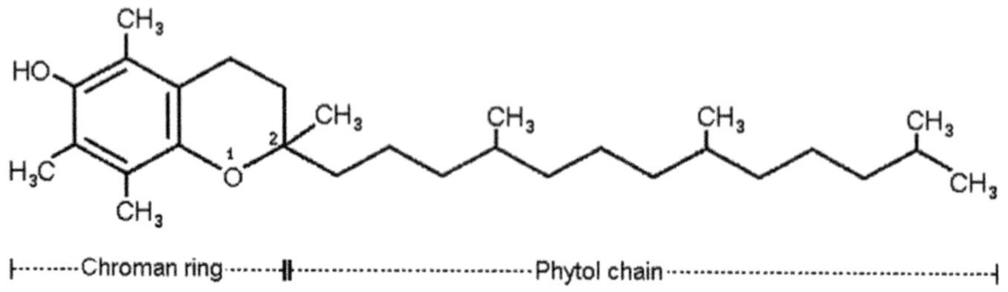

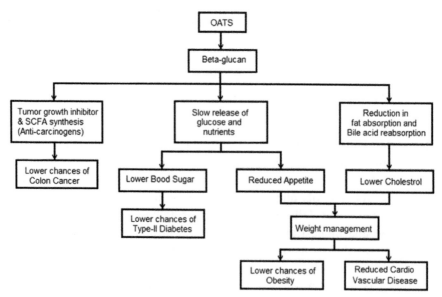

FIGURE 3.1   Chemical structure of α-tocopherol.

3.4   HEALTH BENEFITS OF OAT CEREAL

Oats have gained significant importance over the time compared to wheat, rice, or barley due to several nutritional and health beneficial properties. These are a good source of several nutrients like starch, proteins as a source of essential amino acids, lipids for essential fatty acids, minerals, and vitamins. These also consist of high proportions of several phytochemicals, such as β-glucans, phenolic acid, flavonoids, avenanthramides, lignans, carotenoids, and tocopherols [16] that play a vital role in our physical health (Figure 3.2).

FIGURE 3.2   Health benefits of oats.

Whole grains are rich in phenolic compounds that are well known for their antioxidant activities and contribution in reducing the adverse effects of hypertension, arthrosclerosis, hypercholesterolemia, type 2 diabetes, obesity, and cancer. The valuable physiological and nutritional attributes of oat β-glucan and other dietary fiber components, high in tocopherols and natural antioxidants, have generated an increased demand for oats in human nutrition [54]. For utilizing the beneficial effects, oats can be utilized as whole grains (porridge) or as components of grains (oat bran) to develop novel health products. Oats may also be added in other food products to enhance their antioxidant properties.

Avenanthramides are the compounds found specifically in oats, which exhibit anti-inflammatory properties along with reduction in problems due to atherosclerosis [22]. Combined with ascorbic acid, avenanthramides are known to reduce the oxidation of LDL-cholesterol. Avenanthramides also exhibit anti-allergic properties (antihistamine) that help in reducing burning sensation or itching due to allergic reactions.

3.4.1 CARDIOVASCULAR DISEASES

Consumption of soluble β-glucan in oats facilitates reduction in total and LDL cholesterol, which in turn reduces the chances of cardiovascular diseases [38]. The specific mechanism probably involves hindrance to absorption of dietary cholesterol and reabsorption of bile acids by β-glucan in the intestine by forming a thick viscous mass of fibers [14]. This viscous mass holds the cholesterol and bile acids in it, which are then carried out of the body through feces. The reduced reabsorption of bile acids in the intestine increases their production, which in turn helps to further reduce the levels of cholesterol in our body [3].

3.4.2 GLYCEMIC CONTROL

Consumption of oats and its products helps in sustenance of blood sugar levels, which in turn reduces the chances of type 2 diabetes and glycemia that are affecting a large number of people due to reasons associated with heredity, dietary habits, lifestyle, smoking, and age-related factors. High blood glucose level, if not treated, can have many damaging effects on the vital organs that may lead to kidney failure, retinal damage, high blood pressure, slow healing of wounds, and so on [17]. Consumption of β-glucan slows down intermixing of gastric juices with food particles in stomach, which in turn restricts the breakdown of complex carbohydrates, like starch, to release glucose. β-Glucans form a viscous mass of soluble fibers that slows down the movement and assimilation of digested carbohydrates or sugars in the intestine. Hence, these sugars are released at a much slower rate into the bloodstream that attenuates the response of insulin. Therefore, oat β-glucan can be highly beneficial for controlling glycemic response either through increasing the gastric retention time or through reduced digestion of carbohydrates and absorption of glucose [50].

3.4.3   CELIAC DISEASE

Celiac disease is characterized by an immune reaction that arises due to consumption of gluten, a protein found in wheat, barley, and rye. The intake of wheat and related products results in an immunological response in the small intestine that destroys the absorptive villi and epithelial cells on the internal lining of small intestine [6]. Gluten formation occurs as a result of combination of two insoluble protein fractions, called gliadins and glutenin, which on hydration provide viscoelastic properties to dough. The symptoms may include severe abdominal pain and intestinal problems, such as diarrhea or simply as constipation, bloating, and weight loss due to malnourishment. The patient may also suffer from anemia, chronic fatigue, migraine, vitamin deficiency, depression, and anxiety. A strict life-long gluten-free diet can curb this disease, whereas oats can be consumed in small quantities by the celiac patients as they can be tolerated in moderate amounts [43]. Oats bring variety and nutrition in the diet of people suffering from celiac disease.

3.4.4   ANTICANCER ACTIVITY

Oats consist of biologically active compounds that act as anticarcinogens. Oat β-glucan shows capability of inhibiting tumor growth, although the effectiveness is largely dependent on factors, such as dosage, type and size of the tumor, and genetic background of the subject [36]. Oat β-glucan attenuates the functioning of anaerobic bacteria found in colon region of large intestine and suppresses the formation of cancer-causing secondary bile acids. In addition, β-glucan encourages formation of short-chain fatty acids, which are well-known anticarcinogenic compounds, by colonic anaerobic bacteria and facilitates tumor cell apoptosis [10].

3.4.5   INCREASED SATIETY AND WEIGHT MANAGEMENT

Obese people are considered more vulnerable to diet-related chronic noncommunicable diseases. Food products from oats can play significant role in lowering obesity or weight gain as they provide higher satiety than foods having wheat or rice as ingredients. In comparison to foods containing refined wheat flour, consumption of oats leads to lowered hunger necessities.

The soluble fiber present in oats and oat products probably contributes to the reduction of body weight due to lowering of calorie intake and enhanced satiation. Oat β-glucan slows down the digestion process in stomach and further in the intestine, which leads to lowered glycemic and hunger response. Majority of the studies suggest that oat β-glucan accounts for higher satiety probably through the stimulation of gut hypothalamic axis [32]. Oat β-glucan, when consumed, form a viscous mass of fibers that slow down the action of gastric juices in stomach and hence slow down the digestion process. This continues in the intestine, where absorption of nutrients into the bloodstream is also delayed and ultimately inhibits weight gain in the body.

3.5   OAT CEREAL: FOOD AND NONFOOD APPLICATIONS

3.5.1   FOOD APPLICATIONS

Utilization of oats as a food source varies widely owing to their availability in different forms, for example, whole groats, rolled oats, oat flakes, oat flour, and oat bran. Oats can be added into a wide range of bakery products in the form of flour or bran to produce high fiber oat-based breads, cookies, and cakes (Table 3.1).

The composition of oats is quite distinctive, which makes it an important ingredient for the development of health-based products in the food industry (Table 3.1). Oats usually require minimal processing prior to their consumption or utilization into manufacturing of other food products. In comparison to other cereals, they possess a superior composition profile, which makes them suitable for formulating health foods containing high proportion of proteins especially with higher levels of lysine content. Oats also contain balanced proportion of saturated and unsaturated fatty acids, which include essential fatty acids, antioxidants, vitamins, and minerals. Such nutritional characteristics make them ideal choice for the manufacture of nutritional supplements and nutraceutical food products.

TABLE 3.1 Popular Food Applications of Oat Cereal and Its Components

Oats or Its Components	Processing	Product	Popular Food Applications	Ref.
Oat groat, groat fractions, oat flour	Stabilization and size reduction (for groats) and sieving	Oat bran [≥5.5% of β-glucan (dry basis)]	Hot cereals, RTE cereals, oat-based breads, cookies, and oatcakes, granola bars, muffins, pretzels, and other snack foods	[31, 46]
Oat groat	Stabilization and cutting into ~2 to 4 pieces.	Steel-cut oats.	Oat flour, oatmeal, hot cereals, granola bars, muesli, multi grain breads	[46]
Oat groat	Stabilization, cutting (optional), rolling, and flaking	Rolled oats (0.508–0.762 mm).	Oatmeal (porridge), breakfast hot cereals, granola, muesli, RTE cereals, cookies, and oatcakes, infant foods	[24, 46]
Oat groat, rolled oats	Stabilization (for groats) and size reduction	Whole-oat flour, oat flour	RTE cereals, oat-based breads, cookies and oatcakes, infant foods, oat-based beverages, thickeners in soups and gravies	[37]
Steel-cut oats	Steaming and flaking.	Oat flakes (0.279–0.457 mm)	Quick oats, instant oat flakes, and instant oat-based foods.	[46]
Whole grain of oats	Dehulling	Oat groats	Steel-cut oats, rolled oats, whole-oat flour/oat flour, oat bran	[46]

Note: *RTE*, ready-to-eat.

3.5.1.1 BREAKFAST CEREALS

Breakfast cereals from oats can be of different types depending on the raw materials or cereal used, the type of processing, the form or shape of the end-product, and the cooking time required prior to consumption. Oat groats obtained after dehulling can be processed into different shapes and sizes to produce steel-cut oats or oatmeal, rolled oats, and oat flakes, which are served as hot cereals as they are cooked prior to eating to gelatinize the starch. Since these products have their bran intact and are minimally processed, they form a nutritious source of vitamins, minerals, and phytochemicals in the breakfast foods.

Manufacture of rolled oats involves partial cooking of oat groats by steaming and then followed by flattening in flaking rolls. Flaking reduces the thickness of groat pieces and simultaneously increases the surface area, which increases the rate of hydration and reduces the cooking time significantly. The thickness of flakes may vary from ~0.5 to 0.8 mm depending on the final texture [46]. Thickness of flakes provides a chewy texture and maintains the shape of flakes even after longer period of cooking.

Steel-cut oats are produced after stabilization process for inactivation of enzymes and followed by cutting the groats into 2–4 pieces. They are further used for processing into quick cooking flakes or for producing oat flour. Since the pieces are relatively smaller in size, the flakes produced are thinner than the rolled oats and require lesser time for cooking. Quick-cooking oats or instant oat flakes have much thinner, yet firmer flakes consisting of pregelatinized starch to further shorten the cooking time. They, however, do not retain their shape and provide a smooth texture after cooking. The instant oats may also contain flakes from wheat, rice, or corn to form multigrain cereal for further improving the product variety and consumer acceptability.

3.5.1.2 READY-TO-EAT CEREALS

Oats are also utilized to produce ready-to-eat (RTE) food products that can be consumed readily and do not require any household preparation or cooking. Some popular RTE products manufactured using oats are granola, muesli, and snack foods, which include extruded, flaked, shredded, and puffed products [46]. These products are available in a wide range of variety and contain cereals like corn, wheat, rice, oats, or barley along with a number of other ingredients like flaxseed, nuts, or dried fruits. They all may be blended in different forms, such as flour, bran, flakes, extrusion cooked, shredded or puffed, and may further be coated with sweeteners, like honey, corn syrup, sugars, or malt extract.

Extrusion cooking has gained high popularity due to its versatility and high productivity in producing snack foods and breakfast cereals. Flours or grits oats can be mixed and used for manufacturing extruded RTE products. However, these products may undergo oxidative rancidity if the moisture content is kept below a critical level. Extrusion cooking results in several modifications in physical, chemical, and textural properties along with starch gelatinization and flavor generation in oats [49].

3.5.1.3   BAKERY PRODUCTS FROM OATS

Owing to health benefits of oats and oat products, they are widely used in the manufacture of breads and other bakery products to form an important part of our staple diets. Bread made from wheat flour gets its unique texture primarily due to the availability of monomeric and polymeric proteins, which combine to form gluten. Since oats lack these proteins, they cannot be used exclusively for making a good quality bread.

Among various bakery products, bread production using oats and oat products exclusively would offer a great challenge to the researchers, however if accomplished, it would offer a better alternative to people suffering from celiac disease or gluten-intolerance rather than not to consume wheat-based bakery products. This will not only boost the utilization of oats but will also provide novel and healthy food availability to the consumers.

Incorporation of oats into wheat-bread systems has been carried out by many researchers either to increase the protein content or to enhance the soluble fiber content of the breads. Incorporation of high-protein oat flours (~52% protein) in wheat breads up to 3%–6% of total flour weight resulted in an improved loaf volume of bread [19]. Similarly, oats or oat fractions can be used for the preparation of cookies, oatcakes, pancakes, and pretzels, wherein they contribute to the texture and their unique flavor. Oat flour when added into cookie dough improves water absorption and spreading factor during baking [28].

3.5.1.4   OTHER OAT-BASED FOODS

Apart from utilization in breakfast foods and bakery products, oats and their oat fractions are also used in the manufacture of infant foods; oat-based nondairy milk beverages; ice creams and yogurts; thickeners for soups and gravies; and many specialty flours. Owing to the better nutritional quality and lower allergenicity, oat flour finds its use as one of the major ingredients in infant foods. It also provides viscous properties to infant foods [31].

Innovative processes have been developed to manufacture high moisture foods and nondairy products such as milk, ice cream, and yoghurt using oats. A product similar to yoghurt can be prepared by fermentation of flour slurry from oats and can be utilized by people allergic to products of dairy origin. Such products can also be extremely helpful for patients suffering from milk intolerance or lactose intolerance.

Numerous drinks like oat milk or oat-berry beverages [42] are now available. Stabilization of fats during storage can be achieved in milk and meat products due to the natural antioxidants and soluble fiber present in oat flour.

Gums obtained from β-glucan in oats can be used to stabilize frozen desserts and ice cream, whereas proteins from oats find a wide usage in food products for their fat emulsification properties. Due to the presence of higher proportion of lipids, oat starch exhibits stability toward retrogradation and finds its usage as a thickening agent for sauces, dressings, and gravies. Higher lipids also favor its usage as fat replacers in ice creams and frozen desserts, whereas oat starch can be used as a fat substitute in cheese making [52].

3.5.2   OATS AS ANIMAL FEED

Utilization of oats as an animal feed has been carried out since ages, however recent developments in the usage of feed grains from noncereal sources like soybean or canola has impacted the overall production of oats for the feed purpose. In the United States alone, usage of oats as animal fodder has declined from 120 million bushels in 2007 to about 82 million bushels in 2016 [2]. Compared to corn, oats still remain a grain of choice for feeding racehorses due to their suitability in terms of balanced nutritive value and better digestibility of starch and proteins. The lipids present in oats form a vital component of the horse feeds as they contribute to the total energy content of the feed.

Majority of the dairy farmers in the United States prefer corn grains for feeding the dairy cattle; however, a comparative study between oats (regular and high protein oats) and corn grain revealed that the fat-corrected milk yields from the animals were almost the same and the high protein oats could easily substitute soybean meal supplementation in their rations [35]. Oat feeds are highly recommended for the beef cattle during the growth period, but due to their lower calorific values, they are not as efficient as other feeds for the finishing cattle [9].

Oats are usually rolled or ground for effective feeding purpose especially for younger animals. Both covered as well as naked type of oats are used in the animal feed industry. Covered oats contain fibrous hull, which forms a suitable component in feeds for ruminants; however, it dilutes the nutritive fractions in the feed. Hence, oat dust or hulls obtained during oat milling can be added as diluents in the feeds containing corn or other cereals.

The naked or dehulled oats have higher protein and starch but lower fiber content compared to covered oats; and therefore, can partially replace the feeds prepared from corn or wheat. In an experimental diet for egg laying birds, naked oats were used to partially replace (up to 60%) both corn and soybean meal, without affecting the egg yield or performance of the birds [7]. Naked oats are also suitable for feeding of swine, poultry birds, and for pets like dogs and cats as they improve the fur sheen, prevent allergies, and reduce diarrhea.

3.5.3   NONFOOD APPLICATIONS OF OAT CEREAL

Hull obtained from oat grains finds number of applications for the manufacture of various nonfood products. Oat hull contains high amounts of pentosans, which finds its usage in the production of furfural. Furfural is further utilized as a raw material for the manufacture of different types of polymers, solvents, and many other useful chemicals. Furfural is also used for producing furan derivatives, which form as an ingredient of several adhesives, composites, and coatings [5]. As a by-product of oat milling industry, oat hull was the only material used initially for furfural production; however, with the increasing demand, other sources such as rice husk, bagasse, or corncob also became useful.

Other uses of hull include as cleanup material for oil-spills, as a biomass fuel for the replacement of coal and as a filter-aid in breweries during mashing process. During wort extraction, oat hull helps to form a porous bed in the mashing "tun" for better filtration of wort. The abrasive properties of oat hull are exploited by using it as an abrasive in air blasting to remove old paint, corrosion, rust, or oil from any hard surface.

Oats in the form oatmeal and oat flour are also used for the manufacture of cosmetic products. They have been used since long for relieving skin irritations or rashes as soap replacement or as facial masks. They are commonly used in cosmetics, soaps, and antiaging creams for their hypoallergic properties.

Usage of oatmeal for treatment of skin irritations due to allergies or insect bites has also been acknowledged. The bioactive components and antioxidants present in oats are utilized for manufacturing skin-care products that provide anti-inflammatory, antiallergic, and anti-itching properties. The oil extracted from oats also forms an important component of skin-care products. It contains rich proportions of glycolipids and phospholipids, which impart moisturizing and skin hydrating properties to the products. Similarly, concentrates of oat β-glucan are also used in the manufacture of several skin-care soaps, shampoos, and shaving products [46].

3.6   FUTURE ASPECTS

High-calorie diet intake and improper living styles are contributing largely to the rise in rate of obesity, metabolic syndrome, and cardiovascular diseases. Prevention of these harmful diseases through beneficial diet plans and modifications in the lifestyles will play an important role to promote public health. For better management of disease risk factors and for promoting overall health, the consumer preferences are likely to shift toward functional foods and dietary solutions.

Oat consumption has increased over the years due to its health benefits. Oats offer many opportunities for future functional food development as they deliver the benefits of macro- and micronutrients, along with fiber and many bioactive components. The bioactive components present in oats are largely responsible for factors that contribute to the health benefits of oats and help in the prevention of many life-threatening diseases. Besides these, oats also help in controlling calorie intake by slowing down the digestion and assimilation process, which helps in weight management and prevention of obesity. Hence, developments in the varieties of oats with beneficial health factors like phytochemicals, soluble fiber, and antioxidants will enhance the uses of oats and perhaps may open new potential uses.

Active involvement from the consumers and the production houses involved in marketing of oat-based products will push the needs for development and introduction of more health beneficial oat products that are far tastier and attractive to look than some of the high-calorie foods available in the food chain restaurants or café; as it will help largely in improving the overall health of our younger generation. These products may counter the increasing menace of obesity or overweight problems in the population all over the world.

Greater utilization will further boost developments in the cultivation techniques for oats and will further improve the production in agriculture sector. A shift has been noticed in the utilization of oats from animal feed sector toward human food consumption and this shift is likely to increase in the coming future due to increased awareness among the consumers for healthier foods. With the increasing number of people that are allergic to wheat or those suffering from gluten intolerance or celiac disease, oat-based products can offer better dietary choices and thus may even pose a challenge to the production of wheat-based products in future.

3.7 SUMMARY

Like all other cereals, starch forms the largest component in oats with amylose and amylopectin as the major constituents of oat starch. The granules of oat starch are similar in size to that of rice but quite smaller than the starch granules of other cereals like wheat or barley. As compared to other cereals, the proportion and the lysine content of proteins are quite larger in oat groats. Lipid content in the oat grain ranges from 2% to 11%, depending on the variety. Oats contain several phenolic compounds, β-glucan, avenanthramides, and tocopherols that contribute to numerous health benefits of oats and help in prevention of diabetes, cardiac and gastrointestinal complications, and cancers. Food uses for oats include breakfast cereals, breads, biscuits, and instant foods prepared using oat bran, oatmeal, oat flour, or oat flakes.

KEYWORDS

- *Avena*
- cereals
- composition
- fiber
- health benefits
- *β*-glucan

REFERENCES

1. Adom, K.K.; Liu, R.H. Antioxidant Activity of Grains. *Journal of Agricultural and Food Chemistry*, **2002**, *50*, 6182–6187.
2. Agricultural Statistics; National Agricultural Statistics Service (NASS), USDA; Washington, D.C.: U.S. Government Printing Office; **2017**; online; https://www.nass.usda.gov/; Accessed on December 31, 2019.
3. Bae, I.Y.; Kim, S.M.; Lee, S.; Lee, H. G. Effect of Enzymatic Hydrolysis on Cholesterol-Lowering Activity of Oat β-Glucan. *New Biotechnology*, **2010**, *27*, 85–88.
4. Bishop, J. Oat Extract Called New Fat Fighter. *Wall Street Journal*, **1990**, *85* (80: April), B1–B10.
5. Brydson, J.A. Furan Resins. In: *Plastics Materials*; 7th edition; Brydson, J.A. (Ed.); Oxford: Butterworth-Heinemann; **1999**; pp. 810–813.
6. Catassi, C.; Fasano, A. Celiac Disease. In: *Gluten-Free Cereal Products and Beverages*; Arendt, E. K. and Dal Bello, F. (Eds.); New York, USA: Academic Press; **2008**; pp. 1–22.
7. Cave, N.A.; Hamilton, R.M.G.; Burrows, V.D. Evaluation of Naked Oats (*Avena Nuda*) as a Feedstuff for Laying Hens. *Canadian Journal of Animal Science*, **1989**, *69*, 789–799.
8. Coffman, F. A. *Oat History, Identification and Classification*; Washington, D.C.: US Government Printing Press; **1977**; p. 365; online; https://oatnews.org/oatnews_pdfs/2018etc/Oat_History_Identification_and_Classific.pdf; Accessed December 31, 2019.
9. Comerford, J. *Feeding Small Grains to Beef Cattle*, **2017**; Penn State University Extension; http://extension.psu.edu/animals/beef/nutrition/articles/feeding-smallgrains-to-beef-cattle; Accessed July 19, 2019.
10. Daou, C.; Zhang, H. Oat Beta-Glucan: Its Role in Health Promotion and Prevention of Diseases. *Comprehensive Reviews in Food Science and Food Safety*, **2012**, *11*, 355–365.
11. Decker, E.A.; Rose, D.J.; Stewart, D. Processing of Oats and the Impact of Processing Operations on Nutrition and Health Benefits. *British Journal of Nutrition*, **2014**, *112*, S58–S64.
12. Dimberg, L.H.; Molteberg, E.L.; Solheim, R.; Frølich, W. Variation in Oat Groats Due to Variety, Storage and Heat Treatment. I: Phenolic Compounds. *Journal of Cereal Science*, **1996**, *24*, 263–272.
13. Draper, S.R. Amino Acid Profiles of Chemical and Anatomical Fractions of Oat Grains. *Journal of Science Food and Agriculture*, **1973**, *24*, 1241–1250.
14. Erkkila, A.T.; Lichtenstein, A.H. Fiber and Cardiovascular Disease Risk: How Strong Is the Evidence? *Journal of Cardiovascular Nursing*, **2006**, *21*, 3–8.
15. FAOSTAT (Food and Agriculture Organization of the United Nations), Rome, Italy. http://www.fao.org/faostat/en/#data/QC; Accessed June 16, 2019.
16. Gangopadhyay, N.; Hossain, M.B.; Rai, D.K.; Brunton, N.P. Review of Extraction and Analysis of Bioactives in Oat and Barley and Scope For Use of Novel Food Processing Technologies. *Molecules*, **2015**, *20*, 10884–10909.
17. He, L.X.; Zhao, J.; Huang, Y.S.; Li, Y. The Difference between Oats and Beta-Glucan Extract Intake in the Management of Hba1c, Fasting Glucose and Insulin Sensitivity: A Meta-Analysis of Randomized Controlled Trials. *Food & Function*, **2016**, *7*, 1413–1418.
18. Kent, N.L.; Evers, A.D. *Kent's Technology of Cereals.* 4th edition; Oxford: Pergamon; **1994**; pp. 73–75.
19. Lapvetelainen, A. Barley and Oat Protein Products From Wet Processes: Food Use Potential. PhD Dissertation; Finland: University of Turku, **1994**; p. 229.

20. Lasztity, R. Oat Grain-A Wonderful Reservoir of Natural Nutrients and Biologically Active Substances. *Food Reviews International*, **1998**, *14*, 99–119.

21. Lippi, M. M.; Foggi, B.; Aranguren, B.; Ronchitelli, A.; Revedin, A. Multistep Food Plant Processing at Grotta Paglicci (Southern Italy) Around 32,600 Cal B.P. *Proceedings of the National Academy of Sciences*, **2015**, *112*, 12075–12080.

22. Liu, L. P.; Zubik, L.; Collins, F.W.; Marko, M.; Meydani, M. The Antiatherogenic Potential of Oat Phenolic Compounds. *Atherosclerosis*, **2004**, *175*, 39–49.

23. Makela, M. J.; Laakso, S. Studies on Oat Starch with a Celloscope-Granule Size and Distribution. *Starch*, **1984**, *36*, 159–163.

24. Mckechnie, R. Oat Products in Bakery Foods. *Cereal Foods World*, **1983**, *28*, 635–637.

25. Menon, R.; Gonzalez, T.; Ferruzzi, M.; Jackson, E.; Winderl, D.; Watson, J. Oats-From Farm to Fork. *Advances in Food Nutrition Research*, **2016**, *77*, 1–55.

26. Morrison, W.R.; Milligan, T.P.; Azudin, M.N. A Relationship between the Amylose and Lipid Contents of Starches from Diploid Cereals. *Journal of Cereal Science*, **1984**, *2*, 257–271.

27. Murphy, J.P.; Hoffman, L.A. *The Origin, History, and Production of Oat*. In: *Oat Science and Technology*; Sorrells, M. E. and Marshall, H. G. (Eds.); Madison, WI: American Society of Agronomy and Crop Science Society of America; **1992**; pp. 1–28.

28. Oomah, B. D. Baking and Related Properties of Wheat-Oat Composite Flours. *Cereal Chemistry*, **1983**, *60*, 220–225.

29. Panfili, G.; Fratianni, A.; Irano, M. Normal Phase High-Performance Liquid Chromatography Method for the Determination of Tocopherols and Tocotrienols in Cereals. *Journal of Agricultural and Food Chemistry*, **2003**, *51*, 3940–3944.

30. Peterson, D.M.; Brinegar A.C. *Oat Storage Proteins*. In: *Oats: Chemistry and Technology*; F. H. Webster and Wood, P. J. (Eds.); St Paul, MN: American Association of Cereal Chemists; **2011**; pp. 123–139.

31. Ranhotra, G. S.; Gelroth, J. A. Food Uses of Oats. In: *The Oat Crop*; Welch, R. W. (Ed.); Chapman and Hall, London; **1995**; pp. 409–432.

32. Rebello, C.J.; O'Neil, C.E.; Greenway, F.L. Dietary Fiber and Satiety: The Effects of Oats on Satiety. *Nutrition Review*, **2016**, *74*, 131–147.

33. Sampson, D. R. *The Origin of Oats*. In: *Botanical Museum Leaflets*; Boston, MA: Harvard University; **1954**; volume *16*; pp. 265–303.

34. Sayar, S.; White, P.J. Oat Starch: Physicochemical Properties and Function. In: *Oats: Chemistry and Technology*; 2nd edition; Webster, F. H. (Ed.); St Paul, MN: American Association of Cereal Chemistry Inc.; **2011**; pp. 110–111.

35. Schingoethe, D.J.; Voelker, H.H.; Ludens, F. C. High Protein Oats Grain for Lactating Dairy Cows and Growing Calves. *Journal of Animal Science*, **1982**, *55*, 1200–1205.

36. Shen, R.L.; Wang, Z.; Dong, J.L.; Xiang, Q.S.; Liu, Y.Q. Effects of Oat Soluble and Insoluble B-Glucan on 1,2-Dimethylhydrazine Induced Early Colon Carcinogenesis in Mice. *Food & Agricultural Immunology*, **2016**, *27*, 657–666.

37. Shukla, T. P. Chemistry of Oats: Protein Foods and Other Industrial Products. *Critical Review of Food Science & Nutrition*, **1975**, *6*, 383–446.

38. Sing, C. F.; Stengard, J. H.; Kardia, S. L. R. Genes, Environment, and Cardiovascular Disease. *Arteriosclerosis, Thrombosis, and Vascular Biology*, **2003**, *23*, 1190–1196.

39. Singh, N.; Singh, J.; Kaur, L.; Sodhi, N. S.; Gill; B. S. Morphological, Thermal and Rheological Properties of Starches from Different Botanical Sources Review. *Food Chemistry*, **2003**, *81*, 219–231.

40. Singh, S.; Singh, N.; Isono, N.; Noda, T. Relationship of Granule Size Distribution and Amylopectin Structure with Pasting, Thermal, and Retrogradation Properties in Wheat Starch. *Journal of Agricultural and Food Chemistry*, **2010**, *58*, 1180–1188.

41. Smeds, A.I.; Jauhiainen, L.; Tuomola, E.; Peltonen-Sainio, P. Characterization of Variation in the Lignan Content and Composition of Winter Rye, Spring Wheat, and Spring Oat. *Journal of Agricultural and Food Chemistry*, **2009**, *57*, 5837–5842.

42. Sontag-strohm, T.; Lehtinen, P.; Kaukovirta-norja, A. Oat Products and Their Current Status in the Celiac Diet. In: *Gluten-Free Cereal Products and Beverages*; Elke, K. A. and Fabio Dal, B. (Eds.); San Diego, CA: Academic Press; **2008**; pp. 191–199.

43. Storsrud, S.; Olsson, M.; Arvidsson Lenner, R. Adult Coeliac Patients Do Tolerate Large Amounts of Oats. *European Journal of Clinical Nutrition,* **2003**, *57*, 163–169.

44. Valentine, J. *The Oat Crop: Production and Utilization*; London: Chapman and Hall; **1995**; pp. 168–172.

45. Valentine, J.; Cowan, A. A.; Marshall, A. H. *Oat Breeding.* In: *Oats: Chemistry and Technology*; 2nd edition; Webster, F. H. and Wood, P. J. (Eds.); St Paul, MN: AACC International; **2011**; pp. 11–30.

46. Webster, F. H. *Oat Utilization: Past, Present, and Future.* In: *Oats: Chemistry and Technology*; 1st edition; Webster, F. H. (Ed.); St Paul, MN: American Association of Cereal Chemistry International; **1986**; pp. 413–426.

47. Webster, F. H. Oats. In: *Cereal Grain Quality*; Henry, R and Kettlewell, P. (Eds.); London: Chapman and Hall; **1996**; pp. 179–203.

48. Welch, R. W. (Ed.) *The Oat Crop: Production and Utilization*; Chapman & Hall, London; **1995**; pp. 516–518.

49. Yao, N; Jannink, J.L.; Alavi, S.; White, P.J. Physical and Sensory Characteristics of Extruded Products Made from Two Oat Lines with Different β-Glucan Concentrations. *Cereal Chemistry*, **2006**, *83(6)*, 692–699.

50. Zhang, Y.; Zhang, H.; Wang, L.; Qian, H.; Qi, X.; Ding, X.; Hu, B.; Li, J. The Effect of Oat B-Glucan on In Vitro Glucose Diffusion and Glucose Transport in Rat Small Intestine. *Journal of Science of Food and Agriculture*, **2016**, *96*, 484–491.

51. Zhou, M. X.; Robards, K.; Glennie-Holmes, M.; Helliwell, S. Oat Lipids. *Journal of American Oil Chemistry Science*, **1999**, *79*, 585–592.

52. Zhu, F. Structures, Properties, Modifications, and Uses of Oat Starch. *Food Chemistry*, **2017**, *229*, 329–340.

53. Zwer, P. K. Oats. In: *Encyclopedia of Grain Science*; Wrigley, C., Corke, H., and Walker, C. (Eds.); Oxford: Elsevier; **2004**; pp. 153–158.

54. Zwer, P. K. Oats: Overview. In: *Encyclopedia of food grains.* 2nd edition; Wrigley, C., Corke, H., Seetharaman, K., and Faubion, J. (Eds.); Oxford: Elsevier, **2016**; pp. 173–183.

CHAPTER 4

MAIZE: A POTENTIAL GRAIN FOR FUNCTIONAL AND NUTRITIONAL PROPERTIES

RAMANDEEP KAUR, AKANKSHA PAHWA, and SURESH BHISE

ABSTRACT

The rich nutritional profile of maize makes it unique among other cereal grains. It has a number of health benefits, such as high levels of antioxidants, dietary fiber, lipids with unsaturated fatty acid, antioxidants, vitamins, and minerals. The chemical characteristics of maize kernels are affected by variety and age of maize plant, heredity conditions, surrounding conditions, geographic location, and others. The maize protein has numerous functional properties, which make it suitable to food processors and for human nutrition. This chapter presents detailed information about the bioactive and nutritional properties of maize.

4.1 INTRODUCTION

Maize (*Zea mays* L.) is mainly used as food, feed, and raw material for various industrial purposes. It is consumed as staple diet by people in Asian and African countries [19]. It is the only grain which is 100% utilized after wet milling process (Figure 4.1). The United States shares about 35% of the total maize production holding first position in production of maize. Bihar, Haryana, Rajasthan, Uttar Pradesh, Madhya Pradesh, Maharashtra, Jammu, Himachal Pradesh, Punjab, Karnataka, West Bengal, and Kashmir are contributing more than 95% of maize production in India [21, 34]. Functional foods are rich source of physiologically-active food components with health-promoting role. Maize kernels can be consumed in the form of maize cob, roasted, fried, boiled, dried, and fermented, which is further utilized in the production of breads, gruel, porridges, cakes, and alcoholic drinks. Starch derived from maize is used as a food thickener, sweetener, oil, and produce nonedible products [6].

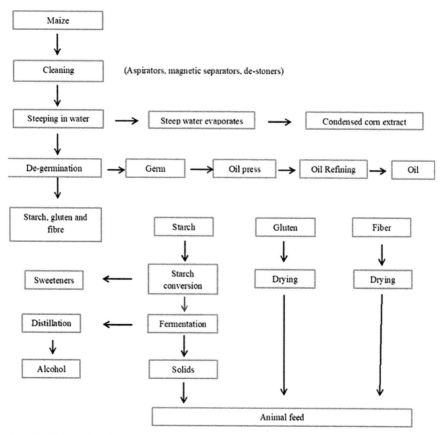

FIGURE 4.1 Wet milling of maize.

Maize has many health-promoting benefits due to the presence of large amounts of phytochemicals having antioxidant activity (AO). The risk of chronic diseases is reduced due to consumption of different varieties of maize providing optimum nutritional benefits. Bran and germ fractions of maize are significant source of bioactive compounds, such as phenolic acids and antioxidants. Besides these, different varieties of maize are rich sources of vitamins, dietary fiber, minerals, flavonoids, plant sterols, and other phytochemicals.

The various scientific evidences and studies have concluded that more and regular consumption of maize is beneficial to well-being and can decrease the risk of cardiovascular diseases (CVDs), cancer, obesity, type II diabetes, and improve digestive health. Pigmented corn, sweet corn, high amylose corn, and quality protein maize are different types of maize. Different value-added products are prepared by incorporating maize, such as cornmeal, corn starch, corn grits, corn flour, snacks, and breakfast cereals, *chapattis*, or flatbreads because of its nutritional and health-promoting effects.

This chapter discusses various types of maize, their chemical composition, bioactive compounds, functional properties, and health benefits.

4.2 HIGH AMYLOSE CORN

Maize with high amylose content makes maize starch as type 2 resistant starches. Resistant starch (RS) is a portion of starch not broken down in the small intestine by pancreatic α-amylase due to its passage as raw to the large intestine, where it acts as a substrate for microbial fermentation [27]. In the small intestine, incomplete digestion and absorption of starch is a normal phenomenon, and such nondigestible starch fractions are known as "resistant starches." Physiological functions of resistant starches are similar to dietary fiber. Chemically, starches are homopolysaccharides, which consist of a number of monosaccharides. On the basis of the nature of bond present, they are classified as amylose and amylopectin. The amylose consists of glucose monomer linked to each other by α-(1-4) bond while the amylopectin is a highly branched molecule having α-(1-4) and β-(1-6) linkages and is a major constituent of starch.

Starches could be classified as slowly digestible starch, rapidly digestible starch, and RS based on the behavior under the action by enzymes [21]. The rapidly and slowly digestible come under digestible starch, whereas the resistant starch is of great importance due to its many physiological health benefits and it remains indigestible by human body enzymes. Resistant starch (classified as RS_1, RS_2, RS_3, and RS_4) is physically protected in the whole cereal grain. RS_2 is packed in a compact radial pattern, which confines the convenience of digestive enzymes and therefore reveals the resistant nature of RS_2 (ungelatinized starch). RS_3 is categorized as retrograded starch and RS_4 has structure of starches found by chemical treatment of starches, which are referred as modified starches. High-amylose maize has various physiological beneficial effects on blood sugar, colon health, as it is a rich source of RS2 type resistant starch, which is not digested in the human intestine.

4.3 PIGMENTED MAIZE

The colorful maize is uncommon, whereas white maize and yellow maize are most common. All parts of maize or corn (such as husk, silk, kernel, cob, and leaves of maize) have been utilized by consumers. The pigmented maize comes in several colors (such as blue, red, or purple-colored maize), which are same species botanically as white and yellow maize. This maize is utilized in the development of colorful foods and beverages. The importance of pigmented maize is because of its bioactive compounds (such as anthocyanins, flavonoids, p-hydroxycinnamic acids, carotenoids, lignans, vitamin E, policosanols, phytosterols, and xylans).

Anthocyanins, which are present in all parts of pigmented maize, are most commonly present in pericarp and aleurone layers. Pericarp layer can be transparent or red, orange-brown in color, whereas aleurone layer can be transparent, red, or purple. Recently, various studies have focused on anthocyanin content of pigmented maize. One study showed that Mexican maize contains significant amounts of anthocyanins. The major anthocyanins present in the maize are cyanidin-3-glucoside (C3G) and cyanidin-3-(6"-malonyl) glucoside. In the red maize, peonidin-3-glucoside (P3G) and pelargonidin-3-glucoside and their derivatives have major variability, whereas blue corn does not contain pelargonidin-3-glucoside and peonidin-3-glucoside. Moreover, blue corn has more cyanidin-3-(6"-malonyl) glucoside than the purple corn [29].

The risk of chronic diseases (such as CVDs, cancer, and obesity) can be reduced by consumption of anthocyanins rich foods [9, 29, 30]. The concentration and composition of phytochemicals can affect the color of maize kernel. Moreover, high amount of antioxidants is present in pigmented corn than the nonpigmented types. The waxy corn is harvested and consumed at immature stage, as it is rich in amylopectin. In recent times, much research has been carried out in production of different color of maize kernels (such as white, purple, red, black, yellow, etc.) to enhance the functional properties of corn. The compounds responsible for these pigments are mainly anthocyanins, which mainly include cyanidin-3-glucoside, cyanidin-3-(6"-malonyl glucoside), cyanidin-3-(3",6"-dimalonyl glucoside), pelargonidin-3-glucoside, and peonidin-3-glucoside.

4.4 SWEET CORN

Sweet corn stands sixth in the United States among all vegetables. It is consumed in bakery products, canned sweet corn, tortillas, and snack foods. It is source of many vitamins [thiamine (B1), riboflavin (B2), niacin (B3), pyridoxine (B5), folate (B9), vitamin-E, vitamin-K and vitamin-C], carotenoids, minerals (Ca, Mg, Ph, K, Na, and Zn), and resistant starches. Moreover, sweet corn consists of 30% monosaturated and 50% polyunsaturated and a smaller portion of saturated lipids (20%) [17].

The particular sweet taste, unique carotenoid, and tocopherol profiles are some of the important properties of sweet corn compared to the common maize. The primary carotenoids in sweet corn are lutein and zeaxanthin [lutein: 0–27.6 μg/g dw (dry weight), zeaxanthin: 0–7.7 μg/g)] and primary tocopherol is γ-tocopherol (2.4 to 63.3 μg γ-tocopherol/g dw) [17].

The thermal processing of sweet corn increases the AO due to the release of bound form phytochemicals [4]. Dewanto et al. [4] found that total AO in sweet corn was 44% and phytochemicals (such as ferulic acid) were increased by 550%, total phenolics content was increased by 54%, and AO was increased to 210 mg of. vitamin C/100 g.of corn under thermal treatment of sweet corn at 115 °C for 25 min.

4.5 QUALITY PROTEIN MAIZE

The amino acid profiles of each corn protein fractions are exceptional. The overall quality of protein in maize is represented by zein (prolamin) protein, which is 50% of the total nitrogen in maize kernel. The biological value of maize protein is deprived as zein protein, which is deficient in essential amino acids (EAA), such as tryptophan and lysine. Maize is poor source of high-quality protein, though germ of maize kernel consists of these two EAA. The total protein present in maize kernels ranges from 7 to 11 g of protein/100 g. To improve the quality of protein in maize, many breeding programs began in the mid-1960s. The opaque-2 (o-2) is a mutant to increase the level of lysine and tryptophan in maize protein, which makes it more nutritious compared to common maize. The level of amino acids (such as histidine, arginine, aspartic acid, and glycine) is increased while the level of glutamic acid, alanine, and leucine was decreased [8, 20]. Lower quantity of leucine is desirable to make leucine–isoleucine ratio advantageous aiding to release more tryptophan, which is used for biosynthesis of niacin helping to fight against pellagra [33].

4.6 GROSS CHEMICAL COMPOSITION

The chemical composition of maize kernels is affected by variety, genetic background, plant age, geographic location, and environmental conditions.

Cortez et al. [3] studied the chemical composition of different types of maize with highest carbohydrate content in Salpor and black maize (75.90%) compared to pop type of maize (66%). Pop maize had maximum amount of protein (13.7%) compared to Black maize (5.7%) and ash content in all maize varieties ranges from 1.2% to 2.9%, moisture content varies from 9.5% to 12.3% and ether extract was 5.7% in pop maize and 3.9% in sweet maize.

4.6.1 STARCH

Maize kernel consists of 72%–73% of starch on a percentage weight basis. Level of simple sugars (such as glucose, sucrose, and fructose) ranges from 1% to 3% of the kernel weight. The linear amylose and a branched-chain amylopectin polymer are present in maize starch. The endosperm in the maize present either in dent or flint form. The amylose and amylopectin each contribute up to 25%–30% and 70%–75% of the starch, respectively. The waxy maize contains 100% of amylopectin of total starch. An amylose extender is an endosperm mutant that could increase the portion of amylose up to 50% and higher of the total starch. The genetic alteration can change the amylose-to-amylopectin ratio in the maize starch [14].

4.6.2 PROTEIN

The protein content varies from 8% to 11% and most of it present in endosperm of maize. Maize protein has five different protein fractions [33, 34]. About 18% of total nitrogen in maize protein is present in the form of albumins (7%), globulins (5%), and nonprotein nitrogen (6%). The prolamine is soluble in 55% solution of isopropanol and isopropanol with mercaptoethanol, which donates 52% of the nitrogen in the kernel. The concentration of prolamine-1 or zein-1 is about 42%, which is soluble in 55% isopropanol, whereas prolamine-2 or zein-2 contributes only 10%.

Lysine and tryptophan are absent in maize proteins. In a study conducted on children, the essential amino acid contents of lime-treated maize supplemented with 5% maize gluten to increase the protein content [35] were compared with the reference protein amino acid. It was observed that the addition of 148 mg of DL-tryptophan per gram of nitrogen was enhanced by the concurrent adding of tryptophan and lysine (243 mg/g nitrogen). It was found that the addition of methionine decreased the nitrogen retention. The efficiency of protein utilization of maize was affected due to the excess presence of certain essential amino acids.

The average of protein digestibility of o-2 maize protein was 92% (ranges from 67% to 106%) compared to 96% of egg protein (ranges from 78% to 103%). The average biological value for o-2 maize and egg was 80% and 96%, respectively.

4.6.3   OIL AND FATTY ACIDS

The oil content of maize germ varies from 3% to 18%, which can be affected by genetic makeup and variety of maize. Maize oil contains 11% of palmitic acid and 2% of stearic acid, that is, saturated fatty acids. The high level of polyunsaturated fatty acids (such as α-linolenic and arachidonic acids) were present in maize oil. The small amounts of 0.7% of linoleic acid and excellent levels of natural antioxidants make the maize more stable during storage. The presence of oleic and linoleic acids makes the maize oil valuable.

4.6.4   DIETARY FIBER

Maize is a rich source of dietary fiber. The complex polysaccharides are found in the maize kernel mainly present in pericarp and tip cap, followed by endosperm cell walls to a lesser extent from cell-walls of germ. Maize bran consists of 75% hemicellulose, 25% cellulose, and 0.1 % lignin on dry weight basis [25]. The whole maize kernels had more amount of dietary fiber compared to that of dehulled maize.

4.6.5   OTHER CARBOHYDRATES

Small amount of carbohydrates is present in mature maize. Maize kernel contains 1% and 3% of total sugars. The major sugar in maize germ is sucrose. Mature kernels of maize are rich source of monosaccharides, disaccharides, and trisaccharides. As the maturity of maize kernel progresses, the level of sugars is declined with increasing level of starch. High level of reducing sugar occurs in maize at immature stage, therefore sweet maize was more liked by consumers.

4.6.6   MINERALS

The ash content of maize kernel is about 1.3%. The crude fiber content is slightly higher than ash content of maize kernel. The germ constitutes 78% of minerals present in whole kernel. The phosphorus is present as phytate of potassium and magnesium in maize. Mainly, phosphorus is present in the embryo of maize. Generally, maize contains 0.90% of phosphorus compared to 0.92% in o-2 maize. Phosphorus (P), magnesium (Mg), and potassium (K) are most predominant minerals that are present in maize, contributing to 85% of kernel mineral content. Sulfur is available in an organic form as a part of cysteine and methionine. Low level of calcium and iron are present in maize. The presence of phytate in germ decreases the bioavailability of calcium and iron. Manganese, copper, selenium, and iodine are present as trace minerals [37].

4.6.7 VITAMINS

Pro-vitamin A (carotenoids) and vitamin-E are rich fat-soluble vitamins present in maize kernel. Genetically modified yellow maize has more carotenoids compared with white maize with negligible carotenoid content. Endosperm of the kernel is rich source of carotenoids, whereas little amount present in germ. Yellow maize has 22% of beta-carotene of total carotenoids (6.4–11.3 µg/g) [31], whereas 51% of cryptoxanthin of total carotenoids are present.

The vitamin-A in yellow maize ranges from 1.5 to 2.6 µg/g. The germ of maize is a good source of vitamin-E. The biologically active form of vitamin-E in maize is alpha-tocopherol out of four forms of vitamin-E in maize. The AO of gamma-tocopherol is high compared to alpha-tocopherol.

Aleurone layer of maize kernel is a rich source of water-soluble vitamins, followed by germ and endosperm. The high consumption of maize causes pellagra in populations due to deficiency of nicotinic acid/niacin vitamin [20]. The average content of niacin in different varieties is 20 µg/g. The availability of niacin to human was increased by processing method, such as hydrolysis [34]. Maize is deficient in vitamin-B_{12}. Pyridoxine content of maize is 2.69 mg/kg. Folic acid (choline) and pantothenic acid are present in very low concentrations in maize.

4.6.8 BIOACTIVE COMPONENTS

Bioactive compounds are naturally found in maize that deliver several human health benefits to decrease the risk of main chronic diseases. Oxidative stress results in many health problems, such as cancer, obesity, and CVDs. Oxidative damage of cells is prevented by consumption of cereals containing phytochemicals having antioxidant properties. Maize is a rich source of antioxidants [such as vitamin E (tocols), beta-cryptoxanthin], which can prevent lung cancer, while lutein can prevent age-related vision loss. The horny and floury endosperm of yellow maize grains is rich source of carotenoid pigments [37]. The pigments in maize are classified as carotenes and xanthophylls (lutein and zeaxanthin). The major phenolic compounds in maize bran are ferulic acid and anthocyanins. The amount of anthocyanin content decides the color of maize kernels. The anthocyanin content of blue maize is 62.7 mg/100 g while for purplish-red maize it varies from 8.7 to 61.0 mg/100 g of flour. Corn germ oil is rich in β-sitosterol (62%–69%), campesterol (11%–18%), and stigmasterol (5%–13%) [21]. Table 4.1 summarizes the bioactive compounds in various types of maize and their functional properties.

TABLE 4.1 Major Bioactive Components in Different Types of Maize and Their Functional Properties

Type of Maize	Bioactive Components	Functional Properties	Ref.
High amylose maize	Resistant starch 2	Prevention of colon cancer; Improvement in colon health; As a prebiotic.	[27]
		Reduction in gall stone formation.	[21]
		Hypocholesterolaemic effects; Mineral absorption; Inhibition of fat accumulation.	[14]
Pigmented corn	Anthocyanins β-carotene	Cardio protection; Reduction in the risk of CVDs.	[30]
		Prevent weight gain and obesity.	[29]
		Retard diabetic and nephropathy.	[31]
		Antitumor agents.	[9]
		Xerophthalmia and night blindness.	[5]
Sweet corn	Thermal treatment of sweet corn at 115 °C for 25 min enhances antioxidant activity.	Reduced risk of colon cancer; Protection against colorectal and gastric cancers;	[4]
	High antioxidant capacity due to carotenoid and tocopherol, ferulic acid, total phenolic content.	Reduce the risk of developing chronic diseases, including CVD, type-II diabetics, overweight and obesity and digestive problems.	[16]
Quality protein maize	High level of essential amino acid (like lysine and tryptophan)	Fight against hunger and protein malnutrition.	[20]
		Improve child growth.	[8]
	Bioavailability of niacin.	Fight against pellagra.	[33]

Note: *CVDs*, cardiovascular diseases.

Maize also has anti-HIV activity owing to the presence of *Galanthus nivalis* agglutinin (GNA) lectin, also known as GNA-maize. Lectins present in maize have the ability to bind with carbohydrates or carbohydrate receptors, which are present on cell membranes. The binding of lectins with sugars can inhibit activity of the HIV virus. Zein is an alcohol-soluble prolamine protein, which is present in maize endosperm. Zein is found nontoxic, biodegradable, and also generally recognized as safe. It has potential to offer significant health benefits to humans. It is also used as nanoscale biopackaging material having unique solubility and film-forming properties. Zein also has applications in pharmaceutical and nutraceutical fields to develop promising nanocomposite antimicrobial materials, coat nanoparticles for novel food packaging, encapsulation of nutrients, and target delivery with controlled release [36].

Phytosterols in maize provide several health benefits. However, dietary consumption of phytosterol is also negatively associated with total serum, cholesterol absorption, and low-density cholesterol. The main mechanism of dietary phytosterols is the prevention of cholesterol absorption by the intestine and prompting of cholesterol synthesis. Hence, there is increased removal of cholesterol in stools. One study showed that the phytosterols have cholesterol-lowering effect when human consumed a diet with or without phytosterol. The study concluded that the cholesterol absorption was 38% more in persons, which consumed without-phytosterol diet than the humans consuming the phytosterol-diet for two weeks. When corn oil phytosterols were added again to diet, the cholesterol absorption was dropped significantly again. Therefore, the ingestion of maize oil for a long-term period can decrease cholesterol concentration and can avoid atherosclerotic disease [18].

4.7 PROTEIN QUALITY FOR CHILDREN

The intake of nitrogen affects the retention of nitrogen. A recent study showed that the retention of nitrogen for maize protein was significantly lower than the nitrogen retention at high nitrogen intake (469 mg/kg body weight per day) as compared to the same level of milk protein. The nitrogen availability varied from 72% to 78%, which was fairly similar for different nitrogen intakes as indicated by protein digestibility. Nitrogen retention of milk was significantly higher compared to maize at the same level of protein intake. Protein digestibility was 80% for milk compared to 75% for maize [7]. Nitrogen balance was lower for common maize endosperm and whole kernel compared to the nitrogen balance for the reference protein, that is, casein. To match the nitrogen retention of maize protein same as with casein protein, the children have to gain 203.9% of energy from maize, which is impossible [27]. The protein from maize germ is a rich source of EAAs, therefore maize products deficient in germ are always lower in protein quality than the whole kernel. More consumption of maize having high zein protein causes in the deficiency of lysine and high imbalance of leucine and isoleucine essential amino acids.

4.8 HEALTH BENEFITS OF MAIZE

Maize grain is abundant in nutrients and bioactive compounds, which include fiber, vitamins, minerals, and phytochemicals. More scientific studies have proved that the regular intake of whole maize grain reduces the threat of rising chronic diseases, such as type-II diabetics, CVDs, overweight and obesity, and digestive disorders. Maize is believed to have several health benefits.

The B-complex vitamins in maize provide excellent benefits for hair, skin, brain, heart, and proper digestion. Vitamins also avert the rheumatism symptoms, since they are reported to maintain the joint motility. The other vitamins present in maize are A, C, and K, which together with beta-carotene and selenium support to enhance the functioning of immune system and thyroid gland. Potassium is the other main nutrient found in maize, which possesses diuretic properties. Maize silk has also several benefits linked with it. In India, Spain, China, Greece, and France, maize is utilized to treat urinary tract infections, kidney stones, jaundice, and fluid retention. Maize also supports liver functioning, improves blood pressure, and produces bile. It acts as a soothing medicine for swelling, wounds, and ulcers. Moreover, silk, roots, and leaves decoction of maize are used for nausea, bladder problems, and vomiting, whereas cob decoction is utilized for stomach complaints.

Maize RS is also known as high-amylose maize and has numerous health benefits. Maize endosperm contains 39.4 mg/100 g of RS. The RS intake aids in the reduction of cholesterol, changing microbial populations and improving its fecal excretion, increasing the fermentation, and short-chain fatty acid production in large intestine, decreasing diarrhea problems, which together diminishes the hazard of atherosclerosis, cecal cancer, and obesity problems. RS also improved the required composition of colonic bacteria in mice, hence might have potential prebiotic properties. Its ingestion drops body fat storage and affects cholesterol metabolism, thus lessens the risks of atherosclerosis, diabetes, hyperlipidemia and obesity. It can significantly reduce the intestinal transit time that causes to eradicate the waste material through feces in a faster time [10].

Maize is a vital source of several phytochemicals that have important role in our body. There is an inverse correlation between the consumption of phytochemicals and the development of chronic diseases. The phytochemicals in maize have not received much attention and most of the times are underestimated. The various studies reported the bioactive compounds in maize owing to their high AOs exhibit significant beneficial impact in dropping the chance of various diseases. Maize grain, especially yellow maize, contains high quantities of the carotenoid pigments and have vital significance in the diet as human beings are not able to biosynthesize carotenoids. These pigments are also beneficial in preventing cancer [21].

Carotene has also several health benefits due to its high AO. Alpha and beta carotenes have provitamin-A activity. The high concentration of β-carsotene in maize has been proved to act as a proantioxidant and persuades apoptosis of colon cancer cells, melanoma cancer cells, leukemia cells, and gastric cancer cells, thus rendering potent chemopreventive effect. Whereas, a diet with a high amount of β-carotene might not be suitable for smokers due to higher risks of lung cancer occurrences.

Xanthophylls in maize have many critical and specific biological functions. Lutein supplementation in food at dose-dependent manner elevates tumor latency, lymphocyte proliferation, inhibits mammary tumor growth, decreases the chance of palpable tumor, and significantly protects cells against oxidant-induced damages. Zeaxanthin and Lutein are proved to be the only carotenoids that are required for sharp and proper vision. They also proved to defend humans against phototoxic damage and play a role in protection against age-related macular degeneration and cataract formation. Supplementing lutein in daily diets for a period showed a significant improvement in macular pigment optical density and notable protection of the macula from light damage. Lutein also plays role as a cancer chemopreventive suppressing agent by offering inhibitory actions during elevation of disease [16].

Fatty acid in maize has strong antioxidant properties, hence it shields the cell membranes against oxidation. The numerous benefits of fatty acids in maize include anti-inflammatory, antidiabetic, anticancer preventive effects against bone loss, and hepatoprotective activities.

Anthocyanins have been known for their health benefits, such as antiatherogenic, anticarcinogenic, lipid-lowering, antimicrobial, antidiabetic, and anti-inflammatory properties. Owing to the effective antioxidant properties, anthocyanins are capable to reduce the immune system stimulation, capillary permeability, and prevent platelet aggregation. The utilization of anthocyanins in purple maize for the 36-week administration period proved to be a distinct inhibition of colorectal cancer in male rats showing that the lesion of colon was significantly suppressed. The dietary administration of purple maize pigment has antihypertensive effects on spontaneously hypertensive male rats through lowering the systolic blood pressure. The pigments from black glutinous maize cob have shown to possess potent antihyperlipidemic effects in high-fat-fed mice by improving the serum lipids profile and reducing the atherogenic index [36].

4.9 SUMMARY

Maize grain is a good source of protein, fat, minerals, antioxidants, and dietary fiber. The genetic background, variety, environmental and storage conditions, and age of plant can affect the bioavailability of the nutrients in maize. The maize is incorporated in many food products due to its important nutritional and functional properties. The rheological and textural properties of finished products can be improved by blending maize flour with other cereal flours without affecting their sensory attributes. Biofortification technique is used to enrich maize grain with macronutrients and micronutrients to overcome the problems of undernourishment. Maize is a healthy food owing to its nutrients and phytochemicals. Due to the health benefits of maize, it can be a part of our daily diet.

KEYWORDS

- **antioxidant**
- **bioactive**
- **colored maize**
- **functional**
- **maize**
- **sweet corn**

REFERENCES

1. Boyer, C. D.; Shannon, J. C. Carbohydrates of the Kernel. In: *Corn Chemistry and Technology*; Watson, S. A. and Ramstad (Eds.); American Association of Cereal Chemist Press, Eagan, MN; **1987**; pp. 253–268.
2. Christianson, D.D.; Wall, J.S.; Dimler, R.J.; Booth, A.N. Nutritionally Available Niacin in Corn. Isolation and Biological Activity. *Journal of Agriculture and Food Chemistry,* **1968**, *16,* 100–104.
3. Cortez, A.; Wild-Altamirano, C. Contributions to the Lime Treated Corn Flour Technology. In: *Nutritional Improvement of Maize*; Bressani, R., Braham, J. E. and Behar, M. (Eds.); Institute of Nutrition of Central America and Panama (INCAP), Guatemala; **1972**; INCAP Pub. L4; pp. 99–106.
4. Dewanto, V.; Wu, X.; Liu, R. H. Processed Sweet Corn Has Higher Antioxidant Activity. *Journal of Agriculture and Food Chemistry,* **2002**, *50,* 4959–4964.
5. Fukamachi, K.; Imada, T.; Ohshima, Y.; Xu, J.; Tsuda, H. Purple Corn Color Suppresses Ras Protein Level and Inhibits 7,12-Dimethylbenz[a]anthracene-Induced Mammary Carcinogenesis in the Rat. *Cancer Science,* **2008**, *99,* 1841–1846
6. Gardner H.W.; Inglett G.E. Food Products from Corn Germ: Enzyme Activity and Oil Stability. *Journal of Food Science,*1971, *36,* 645–648.
7. Graham, G.G.; Glover, D.V.; de Romana, G.L. Nutritional Value of Normal, Opaque-2 and Sugary-2 Opaque-2 Maize Hybrids for Infants and Children, I: Digestibility and Utilization. *Journal of Nutrition,* **1980**, *110,* 1061-1069.
8. Gunaratna, N.S.; Groote, H.D.; Nestel, P.; Pixley, K.V., McCabe, G.P. A Meta-Analysis of Community-Based Studies on Quality Protein Maize. *Food Policy,* **2010**, *35,*202–210
9. Kang, M.K.; Lim, S.S.; Lee, J.Y.; Yeo, K.M.; Kang, Y.H. Anthocyanin Rich Purple Corn Extract Inhibit Diabetes-Associated Glomerular Angiogenesis. *PLoS One,* **2013**, *8,* e79823.
10. Kim, W. K.; Chung, M. K.; Kang, N. E.; Kim, M. H.; Park, O. J. Effect of Resistant Starch from Corn or Rice on Glucose Control, Colonic Events and Blood Lipid Concentrations in Streptozotocin-Induced Diabetic Rats. *Journal of Nutritional Biochemistry,* **2003**, *14,* 166–172.
11. Landry, J.; Moureaux, T. Distribution and Amino Acid Composition of Protein Groups Located in Different Histological Parts of Maize Grain. *Journal of Agricultural and Food Chemistry,* **1980**, *28,* 1186–1191.
12. Landry, J.; Moureaux, T. Physicochemical Properties of Maize Glutelin as Influenced by Their Isolation Conditions. *Journal of Agricultural and Food Chemistry,* **1981**, *29,* 1205–1212.
13. Liu, R. H. Whole Grain Phytochemicals and Health. *Journal of Cereal Science,* **2007**, *46,* 207–219.
14. Maki, K. C.; Pelkman, C. L.; Finocchiaro, E. T. Resistant Starch from High-Amylose Maize Increases Insulin Sensitivity in Overweight and Obese Men. *Journal of Nutrition,* **2012**, *142,* 717–723.
15. Milind, P.; Isha, D. *Zea mays*: A Modern Craze. *International Research Journal of Pharmacy,* **2013**, *4,* 39–43.
16. Moreno, F. S.; Toledo, L. P.; de Conti, A.; Heidor, Jr. R. Lutein Presents Suppressing but not Blocking Chemopreventive Activity During Diethylnitrosamine-Induced

Hepatocarcinogenesis and This Involves Inhibition of DNA Damage. *Chemico Biological Interactions*, **2007**, *168*, 221–228.

17. Nuss, E. T.; Tanumihardjo, S. A. Maize: A Paramount Staple Crop in the Context of Global Nutrition. *Comprehensive Reviews of Food Science*, **2010**, *9*, 417–436.

18. Ostlund, Jr. R. E.; Racette, S. B.; Okeke, A.; Stenson, W. F. Phytosterols that are Naturally Present in Commercial Corn Oil Significantly Reduce Cholesterol Absorption in Humans. *American Journal of Clinical Nutrition*, **2002**, *75*, 1000–1004.

19. Patterson, J.I.; Brown, R.R.; Linkswiler, H.; Harper, A.E. Excretion of Tryptophan Niacin Metabolites by Young Men: Effects of Tryptophan, Leucine and Vitamin B6 Intakes. *American Journal of Clinical Nutrition*, **1980**, *33*, 2157–2167.

20. Prasanna, B. M.; Vasal, S. K.; Kassahun, B.; Singh, N. N. Quality Protein Maize. *Current Science*, **2001**, *2001*, 1308–1319.

21. Rouf Shah, T.; Prasad, K.; Kumar, P. Maize: A Potential Source of Human Nutrition and Health: A Review. *Cogent Food and Agriculture*, **2016**, *2* (1), article I.D. 1166995; doi. org/10.1080/23311932.2016.1166995

22. Sajilata, M. G.; Singhal, R. S.; Kulkarni, P. R. Resistant Starch–A Review. *Comprehensive Review of Food* Science, **2006,** *5*, 1–17

23. Salinas-Moreno, Y. S.; Sanchez, G. S.; Hernandez, D. R.; Lobato, N. R. Characterization of Anthocyanin Extracts from Maize Kernels. *Journal of Chromatographic Science,* **2005**, *43*, 483–487.

24. Sandhu, K. S.; Singh, N.; Malhi, N. S. Some Properties of Corn Grains and Their Flours, I: Physicochemical, Functional and Chapati-Making Properties of Flours. *Food Chemistry*, **2007**, *101*, 938–946.

25. Sandstead, H. H.; Munoz, J. M.; Jacob, R. A.; Kelvay, L. M. Influence of Dietary Fiber on Trace Element Balance. *American Journal of Clinical Nutrition*, **1978**, *31*, 5180–5184.

26. Scrimshaw, N. S.; Behar, M.; Wilson, D.; Viteri, F.; Arroyave, G.; Bressan, R. All-Vegetable Protein Mixtures for Human Feeding, V: Clinical Trials with Corn and Beans. *American Journal of Clinical Nutrition*, **1961**, *9*, 196–205.

27. Siyuan, S.; Tong, L; Liu, R. H. Corn Phytochemicals and Their Health Benefits. *Food Science and Human Wellness*, **2018**, *7*, 185–195.

28. Squibb, R.L.; Bressani, R.; Scrimshaw, N.S. Nutritive Value of Central American Corns, V: Carotene Content and Vitamin A Activity of Three Guatemalan Yellow Corns. *Food Research*, **1957**, *22*, 303–307.

29. Toufektsian, M.C.; De Lorgeril, M.; Nagy, N.; Salen, P. Chronic Dietary Intake of Plant-Derived Anthocyanins Protects the Rat Heart Against Ischemia-Reperfusion Injury. *Journal of Nutrition*, **2008**, *138*, 747–752.

30. Toufektsian, M. C.; Salen, P.; Laporte, F.; Tonelli, C.; De Lorgeril, M. Dietary Flavonoids Increase Plasma Very Long-Chain (n-3) Fatty Acids in Rats. *Journal of Nutrition*, **2011**, *141*, 37–41

31. Tsuda, T. Dietary Anthocyanin-Rich Plants: Biochemical Basis and Recent Progress in Health Benefits Studies. *Molecular Nutrition and Food Research*, **2012,** *56*, 159–170.

32. Tsuda, T.; Horio, F.; Uchida, K.; Aoki, H.; Osawa, T. Dietary Cyanidin 3-O-b-D-Glucoside-Rich Purple Corn Color Prevents Obesity and Ameliorates Hyperglycemia in Mice. *Journal of Nutrition*, **2003**, *133*, 2125–2130.

33. Wright, K.N. Nutritional Properties and Feeding Value of Corn and its By-Products. In: *Corn: Chemistry and Technology*; Watson, S. A. and Ramstad, P. E. (Eds.); St. Paul, MN: *American Association of Cereal Chemistry*; **1987**; pp. 447–478.

34. Yen, J.T.; Jensen, A.H.; Baker, D.H. Assessment of the Concentration of Biologically Available Vitamin B6 in Corn and Soybean Meal. *Journal of Animal Sciences,* **1976**, *42,* 866–870.

35. Young, V.R.; Ozalp, I.; Chokos, B.V.; Scrimshaw, N.S. Protein value of Colombian Opaque-2 Conn for Young Adult Men. *Journal of Nutrition,* **1971**, *101,* 1475–1481.

36. Zhang, Z.; Yang, L.; Ye, H.; Du, X. F.; Gao, Z. M.; Zhang, Z. L. Effects of Pigment Extract from Black Glutinous Corncob in a High-Fat-Fed Mouse Model of Hyperlipidemia. *European Food Research and Technology,* **2010**, *230*, 943–946.

37. Zhao, Z.; Egashira, Y.; Sanada, H. Phenolic Antioxidants Contained in Corn Bran Are Slightly Bioavailable in Rats. *Journal of Agriculture and Food Chemistry,* **2005**, *53,* 5030–5035.

PART II

Technological Advances in Processing of Cereals for Healthcare

CHAPTER 5

CURRENT TRENDS IN MULTIGRAIN FOODS FOR HEALTHCARE

KAMALJIT KAUR, JASPREET KAUR, and AMARJEET KAUR

ABSTRACT

Extension of the cereal-based industry has generated a need for development of wholegrain-based multigrain products that can cater the need of consumers due to several health benefits. Consumption of wholegrain-based multigrain food products has been connected with reduction of developing chronic disorders, such as type 2 diabetes, cardiovascular diseases, some cancers, and so on. Food scientists are continuously developing new multigrain products by keeping the balance between nutrition and sensory parameters. This chapter reviewed the market trends and research advances in multigrain products and their health benefits.

5.1 INTRODUCTION

Expanding world population and increasing prices of rice, wheat, and corn have led to the expansion of multigrain food products particularly in developing countries. Multigrain or composite flour and products prepared from it are more nutritive with potential health benefits compared to single grain or refined flour [37]. Consumers are getting aware of the relationship between the diet and disease and there is a constant shift from animal-derived to plant-derived meals.

Phytochemicals in plant-based foods have beneficial impacts on the disease prevention and healthcare. Phenolic compounds, tocopherol, ascorbic acid, glutathione, beta-carotene, copper, and selenium are vital antioxidants to avert oxidation reactions in our body [46]. Whole grain (WG) cereals have received substantial attention during the last decades due to the presence of antioxidants and phytochemicals. Major cereal grains are wheat, maize, rice; and barley, oats, rye, buckwheat, millets, and sorghum are minor grains. About 50% of the calories taken by world population originate from three main cereals: wheat (17%), rice (23%), and maize (10%) [25]. These cereals are grown for their nutritious seeds, which are often named as grains. These three grains are staple food among diets of people throughout the world. These grains are composed of endosperm rich in starch, germ, and bran including aleurone. Cereals are the main source of calories, besides these supply a number of nutrients and other phytochemicals [30].

Among prime cereals, wheat is the predominant food accounting for 20%–80% of the entire food utilization in various parts of the world. Indians are mainly dependent on wheat, maize, and rice as a staple food to provide calories and basic nutrition. Traditionally, whole wheat meal (atta) has been used for preparation of chapattis and refined wheat flour (maida) for preparation of bread [36]. Regular consumption of a single item affects health directly, for example, people who regularly feed on wheat are deficient in lysine; moreover, gluten protein in wheat causes allergic reactions to some persons.

Diet must be balanced besides being appetizing, wholesome, satisfying, and palatable. With an increasing consumer perception, improved standards of living, educational status, change in food habits, knowledge about natural foods, and rising cost of medicines, there is an increased shift in consumption of healthy eating. Although people are becoming health conscious, yet they are not having enough time to consume every single grain at a time, hence alternate flour mixes serve as excellent source to provide functional ingredients from other grains in the diet. Multigrain flours and its products feature a combination of grains, such as wheat, barley, oats, maize, rice, pseudocereals, and others. [18]. WGs must be used to prepare multigrain mixes to offer maximum nutritional properties. Apart from balanced nutrition, multigrains provide variety of bioactive compounds, flavors, and improve the texture and sensory quality of products [33].

Multigrain products must be prepared from WGs for better nutritional benefits. In bakery and breakfast cereals, multigrains are well established contributing to the taste and texture of products; and consumers readily accept them due to great health benefits. Products prepared from multigrains reduce the risk of diabetes, help in weight control, reduce the risk of cardiac failure, contribute to healthy digestive system, and prevent the chances of bowl cancer.

Mixing grains is a healthy choice because each grain has its unique nutritional properties and composition. If certain nutrient lacks in one grain, it can be compensated by another grain. The important benefit of combining grains affects mainly the amino acid composition. Grains are not complete proteins. They may lack certain essential amino acids and combination of grains provides complete proteins. However, all multigrain products in market are not WGs, refined grains, because specific ingredient may result in poor nutrition and low quality.

This chapter discusses the role of WGs in multigrain formulations, health benefits of a multigrain diet, and research trends in multigrain food products.

5.2 ROLE OF WHOLE GRAINS IN MULTIGRAIN DIETS

WG has been defined as the grain that is consumed as intact, cracked, ground, or flaked; its principle anatomical portions such as bran, endosperm, and germ are in similar ratios as in the unbroken grain. People are getting aware of the health advantages of WGs and have started reducing the consumption of refined grains. Epidemiological studies have shown that people on WG diet have less possibility of diabetes, cardiovascular diseases (CVDs), and certain cancers [43].

There is a development of hypothesis that the benefit of three anatomical components (endosperm, bran, germ) together have more health benefits compared to any single fraction of cereals. WGs contain physically important nutritional compounds including vitamins (vitamin E and B-complex), minerals (Zn and Mg), and phytochemicals (e.g., phytosterols, phytoestrogens, phenolics, betaine, etc.) that can promote our health [42]. Table 5.1 summarizes the nutritional profiles and potential health benefits of selected grains.

Despite knowledge about the health benefits of WGs, people are falling short of consumption goals. There are many hurdles to increase WG consumption, such as:

- Confusion about mislabeling of products.
- Habitual inclination towards refined diets.
- Lack of disease-resistant variety.
- Limited availability of products made with WGs.
- Poor palatability of WG products.

Processing of cereal-based products can play an important role in helping the consumers to include WG as part of daily diet. WGs can be used for the preparation of wide range of products, such as snacks, pasta, biscuits, buns, bread, and even beverages. Many innovative products are being researched for increasing the WG content in available products. However, such products will only be successful if sensory parameters that are primary consumer necessity are not compromised [42].

TABLE 5.1 Nutritional and Health Benefits of Various Grains Used in Multigrain Diets

Cereal	Nutritional Properties	Health Benefits	Ref.
Barley	Barley grain is used as food, feed and malt. As food, it is consumed in the form of flour, semolina, and as whole dehulled grain.	β-glucan is major soluble fiber that cures hypercholesterolemia, chemically induced colon cancer, and hypoglycemia.	[11]
Buckwheat	• Amino acid composition of buckwheat is nutritionally superior to that of cereal grains. • Rich source of K, Ca, Na, Mg, Zn, Mn, Cu, Se. • Rich in rutin, polyphenols, and catechins that act as potential antioxidants. • Rich source of fibers.	• Buckwheat lowers the plasma cholesterol level; act as neuroprotectant, anti-inflammatory, anticancer, antidiabetic; and improves hypertension conditions. • Prevention of obesity and diabetes.	[11, 12]
Flax seed	Rich in omega-3 fatty acids, lignin, alpha-linolenic acid, fiber, secoisolariciresinol diglucoside, polysaccharides, cyclic peptides, alkaloids, cyanogenic glycosides.	• Flaxseed is a functional food with a significant content of omega-3-fatty acids. • Flaxseed protein is rich in arginine that helps in lowering blood pressure.	[38, 44]
Maize	Corn is rich in vitamins A, B, E and K, and minerals (Mg, P, and K), phytochemicals (phenolic acids: coumaric acid, ferulic acid and syringic acid), flavonoids (anthocyanins), carotenoids, and dietary fiber.	• Quality protein maize (QPM) can be incorporated at 30% level in wheat flour for preparation of bread. • Inclusion of QPM declined the glycemic index (11.86%–15.30%). • Corn bran lowers weight and obesity as it provides satiety feelings.	[1, 49, 56]
Millets (finger millet, foxtail millet, kodo millet, proso millet, barnyard millet, and little millet)	• Millets are major food source in dry-and semidry regions of world. India is the top producer of millets. • Millets provide all essential nutrients; rich in essential amino acids, vitamins and minerals. • Fox tail and proso millets are rich in protein (12.3%–12.5%)	• Multigrain flour: Wheat and millet flour in ratio of 7:3 is simplest flour suitable for making chapatti. • Fortification of millets in wheat improves taste and can control glucose level in diabetic patients. • High fiber content in multigrain flour improves the constipation.	[6, 14, 27]

TABLE 5.1 (Continued)

Cereal	Nutritional Properties	Health Benefits	Ref.
	• Seed coat of finger millet is rich in phytochemicals. • Finger millet is rich in methionine and lysine.	• Millets can cure hypoglycemia, hypercholesterolemia; and act as antiulcerative; • Phytic acid in millets assists in lowering of bad cholesterol.	[11]
Oats	Oats contain high content of β-glucan and compounds rich in antioxidant activity.	Soluble fibers in oats act as prebiotic to maintain the healthy colon wall.	
Rice	Brown rice is a whole grain that contains three parts of grain kernel: bran, germ, and endosperm.	Rice is rich in fiber and antioxidants, gluten free, low glycemic index (GI) that lowers blood glucose level, and cholesterol in the body. Rich in minerals: Mn, Fe, Zn, Mg, P, K, Ca; Rich in vitamins: B1, B2, B3, B6 and K.	[52]
Rye	Rye is rich in fiber content and contains bioactive compounds like lignans, arabinoxylan, benzoxazinoids, phytosterols, folates, phenolic compounds, tocopherols, and tocotrienols.	• Rye bread is rich in antioxidants, dietary fibers, and arabionoxylans that can cure colon, colorectal, stomach, and pancreatic cancers. • Rye stimulates weight loss, lowers the risk of type 2 diabetes, prevents gallstones, improves cardiovascular condition, and promotes gastrointestinal health.	[21, 55]
Sorghum	Important staple crop providing carbohydrates, proteins, fiber, vitamins, minerals, and phytochemicals (such as phenolic acids, tannins, phytosterols, and anthocyanins).	Multigrain flour (Sorghum, wheat, ragi, black gram, fenugreek): TBHQ (tertiary butyl hydro quinine)-rich in protein, vitamin B, Ca, Zn, Fe, Mg, and folic acid.	[39]
Soybean	Soybean peptides (such as soymorphins and lunasin). Major storage proteins glycinin and β-conglycinin contribute 80%–90% of total soybean protein. Isoflavones	• Prevent chronic diseases, e.g., CVDs, obesity, type II diabetes/insulin resistance, immune disorders, and certain type of cancers. • Prevent heart diseases, menopausal symptoms, osteoporosis, diabetes, breast and prostate cancers.	[7]

TABLE 5.1   *(Continued)*

Cereal	Nutritional Properties	Health Benefits	Ref.
	Saponins	• Anticarcinogenic, anti-inflammatory, cardio- and hepatoprotective, antimicrobial.	
Whole wheat	Wheat bran and whole wheat based cereal products are rich source of dietary antioxidants and phenolic acids.	Aleurone layer: rich in protein, minerals, phytates, niacin, B vitamin, folates, and plant sterols; 40%–60% of total wheat mineral lies in aleurone layer	[5]

Note: *CVDs*, cardiovascular diseases.

Holloender et al. [15] performed a meta survey of randomized controlled trials on corn, brown rice, rye; and they concluded that WGs had a tendency to lower 20%–25% CVDs compared with refined grains. In another study, the association between dietary fiber intake and WGs on development of obesity on 74,091 US female nurses was observed. During 12-year period, ladies on high dietary fiber weighed 1.52 kg less compared to those on normal diet. This research concluded that obesity was positively correlated to the intake of refined grains [31]. These studies suggested that the regular consumption of WG and their derived products are linked with less chance of type 2 diabetes. Magnesium, vitamin E and fiber is abundantly available in WGs that are required for insulin metabolism. Also, WGs regulate the insulin level by providing feeling of satiety and lowering body weight index [26, 57].

Aleurone layer of grains contain a number of bioactive compounds (such as tocols, phenolics, phytosterols, and carotenoids) than the inner endosperm. Phenolic compounds present in WGs are phenolic acids, alkyl-resorcinols, and lignans. High intake of WGs is linked with less chances of hypertension and other cardiovascular risks in healthy middle-aged people [51]. The regular intake of three fragments of WG diet can lower cardiovascular risk by observed drop in systolic blood pressure that further decreased the occurrence of stroke and coronary artery diseases. However, consumption of refined cereal products (such as bread, pasta, and rice) have been linked with increased risk of pharynx, larynx, digestive tract, and thyroid cancers. WG are nutrient-dense compared to refined grains regarding the dietary fiber, phytochemicals, and micronutrients; and the mechanism behind health benefits of WG includes improvement in gut transit, prebiotic effect that changes the gut microbiota population and its mechanism, increased supply of phenolics in the body, and improved methyl donor status [40]. Some important health benefits of WGs are as follows [42]:

- Fiber in WG support healthy digestion gives bulk to the stool and feed beneficial gut bacteria.
- Good source of vitamin B: thiamin, niacin, riboflavin, and folate.
- Regular intake of WGs lowers inflammation: a factor responsible for many chronic diseases.
- Rich in RS, which is easily digested and maintains insulin level and keep the blood glucose and cholesterol level down. Oat meal, pearl barley, and brown rice are rich in RS.
- WG acts as prebiotic in diet and lowers the chances of colorectal cancer, breast cancer, and pancreatic cancer.

- WG is rich in nutrients and dietary fiber and more intake is linked with lowered diabetes, obesity, high cholesterol, and high blood pressure.
- WG carry physiologically active compounds, such as vitamins (tocotrienols, B vitamins), minerals, and various phytochemicals (such as phenolics, phytosterols, betaine, and phytoestrogens) that promote health individually, in combination or synergistically.

5.3   HEALTH BENEFITS OF A MULTIGRAIN DIET

Foods formulated from multigrain flours carry higher content of bioactive compounds, such as dietary fiber and phenolics in comparison to products formulated only from refined wheat flour. High antioxidant activity and more phenolic content in multigrain diets could mostly be due to substitution of refined wheat flour with whole wheat flour in multigrain composites [41]. Cereals including millets are essential components of diet throughout the world, especially in subtropical and tropical regions. Cereals contribute to major source of dietary carbohydrates, proteins, minerals, vitamins, particularly for vegetarians throughout the world [13].

A diet was formulated comprising both millets and cereals to examine the role of multigrain diet for controlling lipid and antioxidant metabolisms in the presence of cholesterol. Two diets with- and without cholesterol were used (formulated multigrain diet containing bajra, oats, rice, maize, ragi, jowar) and a commercial diet. The study concluded that formulated multigrain diet was effective in controlling oxidative stress and hyperlipidemia caused by high cholesterol diet [53]. Multigrain products are rich in micronutrients, dietary fiber, and are associated with therapeutic health advantages, such as good gut health, reduction in plasma glucose, bowel transition, and lowering cholesterol levels due to high fiber content [2].

Many parts of the world are suffering from fluorosis. Diets were formulated using millets; and cereals (bajra, ragi, jowar, and maize) are well known in many parts of India. These are the important source of energy and provide iron, calcium, vitamin B-complex, and fiber. Millets are rich in minerals and fiber and are low in fat. These grains are rich in polyphenols having high antioxidant activity.

Vasant et al. [53] investigated the utility of antioxidant and nutrient-rich grains for formulation of high carbohydrate low-protein and low-carbohydrate high protein diet to alleviate fluoride toxicity. Protein-enriched multigrain diets have decreased lipid and plasma glucose levels. This study reported that multigrain diet rich in antioxidants and nutrients fortified with protein is highly effective in mitigating fluoride toxicity.

5.4 MULTIGRAIN FOOD PRODUCTS

Multigrain bread was prepared by replacing wheat flour with oat, maize, barley, and rice flours. Multigrains can decrease the volume, dough expansion, and increase the hardness of bread. This study concluded that multigrain mix up to the level of 30% can be suitable for production of bread [32]. Multigrain composite mixes were developed from different cereals (rice, wheat, and ragi), nuts (almonds, cashew nuts, and sesame), millets (whole chickpea, Bengal gram, and defatted soy flour), and condiments (poppy seeds). These mixes are rich in vitamins, dietary fiber, polyphenols can be used for preparation of food formulations, pancakes, savory products, and snacks [20]. Some of the multigrain products in local market are mentioned in Figure 5.1.

FIGURE 5.1 Some multigrain products available in the market.

Storage studies of multigrain "atta" containing barley meal, gram flour, maize meal, oat fiber, defatted soy flour, garlic powder, psyllium husk, fenugreek powder, and ginger powder in comparison with control stipulated that multigrain "atta" can be kept well above 6 months in high-density polyethylene (HDPE) under ambient conditions. "Chapati" baking quality of multigrain "atta" packed in HDPE was better than low-density polyethylene (LDPE) packaging material [33]. Multigrain flour prepared by blending pulses and millet flour is abundant in vitamins, proteins, minerals, and dietary fiber [17] that encounter the nutritional requirements for social programs and mass feeding. Sorghum-rich multigrain flour improves the nutritional quality and taste of sorghum "roti." This flour was developed using wheat, sorghum, ragi, fenugreek, black gram dhal, karaya gum, and tertiary butyl hydro quinine to lower the lipolytic activity and had a shelf life of 120 days [39].

Multiple scientific studies are accessible for preparation of nutritious bakery products by using combination flours [17, 28]. The most interesting approach is by preparing multigrain mix with composting of pulses, cereals, and oilseeds into biscuits to enhance the nutritive quality [28]. Studies on porridge mix [33], halwa mix [3], and supplementary food [24] revealed that level of 10%–25% multigrain mix improved substantially the nutritional quality, such as dietary fiber content, vitamin–mineral profile, and protein quality of products.

In a recent study [29], multigrain premix prepared by blending of pulses, cereals, and oilseeds at 40% incorporation amount on nutritional profile of multigrain biscuits revealed that protein content was increased more than twofolds, and there was increase in amino acid level, vitamin–mineral profile, and polyunsaturated fatty acids level in biscuits. Multigrain flour prepared by combining finger millet and wheat in ratio of 3:7 is the simplest flour for preparation of chapatti. Gluten content was reduced but flattening was not affected, however color of chapatti turned to slightly dark [22]. Table 5.2 summarizes research trends in preparation of various multigrain products and their corresponding health benefits.

Cukelj et al. [10] developed functional biscuits with combinations of wheat, oat, barley, rye with milled flaxseed flour. Improved bioaccessibility of lignans and omega-3-fatty acids, increased phenolic acids and increased spread factor was observed. In another study, whole wheat meal was replaced with multigrain blend (check pea, barley, soybean, and fenugreek seeds) in combination with dry gluten powder, sodium stearoyl-2-lactylate and hydroxy-propyl methyl cellulose for preparation of North Indian parotta. Multigrains increased the fat, protein, minerals and dietary fiber content of Indian parotta [17]. Mixing grains has multiple applications in the enhancement of nutritive

TABLE 5.2 Multigrain Products and Their Health Benefits

Product	Grains Used	Health Benefits	Ref.
Biscuit	Multigrain premix using a combination of cereals, pulses and oilseeds at 40% incorporation level.	Improved nutritional quality (protein, mineral and fiber), in vitro and in vivo protein digestibility.	[29]
Bread	Wheat flour with oat, maize, barley, rice flour, and flax.	30% multigrain mix increased the fiber content of bread.	[32]
Bread	Wheat flour and quality protein maize.	Inclusion of quality protein maize declined the blood glucose level (glycemic index) of products.	[1]
Bread	Composite flour containing buckwheat, sorghum, chickpea, sprouted wheat, and sprouted barley.	Better nutritional profile with increased protein, fiber, minerals, phenolics, flavonoids and antioxidant compounds, low-fat and moderate carbohydrate content.	[4]
Chapati (roti)	Wheat flour, Bengal gram, dried peas, defatted soy flour, barley, and fenugreek seeds.	Premix supplementation in diet reduced the fasting and postprandial blood sugar level.	[22]
Composite flour bread	Composite four bread using refined wheat flour, high protein soy flour, sprouted mung bean flour, and mango kernel flour in the ratio of 85: 5: 5.	Improved protein content and possibility of using high-fat seed flour-like mango kernel as natural fat replacer.	[35]
Multigrain biscuit	Composite flour biscuit (using wheat flour: barley flour: buckwheat flour in the ratio of 30:20:50) was rich in calcium, iron and zinc.	Economically, the composite biscuits were cheaper compared to commercially available multigrain biscuits.	[16]
Multigrain parotta	Whole wheat flour was replaced with multigrain blend (chickpea split, barley, soybean, and fenugreek seeds flour) at 10, 20, 30, and 40 g/100 g along with Sodium stearoyl-2-lactylate, dry gluten powder and hydroxypropylmethyl cellulose.	Addition of multigrains increased the protein, fat, fiber and mineral content of parotta. Multigrain blend at 30% level of incorporation improved the overall quality of parotta.	[19]
Multigrain muffins	Multigrain flour was prepared by taking oat, maize, pearl millet, ragi, and Bengal gram in the ratio of multigrain flour: refined wheat flour 40:60 was selected best.	Nutritive value of muffins was increased with increase in protein, fat, calcium, iron, and fiber content of multigrain muffins.	[45]

TABLE 5.2 (Continued)

Product	Grains Used	Health Benefits	Ref.
Multigrain *Khakra*	Multigrain *Khakra* prepared using finger millet, pearl millet, maize, sorghum, and whole wheat flour.	Multigrain *Khakra* had high fiber (2.4%), high resistant starch (1.2%), low GI (52) and high protein digestibility (85%). Cereal-millet multigrain combination had high functionality compared to whole wheat *Khakra*.	[8]
Multigrain *Idli*	30% Black gram and rice mix were combined with 35% soybean flour, 10% flaxseed flour, 10% barley flour, 10% gingelly seed powder and 5% curry leaves; and the mix was optimized on the basis of sensory attributes.	Nutritional composition (protein, iron, calcium) was significantly increased.	[47]
Multigrain function biscuits	Wholemeal wheat flour was partially substituted with rye, oat, barley, and milled flaxseed flour.	Lignan concentration was increased 30 times compared to control biscuits. Bioaccessibility of lignans and omega-3-fatty acids improved.	[10]
Multigrain composite mix	Cereals, legumes, millets, nuts, and condiments.	These mixes can be used for preparation of snacks, savory products, and pancakes.	[20]
Multigrain gluten-free biscuit	Flour containing ragi, soybean, jowar, oats, maize, groundnut, and functional food "dink."	Biscuits were found rich in proteins, vitamins especially vitamin A, minerals, and fiber.	[34]
Multigrain unleavened flatbread	Flour contained wheat flour, soy flour, flaxseed flour, barley flour, gingelly seed powder, and curry leaves.	Protein, fiber, calcium, iron, and energy were more in multigrain unleavened bread (*chapatti*); thus, this flour can be included in daily diet to get proper nutrition.	[48]

value and sensory properties of one material with another [50]. Particular ratio of mixing grains increased proteins, dietary fiber, fat, phosphorous, calcium, ash, phytic acid, and tannin in final processed foods [9, 50].

5.5 SUMMARY

Cereal products are staple diet source throughout the world. Cereals contribute a considerable amount of carbohydrates, proteins, fiber, B vitamins, sodium, magnesium, zinc, and selenium. WG cereals consumed in combinations with each other or with legumes cater to the health requirements of mass population. Multigrain concept is not based on random selection of grains, but it requires proper research to select best combination based on nutritional, functional, bioavailability, and sensory parameters. Incorporation of multigrain into traditional products renders improved nutritional quality that opens new dimensions in the development of cereal-based functional foods that can boost the food business to great heights. Future studies must focus on the sensory attributes. Certain combination of grains decreases the bioavailability of essential nutrients and thus upcoming research must focus on bioavailability studies. Although a number of multigrain products are available in market, yet are not cost-effective and people are not aware about their health benefits. There is a need to design the courses and training programs to make consumers aware about health-promoting and functional foods. Further, commercialization of multigrain products will diversify the cropping patterns, improve the health, lead to rural development, and economic well-being of local population.

KEYWORDS

- barley
- health benefits
- millets
- multigrain
- oat
- whole grain

REFERENCES

1. Akanbi, C. T.; Ikujenlola A. V. Physicochemical Composition and Glycemic Index of Whole Grain Bread Produced from Composite Flours of Quality Protein Maize and Wheat. *Croatian Journal of Food Science and Technology*, **2016**, *8*, DOI: 10.17508/CJFST.2016.8.1.01 8.

2. Angioloni, A.; Collar, C. Significance of Lipid Binding on the Functional and Nutritional Profiles of Single and Multigrain Matrices. *European Food Research and Technology*, **2011**, *233*, 141–150.

3. Banu, H.; Itagi, N.; Singh, V.; Indiramma, A. R.; Prakash, M. Shelf Stable Multigrain Halwa Mixes: Preparation of Halwa, Their Textural and Sensory Studies. *Journal of Food Science and Technology*, **2013**, *50*, 879–889.

4. Bhatt, S. M.; Gupta, R. K. Bread (Composite Flour) Formulation and Study of Its Nutritive, Phytochemical and Functional Properties. *Journal of Pharmacognosy and Phytochemistry*, **2015**, *4*, 254–268.

5. Brouns, F.; Hemery, Y.; Price, R.; Anson, N. M. Wheat Aleurone: Separation, Composition, Health Aspects and Potential Food Use. *Critical Reviews in Food Science and Nutrition*, **2012**, *52*, 553–568.

6. Chandra, A.; Singh, A. K.; Mahto, B. Processing and Value Addition of Finger Millet to Achieve Nutritional and Financial Security Case Study. *International Journal of Current Microbiology and Applied Sciences*, **2018**, *7*, 2901–2910.

7. Chatterjee, C.; Gleddie, S.; Xiao, C. W. Soybean Bioactive Peptides and Their Functional Properties. *Nutrients*, **2018**, *10*, 1211–1216.

8. Chauhan, S.; Sonawane, S. K.; Arya, S. S. Nutritional Evaluation of Multigrain *Khakra*. *Food Bioscience* **2017**, *19*, 80–84.

9. Chhavi, A.; Savita, S. Evaluation of Composite Millet Breads for Sensory and Nutritional Qualities and Glycemic Response. *Malnutrition Journal of Nutrition*, **2012**, *18*, 89-101.

10. Čukelj, N.; Novotni, D.; Sarajlija, H.; Drakula, S.; Voučko, B.; Ćurić, D. Flaxseed and Multigrain Mixtures in the Development of Functional Biscuits. *LWT—Food Science and Technology*, 2017, *86*, 85-92.

11. Das, A.; Raychaudhuri, U.; Chakraborty, R. Cereal Based Functional Food of Indian Subcontinent: A Review. *Journal of Food Science and* Technology, **2012**, *49*, 665–672.; doi: 10.3389/fpls.2017.00643.

12. Giménez-Bastida, J. A.; Zieliński, H. Buckwheat as A Functional Food and Its Effects on Health. *Journal of Agricultural and Food Chemistry*, **2015**, *63*, 7896–7913.

13. Gopalan, C.; Sastri, B. V. R.; Balasubramanian, S. C. *Nutritive Value of Indian Foods*. Hyderabad (India): National Institute of Nutrition, ICMR Press; 2004; p. 156.

14. Gupta, S. M.; Arora, S.; Mirza, N.; Pande, A. Finger Millet: A "Certain" Crop for An "Uncertain" Future and A Solution to Food Insecurity and Hidden Hunger Under Stressful Environments. *Frontiers Plant Science*, **2017**, *8*, 643–647.

15. Holloender, P.L.B.; Ross, A. B.; Kristensen, M. Wholegrain and Blood Lipid Changes in Apparently Healthy Adults: A Systematic Review and Meta-Analysis of Randomized Controlled Studies. *American Journal of Clinical Nutrition*, **2015**, *102*, 556–572.

16. Hussain, A.; Kaul, R.; Bhat, A. Development of Healthy Multigrain Biscuits from Buckwheat-Barley Composite Flours. *Asian Journal of Dairy and Food Research*, **2018**, *37*, 120–125.

17. Indrani, D.; Swetha, P.; Soumya, C.; Rajiv, J.; Rao, G. V. Effect of Multigrains on Rheological, Microstructural and Quality Characteristics of North Indian Parotta: An Indian Flat Bread. *LWT—Food Science and Technology*, **2011**, *44*, 719–724.

18. Indrani, D.; Soumya, C.; Rajiv, J.; Venkateswarao, G. Multigrain Bread—Its Dough Rheology, Microstructure, Quality and Nutritional Characteristics. *Journal of Texture Studies* **2010**, *41*, 312–309.

19. Indrani, D.; Swetha, P.; Soumya, C.; Jyotsna, R.; Venkateswara-Rao, G. Effect of Multigrains on Rheological, Microstructural and Quality Characteristics of North Indian Parotta—An Indian Flat Bread. *LWT—Food Science and Technology*, **2011**, *44*, 719–724.

20. Itagi, H. B. N.; Singh, V. Preparation, Nutritional Composition, Functional Properties and Antioxidant Activities of Multigrain Composite Mixes. *Journal of Food Science and Technology*, **2012**, *49*, 74–81.

21. Jonsson, K.; Andersson, R.; Bach Knudsen, K. E. Rye and Health: Where Do We Stand and Where Do We Go? *Trends in Food Science and Technology*, **2018**, *79*, 78–87.

22. Kang, R. K.; Jain, R.; Mridula, D. Impact of Indigenous Fiber Rich Premix Supplementation on Blood Glucose Levels in Diabetics. *American Journal of Food Technology*. **2008**, *3*, 50–55.

23. Kaur, K.; Kaur, A. Storage Studies of Multigrain 'Atta' Containing Various Functional Ingredients. *Processed Food* Industry, **2010**, *13*, 18–20.

24. Khanam, A.; Chikkegowda, R. K.; Swamylingappa, B. Functional and Nutritional Evaluation of Supplementary Food Formulations. *Journal of Food Science and Technology*, **2013**, *50*, 309–316.

25. Khush, G. S. Productivity Improvement in Rice. *Nutrition Reviews*, **2003**, *61*, 114–116.

26. Koh, G.Y.; Rowling, M. J. Resistant Starch as A Novel Dietary Strategy to Maintain Kidney Health in Diabetes Mellitus. *Nutrition* Reviews, **2017**, *75*, 350–360.

27. Kulkarni, D. B.; Sakhale, B. K.; Giri, N. A. A Potential Review on Millet Grain Processing. *International Journal of Nutrition Sciences*, **2018**, *3*, 1018–1022.

28. Kumar, K. A.; Sharma, G. K.; Khan, M. A.; Semwal, A. D. Optimization of Multigrain Premix for High Protein and Dietary Fiber Biscuits Using Response Surface Methodology (RSM). *Food and Nutrition Sciences*, **2015**, *6*, 747–756.

29. Kumar, K. A; Sharma, G. K.; Anilakumar, K. R. Influence of Multigrain Premix on Nutritional, In-Vitro and In-Vivo Protein Digestibility of Multigrain Biscuit. *Journal of Food Science and Technology*, **2019**, *56*, 746–753.

30. Liu, R. H. Whole Grain Phytochemicals and Health. *Journal of Cereal Science*, **2007**, *46*, 207–219.

31. Liu, S.M.; Willett, W.C.; Manson, J.E.; Hu, F.B.; Rosner, B.; Colditz, G. Relation Between Changes in Intake of Dietary Fiber and Grain Products and Changes in Weight and Development of Obesity Among Middle-Aged Women. *American Journal of Clinical Nutrition*, **2003**, *78*, 920–927.

32. Malik, H.; Nayik, G. A.; Dar, B. N. Optimization of Process for Development of Nutritionally Enriched Multigrain Bread. *Journal of Food Processing and Technology*, **2015**, *7*, 544–548; doi:10.4172/2157-7110.1000544

33. Mandge, H. M.; Sharma, S.; Dar, B. N. Instant Multigrain Porridge: Effect of Cooking Treatment on Physicochemical and Functional Properties. *Journal of Food Science and Technology*, **2014**, *51*, 97–103.

34. Mehta, K. R.; Shivkar, S. M.; Shekhar, A. A Study of Multigrain Gluten Free Groundnut and Edible Gum Biscuits. *International Journal of Food and Nutrition Science*, **2014**, *3*, 201–206.

35. Menon, L.; Majumdar, S. D.; Ravi, U. Development and Analysis of Composite Flour Bread. *Journal of Food Science and Technology*, **2015**, *52*, 4156–4165.

36. Nigham, V.; Nambiar, V. S.; Tuteja, S.; Desai, R. Effect of Wheat ARF Treatment on the Baking Quality of Whole Wheat Flours of the Selected Varieties of Wheat. *Journal of Applied Pharmaceutical Science*, **2013**, *3*, 139–145.

37. Noorfarahzilah, M.; Lee, J. S.; Sharifudin, M. S. Applications of Composite Flour in Development of Food Products. *International Food Research Journal*, **2014**, *21*, 2061–74.

38. Parikh, M.; Maddaford, T. G.; Austria, J. A.; Aliani, M.; Netticadan, T.; Pierce, G. N. Dietary Flaxseed as a Strategy for Improving Human Health. *Nutrients*, **2019**, *11*, 1171–1177.

39. Rao, B.D.; Kalpana, K.; Srinivas, K.; Patil, J. V. Development and Standardization of Sorghum-Rich Multigrain Flour and Assessment of Its Storage Stability with Addition of TBHQ. *Journal of Food Processing and Preservation*, **2015**, *39*, 451–457.

40. Ross, A. B.; Bruce, S. J.; Blondel-Lubrano, A. Wholegrain Cereal-Rich Diet Increases Plasma Betaine and Tends to Decrease Total and LDL-Cholesterol Compared with A Refined-Grain Diet in Healthy Subjects. *British Journal of Nutrition*, **2011**, *105*, 1492–1502.

41. Saleh, A. S. M.; Wang, P.; Wang, N.; Yang, S.; Xiao, Z. Technologies for Enhancement of Bioactive Components and Potential Health Benefits of Cereal and Cereal-Based Foods: Research Advances and Application Challenges. *Critical Reviews in Food Science and Nutrition*, **2019**, *59*, 207–227.

42. Schaffer-Lequart, C.; Lehmann, U.; Ross, A. B. Whole Grain in Manufactured Foods: Current Use, Challenges and the Way Forward. *Critical Reviews in Food Science and Nutrition*, **2017**, *57*, 1562–1568.

43. Seal, C. J.; Brownlee, I. A. Whole Grains and Health, Evidence from Observational and Intervention Studies. *Cereal Chemistry*, **2010**, *87*, 167–174.

44. Shim, Y. Y.; Gui, B.; Arnison, P. G.; Wang, Y.; Reaney, M. J. Flaxseed (*Linumu sitatissimum* L.) Bioactive Compounds and Peptide Nomenclature: A Review. *Trends in Food Science & Technology*, **2014**, *38*, 5–20.

45. Shipra, S.; Virginia, P.; Pallavi, S. Sensory Acceptability, Nutrient Composition and Cost of Multigrain Muffins. *International Journal of Food Science and Nutrition*, **2017**,*2*, 38–40.

46. Sidhu, J. S.; Kabir, Y.; Huffman, F.G. Functional Foods from Cereal Grains. *International Journal of Food Properties*, **2007**, *10*, 231–244.

47. Singh, S.; Verma, A.; Bala, N. Nutritional Profiling and Sensory Evaluation of Multigrain Flour Based Indigenous Fermented Food. *International Journal of Food and Fermentation Technology*, **2017**, *7*, 175–179.

48. Singh, S.; Verma, A.; Bala, N. Sensory and Nutritional Evaluation of Unleavened Flat Bread Prepared by Multigrain Flour Mixture. *Journal of Applied and Natural Science* **2016**, *8*, 1168–1171.

49. Siyuan, S.; Tong, L.; Liu, R. H. Corn Phytochemicals and their Health Benefits. *Food Science and Human Wellness*, **2018**, *7*, 185–195.

50. Thapliyal, V.; Singh, K. Finger Millet: Potential Millet for Food Security and Power House of Nutrients. *International Journal of Research in Agriculture and Forestry*, **2015**, *2*, 22–33.

51. Tighe, P.; Duthie, G.; Vaughan, N.; Brittenden, J.; Simpson, W. G. Effect of Increased Consumption of Wholegrain Foods on Blood Pressure and Other Cardiovascular Risk Markers in Healthy Middle-Aged Persons: A Randomized Controlled Trial. *American Journal of Clinical Nutrition*, **2010**, *92*, 733–740.

52. Upadhyay, A.; Karn, S. K. Brown Rice: Nutritional Composition and Health Benefits. *Journal of Food Science and Technology Nepal*, **2018**, *10*, 48–54.

53. Vasant, R. A.; Amaravadi, V. R. L. N. A Multigrain Protein Enriched Diet Mitigates Fluoride Toxicity. *Journal of Food Science and Technology*, **2011**, *50*, 528–534.

54. Vasant, R. A.; Patel, N. D.; Karn, S. S.; Narasimhacharya, A. V. R. L. Physiological Role of a Multigrain Diet in Metabolic Regulations of Lipid and Antioxidant Profiles in Hypercholesteremic Rats. *Journal of Pharmacopuncture*, **2014**, *17*, 34–40.

55. Warechowska, M.; Warechowski, J.; Tyburski, J. Evaluation of Physicochemical Properties, Antioxidant Potential and Baking Quality of Grain and Flour of Primitive Rye (*Secalecereale Var. Multicaule*). *Journal of Food Science and Technology*, **2019**, *56*, 3422–3430.

56. Willis, H.J.; Eldridge, A.L.; Beiselgel, J.; Thomas, W.; Slavin, J.L. Greater Satiety Response with Resistant Starch and Corn Bran in Human Subjects. *Nutrition Research* **2009**, *29*, 100–105.

57. Xi, P.; Liu, R.H. Whole Food Approach for Type-2 Diabetes Prevention. *Molecular Nutrition and Food Research*, **2016**, *60*, 1819–1836.

CHAPTER 6

UTILIZATION OF PSEUDOCEREALS IN BAKERY PRODUCTS AND HEALTHCARE

GURSHARAN KAUR, RITU PRIYA, and SANDEEP SINGH

ABSTRACT

Pseudocereals can replace other cereal crops for improving nutrition, health, and economic status of people. Their regular intake is beneficial for preventing malnutrition in children, immunological disorders in adults due to their unique health benefits. Pseudocereals (amaranth, quinoa, and buckwheat) contain good amount of proteins, calories, bioactive components, which make these suitable for utilization in various bakery products. This chapter focuses on the morphological, nutritional, bioactive composition, health benefits, and utilization of pseudocereals in bakery products.

6.1 INTRODUCTION

Nowadays, human population is dependent on cereal grains, such as wheat, maize, oats, and barley for nutrition. Pseudocereals can be a good substitute for cereals in allergic persons according to the International American Association of Cereal Chemists [3]. Pseudocereals have high amounts of antioxidants, bioactive components, and proteins; and are gluten free, which makes them a component in various food supplementation for manufacturing of new food products. With the increase in demand of healthy foods throughout the world, there is a need of advancements in technology to provide healthy and nutritious food. This objective can be achieved by introducing the pseudocereal crops, which are underutilized but have important biological properties beneficial to human health. The pseudocereals for supplementation in foods are amaranth (*Amaranthus cruentus*), quinoa (*Chenopodium quinoa*), and buckwheat (*Fagopyrum esculentum* and *Fagopyrum tartaricum*).

Pseudocereals are one of the nongrasses that are used in much the same way as cereals (true cereals are grasses). The cereals (such as wheat, rice, and barley) are monocotyledonous plants. Botanically, pseudocereals are dicots; however, they are not categorized as true cereals. They are called as pseudocereals because their seeds or fruits have parity with true cereals in their composition and functional properties. Therefore they can be easily used like cereals in the form of ground flour or as such like whole cereals [18].

This chapter focuses on the history, morphology, chemical composition, bioactive compounds, health benefits, and utilization of pseudocereals in food products.

6.2   ORIGIN AND HISTORY

The genus Amaranthus includes mainly *cruentus*, *caudatus*, and *hypochondriacus*. Amaranthus was thought to be indigenous to Asia due to its vast spread distribution. Nonetheless, domestication of grain Amaranth began in America, based on the data from geographical, morphological, archeological, ethnobotanical, and physiological studies. The important varieties of amaranth are mainly from Peru (*Amaranthus caudatus*), Guatemala (*Amaranthus hypochondriacus*), and Mexico (*Amaranthus cruentus*) [6].

Buckwheat (*Fagopyrum esculentum Moench*) is not a cereal or grass. The origin of buckwheat is from Asia. Now the buckwheat is cultivated in the United States, many regions of Latin America, Africa, and at higher altitudes of India and Nepal [44].

The origin of quinoa (*Chenopodium quinoa* Willd.) is about 7000 years ago in the Andean region of Bolivia and Peru. It is a flowering plant from the family *Amaranthaceae*. Quinoa was the major food crop of the Andean region. It is able to provide nutrition to a larger population due to its high nutrient composition. Its cultivation has spread widespread during the recent decades to Afro-Asian and South American countries with higher yields. Many species of quinoa (bitter or sweet) are present depending on the percentage of saponins present in the seeds [27].

6.3   MORPHOLOGY AND CHEMICAL COMPOSITION

Amaranth seeds are minuscule, lenticular seeds, which vary from pale yellow to red or black in color. The amaranth seed varies from 0.9 to

1.7 mm in diameter with seed weight of 1.0 g per 1000 grains. The seed coat along with globular shaped embryo surrounds the perisperm, which is rich in starch [57]. Figure 6.1 shows the morphological structure of pseudocereals.

FIGURE 6.1   Quinoa, buckwheat, and amaranth seeds.

The buckwheat seeds have a triangular shape, 4–9 mm in length with 15 to 35 g per 1000 seeds. The seed color varies from brown to dark black. Seed consists of two cotyledons and is present in the center of endosperm, which is very thin and mostly consists of starch [52].

Quinoa seeds are 1.5 mm in diameter, spherical, and compact. It consists of pericarp, endosperm, and perisperm. The color of the seeds varies from white to pink, black, or red. Owing to the presence of saponins, the taste of quinoa is bitter. The 60% of the weight of seed is due to the germ portion, which forms a ring around the pericarp [11].

Chemical composition of three pseudocereal grains (i.e., quinoa, amaranth, and buckwheat) is given in Table 6.1. Pseudocereals are rich in nutrients containing a number of nutraceutical components. They are rich source of proteins, starch, bioactive compounds, lipids, and minerals [42]. The biological value of proteins is better or comparable with other protein sources and can contribute to many functions of the human body. The protein content of the pseudocereals is higher than the wheat [27].

6.3.1   PROTEINS

Pseudocereals are considered "ultimate grains" of this century owing to their richness in high protein content. The proteins in these grains are nutritionally safe and have high biocompatibility. Protein content in these pseudocereals (Amaranth, quinoa, and buckwheat) is higher, cheaper, and comparable to most of the cereals (wheat, rice and maize). Amaranth grains contain approximately 16% of proteins followed by quinoa (12%–16%) and buckwheat (10%–14%).

TABLE 6.1   Proximate Composition of Pseudocereal Grains (Amaranth, Buckwheat, and Quinoa)

Constituent	Amaranth	Buckwheat	Quinoa
Ash (g)	–	2.10	–
Carbohydrates (g)	65.25	71.50	64.16
Dietary fiber (g)	6.7	10	7.0
Energy (Kcal)	371	343	368
Lipids (g)	7.02	3.40	6.07
Proteins (g)	13.56	13.25	14.12
Water (g)	11.29	9.75	13.28

The biological value of these proteins is higher [42]. For example, it is around 75 in the case of Amaranth compared to 62 in barley and 57 in wheat. Amaranth grains contain essential amino acids (lysine and tryptophan). The lysine content in wheat is half compared to amaranth. On the other hand, buckwheat has biological value of 93, which provides a proper equilibrium of nutrients in our body due to the presence of amino acids (lysine, threonine, etc.) [5]. The amino acid composition of pseudocereals is summarized in Table 6.2. The biological value of quinoa is 73 that is comparable with many meat proteins like beef, therefore it is considered as one of the most important pseudocereal with utmost safety [15].

The main cause of celiac diseases is the presence of prolamin proteins in cereals, but pseudocereals contain albumins and globulins, which are not toxic in nature [19]. Prolamins, proline, and glutamic acid are present in fewer amounts in albumins and globulins, but they contain higher percentage of essential amino acid lysine. The biological availability of pseudocereal proteins is much higher compared to cereal grains and it is comparable to many animal and milk proteins. Quality of proteins in many cereal grains, milk products, and meat products gets deteriorated due to denaturation during processing but protein quality of pseudicereals gets improved due to increased digestibility. Moreover, the protein quality of pseudocereal is not affected by processing at high temperatures [28]; thus it is beneficial to persons who are gluten intolerant.

TABLE 6.2 Amino acid Composition of Amaranth, Buckwheat, and Quinoa seeds.

Amino Acid	Amaranth	Buckwheat	Quinoa
	g of Amino Acid per 100 g of Seeds		
Alanine	0.799	0.748	0.588
Arginine	1.060	0.982	1.091
Aspartic acid	1.261	1.133	1.134
Cystine	0.191	0.229	0.203
Glutamic acid	2.259	2.046	1.865
Glycine	1.636	1.031	0.694
Histidine	0.389	0.309	0.407
Isoleucine	0.582	0.498	0.504
Leucine	0.879	0.832	0.840
Lysine	0.747	0.672	0.766
Methionine	0.226	0.172	0.309
Phenylalanine	0.542	0.520	0.59
Proline	0.698	0.507	0.773
Serine	1.148	0.685	0.567
Threonine	0.558	0.506	0.421
Tryptophan	0.181	0.192	0.167
Tyrosine	0.329	0.241	0.267
Valine	0.679	0.678	0.594

6.3.2 DIETARY FIBER

Dietary fiber plays an important role in the prevention of diabetes, cardio-vascular diseases, obesity, ulcers, cancer, gastrointestinal diseases, and in improving mineral bioavailability. Dietary fibers are those components of food that are not broken by enzymes and neither are absorbed by small intestine, but are fermented partially or fully in the large intestine by gut microflora. Dietary fiber mainly constitutes both cellulosic and noncellulosic polysaccharides, such as gums, pectin, hemicellulose, ß-glucan, and cellulose.

Pseudocereals are rich source of dietary fibers as it contains many nondigestible constituents present in hulls, seed coat, and others [7]. Dietary fibers play a beneficial and important role in the human body by preventing many chronic diseases. Therefore whole grains are recommended to meet the requirements of daily intake of dietary fibers [26]. In order to maintain a proper dietary fiber level in human nutrition, pseudocereals play an important role. The dietary fiber of different varieties of amaranth seeds varies, for example, the black seeds have 16% of fiber content compared to 8% in pale varieties. Black seeds have more lignin content, which makes it resistant to digestion.

The overall fiber component in the seeds is approximately 27% [4]. The different layers of buckwheat seeds vary in dietary fiber component. The maximum amount of fiber has been reported in seed coat. The buckwheat seeds mainly contain hemicellulose and cellulose as part of fiber content. The quinoa seeds contain more digestible components compared to other cereal grains. The dietary fiber content of quinoa is approximately 16% [35]. The grains of quinoa contain double amount of fiber compared with other grains. The insoluble fiber content of cooked quinoa seeds is 90% that is fermented in gut in a similar fashion as soluble fibers. The recommended daily intake of fiber is 21%, which can be met by eating one serving of cooked quinoa, which contains 21 g of fibers per 100 g.

6.3.4 POLYSACCHARIDES

In the case of cereal grains and pseudocereals, the carbohydrates exist in starch form and nonstarch form (nonsoluble polysaccharides). Starch is the storage polysaccharide and is made up of two components, that is, amylose and amylopectin. The ratio of these components varies according to the type of grain.

In amaranth grains the starch content is less in dark-colored grains compared to the grains that are lighter in color. The amylopectin content varies from 0% to 64%. The starch granules of buckwheat are polygonal, oval, or round in shape. The starch content varies between 60% and 69% on dry weight (d.b.) basis, which is essential for the plant growth. The soluble carbohydrates fagopyritols (A1 and B1) are most significantly active substances, which are used in treating polycystic ovarian syndrome and diabetes by improving glycemic control. The level of fagopyritols in quinoa grains vary from 270 to 465 mg/100 g on dry weight basis [43]. The starch content of quinoa grains is about 70 g/100 g. The starch present in quinoa seeds has better functional properties compared to other grains. The starch granules have polygonal shape (0.4–2 μm). Quinoa amylopectin has significant amounts of short chains and super-long chains [33].

6.3.5   VITAMINS AND MINERALS

Pseudocereals are the grains that have exceptional vitamin and mineral content (Table 6.3). Pseudocereals are rich in B-complex vitamins and fat-soluble vitamin E. Quinoa seeds have the maximum amount of vitamin E (8.7) followed by amaranth (5.7) and buckwheat (5.5 mg/100 g dry weight basis). Quinoa grains are rich in folic acid, thiamine, and riboflavin; buckwheat has good amount of pyridoxine; and amaranth is rich in riboflavin [16, 42]. Many gluten-free products are deficient in important minerals, but pseudocereals are rich source of Mg, Ca, and Fe [2]. The high amount of Ca content prevents celiac diseases and osteoporosis. The problem of lactose intolerance can be eradicated by consuming pseudocereals for long term.

TABLE 6.3   Mineral and Vitamin Contents in Quinoa, Buckwheat and, Amaranth

Mineral	Quinoa	Buckwheat	Amaranth
	per 100 g of Dry Weight		
Calcium (mg)	47	18	160
Copper (mg)	0.60	1.1	0.53
Iron (mg)	4.6	2.2	7.6
Magnesium (mg)	197	231	250
Manganese (mg)	2.0	1.3	3.30
Phosphorous (mg)	457	347	560
Potassium (mg)	563	460	510
Selenium (µg)	8.5	8.3	19.0
Sodium (mg)	5	1.0	4.0
Zinc (mg)	3.1	2.4	2.90
Vitamins	**Quinoa**	**Buckwheat**	**Amaranth**
Alpha-tocopherol (E) (mg)	2.44	0.1	1.20
Ascorbic acid (mg)	–	0.0	–
Folate, total (µg)	184	30	82
Niacin (mg)	1.52	7.02	0.92
Pantothenic acid (mg)	0.77	1.23	1.46
Pyridoxine (B6) (mg)	0.49	0.21	0.60
Riboflavin (mg)	0.32	0.43	0.2
Thiamine (mg)	0.36	0.1	0.12
Vitamin A	1.0	0.0	2.0
Vitamin B-12 (µg)	0.00	0.00	0.00
Vitamin K (Phylloquinone) (µg)	0.0	0.0	0.00

Many antinutritional factors (like saponins and phytic acid) are present in pseudocereals, which binds with minerals and prevents their absorption in our body. These factors can be removed by soaking or sprouting to make the availability of minerals to the body. Some of the antinutritional components are also removed during processing without any loss of minerals [45].

6.3.6   FATS

Fatty acid composition of pseudocereals has been summarized in Table 6.4. They are rich in unsaturated fatty acid (around 77%), which makes them functional grains with improved nutritional value. The lipid content of quinoa and amaranth is higher than cereals and buckwheat [2]. In case of amaranth, plant species and cultivars affect the levels of waxes, terpene, alcohols, and phytosterols. The seed oil is rich in linoleic acid, which varies from 43% to 51% with different varieties. The oleic acid varies from 22% to 31% compared with 18%–21% of palmitic acid [49]. In comparison to amaranth, the quinoa seeds have higher degree of unsaturation (89%). The fat profile of quinoa is superior compared with that in corn, rice, wheat, and barley. It contains good amount of essential fatty acids (2%–10%), which makes it a superfood. The PUFA content is approximately 56% in quinoa seeds [35].

In quinoa seeds, the omega-6 to omega-3 fatty acid ratio (6:1) is very low, which helps in preventing or lowering many life-threatening diseases, such as coronary heart disease, autoimmune diseases, and cancers [54]. About 28% of fatty acids in quinoa is oleic acid, a monounsaturated fat (which is good for heart), and about 5% as alpha-linolenic acid. Quinoa seeds contain natural antioxidant in the form of vitamin E, which maintains quality of seeds during storage [35].

It has been found that maximum fatty acid present in pseudocereals is linoleic acid, which is around 35% in buckwheat and 50% in quinoa and amaranth; while the oleic acid amount to 35% in buckwheat and 25% in quinoa and amaranth grains [42]. The essential amino acid (alpha linolenic acid) makes quinoa a superfood and helps in preventing many chronic diseases [27].

6.4   BIOACTIVE COMPOUNDS

The major bioactive components in pseudocereals are phenolic groups and are present in the outermost layers of amaranth and quinoa seeds. The phenolics lead to physiological effects due to the presence of tannins, flavonoids, and phenolic acids [50]. These compounds are able to prevent oxidative damage to tissue cells. The harmful effects of oxidation lead to cancer, diabetes, and heart diseases. However, the antioxidants in pseudocereals are helpful in the prevention of these diseases [34].

TABLE 6.4   Fatty Acid Composition of Pseudocereals

Lipids	Amaranth	Buckwheat	Quinoa
Cholesterol (mg)	0	0.0	0
Fatty acids, total monounsaturated (g)	1.685	1.040	1.613
Fatty acids, total polyunsaturated (g)	2.778	1.039	3.292
Fatty acids, total saturated (g)	1.459	0.741	0.706
Phytosterol (mg)	24	–	–

Polyphenol compounds play role in the prevention of cancer and cardio-vascular disease [50], due to the higher concentration of functional components (such as phytosterols (plant lipids) in pseudocereals. The flavonoids, tocopherols, and antioxidants in amaranth provide antioxidant properties. Although amaranth grains also contain phytic acid and tannins, yet they have defense mechanism system due to protection of tissues from free-radical mechanism [34]. Due to the unique antioxidant properties, amaranth grains can improve human health.

Buckwheat is considered as super grain because it has better polyphenolic content compared to amaranth [2]. The major antioxidant compounds in buckwheat are protocatechuic acid, vanillic acid, and rutin. Besides these antioxidants, it has other bioactive constituents, such as quercetin, caffeic, vanillic acid, and luteolin [28].

Quinoa grains include phenolics, fagopyritols, carotenoids, phytosterols, and other polyphenolic components [48]. The flavonoid content of these grains is about 144 mg/100 g. The phenolics and polyphenols (such as gallic acid, vanillic acid, and quercetin) have ultimate antioxidant properties. The quercetin content present in quinoa is considered to be even higher than cranberries. Quinoa seeds are of bitter taste due to the presence of high amount of antioxidants.

6.5   HEALTH BENEFITS OF PSEUDOCEREALS

6.5.1   RICH IN PROTEINS

Pseudocereals can reduce muscle and bone mass loss, and help in providing more durability, strength, and physical activity. Pseudocereals contain high-quality protein with almost all amino acids that help to maintain many bodily functions. They help in reducing sarcopenia [40]. Buckwheat contains essential amino acids, which are helpful in maintaining body functions especially for the people who do not eat red meat and are vegetarians or pure vegans.

Superior quality proteins in amaranth grains are required for the growth and formation of new cells and tissues. The amaranth leaves have many important proteins, which are required for metabolic functionality. Protein content of amaranth grains is higher compared with other cereal grains [36].

6.5.2 RICH IN VITAMINS AND MINERALS

Pseudocereals are rich source of iron, calcium, B-complex vitamins, carotenoids, and tocopherols. The grains and leaves of amaranth are rich in iron, which is helpful in increasing levels of hemoglobin, red blood cell count, and enhancing coagulation. Amaranth in daily diet helps to prevent vitamin-A deficiency. It is helpful in eradicating night blindness in children. Pseudocereals also contain good amounts of folates that help in increasing the hemoglobin levels [10]. Other pseudocereals (such as buckwheat and quinoa) are excellent sources of iron content [55]. These pseudocereals are helpful in preventing disorders that occur due to the deficiency of iron.

6.5.3 GLYCEMIC PROPERTIES

The foods have low glycemic index (GI) that can reduce cardiovascular diseases, diabetes, many types of cancers, and insulin resistance. The foods with high GI have score >70 and moderate glycemic range of 56–69 and low GI score <55. The GI of buckwheat is 45, which helps in designing foods of low GI. The GI of amaranth grains and quinoa grains are 40 and 53, respectively. Studies on diabetic rats indicate that amaranth grains can effectively reduce serum glucose levels. Amaranth grains are helpful in preventing hyperglycemia and other diabetic complications [54].

6.5.4 ANTIOXIDANT AGENTS

The peptides in pseudocereals are helpful in enhancing antioxidant characteristics of pseudocereals. The amaranth grains have antioxidant and antithrombotic properties. Among pseudocereals, amaranth grains contain good quantum of lysine and some other peptides along with potassium and iron. These bioactive components in amaranth are helpful in preventing the inflammation and diseases, that is, arthritis and gout. The peptides of amaranth can inhibit free radicals and mutations of healthy cells [2]. Similarly, buckwheat has antioxidant components (such as quercetin and rutin), which help in preventing cancer and heart-related diseases. Moreover, these antioxidants are important for maintaining cardiovascular functioning in human body [29].

Quinoa seeds have significant antioxidant properties due to the presence of lunasin, which can easily get attached with cancer cells and then split the cells away from the healthy cells [20]. Laboratory studies on animals by feeding quinoa seeds indicate that the discrete component lunasin splits the cancerous cells without harming healthy cells [46]. Studies on quercetin in quinoa indicate that it can scavenge the free radicals and therefore it is helpful in preventing many types of cancer, specially lung cancer [39].

6.5.5   IMPROVES BONE QUALITY

Due to calcium in amaranth, it is a fantastic food in the prevention of osteoporosis and maintaining bone strength. Calcium also plays a major role for the proper functioning of heart, muscles, and nerves particularly in the old age. The buckwheat with trace amount of minerals (i.e. zinc and selenium) is essential for human body for strong bones, nails, and teeth. Quinoa seeds are rich in phosphorus, magnesium, and manganese, which help in maintaining and repairing cells and tissues, filtering waste from kidneys, building strong bones and teeth, and producing DNA and RNA—body's genetic building materials [21].

6.5.6   PREVENTS ATHEROSCLEROSIS

Pseudocereal amaranth has high potassium content which helps in the regulation of the body fluid, lowers blood pressure, decreases stress on heart, and preserves bone mineral density. It also helps in preventing the formation of kidney stones and prevents atherosclerosis. Amaranth seeds are rich in phytosterols, a family of molecules related to cholesterol. Therefore low-density lipoproteins can be reduced by consumption of amaranth grains. Quinoa seeds are rich in fatty acid butyrate, which makes a healthy gut [1].

6.5.7 IMPROVES VISION

Carotenoids in pseudocereals protect the healthy cells in the eye and reduce muscle degeneration and prevent cataract formation. With regular intake of pseudocereals, the oxygen stress is reduced in the eye portion. This makes the vision of eyes healthy and strong for many years [53].

6.5.8 ELIMINATES VARICOSE VEINS

Flavonoids like rutin in amaranth strengthen the walls of blood capillaries, which in turn helps in eradicating the varicose veins. Vitamin C plays a key role in the production of collagen and thus improves the functioning of rutin, also it assists in repairing and strengthening the walls of blood vessels [23].

6.5.9 PREVENTS ASTHMA ATTACKS

Buckwheat has high amount of vitamin E and Mg that prevent asthmatic disorders in children [9]. The studies have indicated that children consuming less amounts of cereal grains develop asthmatic disease. Seeds like buckwheat and quinoa have anti-inflammatory properties.

6.5.10 IMPROVES DIGESTION

The dietary fiber content of pseudocereals ranges from 1.1% to 17.3% depending on the variety. The 78% of fiber in pseudocereals is insoluble and is made up of pectic polysaccharides, which help in improving the digestion. Dietary fiber of amaranth has many gastrointestinal benefits (i.e., efficient uptake of minerals) [2]. The dietary fiber of buckwheat is helpful in bowel movements in the digestive tract and improves peristaltic motion.

6.5.11 DIABETES MANAGEMENT

The dietary fibers in pseudocereals are very helpful in maintaining blood sugar levels. Buckwheat grains consist of soluble dietary fibers, which prevent the occurrence of diabetes. It contains the chiroinositol, which is similar in functioning as insulin. It is helpful in controlling type I diabetes. Buckwheat also contains Mg, which is a component of 300 enzymes, which are present in human body and is helpful in regulating type 2 diabetes [52].

6.5.12 GLUTEN FREE

A gluten-free diet does not contain proteins, that is, gliadin and glutenin, which forms gluten. These proteins are available in wheat, rye, and barley and the combination of these proteins form gluten, which results in many celiac conditions. The gliadin and glutenin are not present in pseudocereals and these are proving to be a suitable alternative grain for the people suffering from gluten intolerance. The quinoa is very helpful due to the presence of iron, protein, fiber, and Ca to meet the demands of gluten in intolerant persons [31].

6.5.13 LOWERS CHOLESTEROL LEVEL

Amaranth seed oil contains a bioactive component called squalene ranging from 1.9% to 11.19%. Squalene is also found in quinoa seeds in the range of 3.39%–5.84%. Squalene is a triterpene, which is highly unsaturated and helps in lowering the levels of cholesterol in our body by removing the sterols through feces. Amaranth leaves have an important bioactive component called phytosterol, which is similar in structure to cholesterol. It prevents the absorption of cholesterol in the intestine, thereby lowering the levels of low-density lipoproteins [2]. Amaranth is also rich in phylloquinone, which aids in boosting healthy heart. Moreover, the dietary fibers in amaranth are beneficial in reducing low-density lipoproteins. The vitamin E in amaranth have cholesterol-lowering activity [41].

The buckwheat is superfood because of being a rich source of phytonutrients. Among the phytonutrients in buckwheat, flavonoids are present in major amounts. The flavonoid rutin is beneficial in increasing the levels of high-density lipoproteins (good cholesterol) and reduces coronary heart diseases. The other important role of flavonoid is to prevent the platelets from clotting, which can cause cardiovascular diseases and coronary heart diseases [2]. Quinoa seeds contain 8% of alpha-linoleic acid and oleic acid, which are healthy fatty acids. Diet rich in alpha linoleic acid reduces heart diseases by lowering blood cholesterol levels.

6.5.14   REDUCES RISK OF CANCER

Pseudocereals contain flavonoids and phytosterols, nutraceuticals, amino acids, and peptides (proteins), which can provide many biological functions in prevention of cancer and other diseases [25]. The bioactive peptides in pseudocereals are important in preventing many types of gastric, breast and colon cancer [12]. Besides peptides, good intake of dietary fiber in diet lowers the risk of breast cancer especially in postmenopausal women. Research workers have found that the chances of breast cancer in Swedish women was reduced by half with regular consumption of dietary fiber [32]. The plant lignans are important in the defense mechanism against hormone-based cancers [13]. Buckwheat can play an important role in cancer prevention—colon and gastric [57].

6.6   UTILIZATION OF PSEUDOCEREALS IN BAKERY PRODUCTS

Wheat flour can be replaced with amaranth, quinoa, and buckwheat for the manufacturing of bakery products. This helps in improving the nutritional value of the wheat flour, price reduction of the product, and also to enhance the value of local grains. Moreover, nowadays people are suffering from celiac diseases, which leads to the innovations of gluten-free bakery products from pseudocereals. The patients suffering from celiac diseases rely on gluten-free foods throughout their life and should avoid wheat-based products like snacks, pasta, and other breakfast products [28].

Although the formulation of bread with wheat flour has been standardized many years ago, yet the major challenge today is to make bread from gluten-free flour obtained from pseudocereals. Gluten is a protein formed when gliadin and glutenin form together when water is added to wheat flour. When this mass is kneaded dough better gas holding properties is obtained. and this in return provides good crumb and crust to the bread during baking [14]. However, pseudocereals lack viscoelastic properties due to non-formation of gluten during the dough development [30]. Therefore, different additives (such as starches, gums, emulsifiers and nonstarch polysaccharides) are added. And the starches (such as rice, potato, and cassava) are added to pseudocereal foods.

Dairy ingredients can also be added to enhance the properties of skim milk, caseinate, and dry milk powder. The gums and hydrocolloids like alginates, carrageenan, and carboxymethylcellulose can be added. Similarly, emulsifiers (lecithin and sodium dodecyl lactate) can also be added; and proteins from legumes, eggs, and pulses can be added. Enzymes like cyclodextrin glycosyl transferase, glucose oxidase, laccase, and others can be added. The starches, gums, and hydrocolloids are added as processing aids for enhancing the gluten's viscoelastic properties, which are used for making the network inside the product for better structure of fermented products [58]. Concurrently, different cross-linking enzymes (such as glucose oxidase, transglutaminase, and laccase) are also used to produce a protein network within the flour proteins [47].

Pseudocereals are rich in nutrients and can be used for the manufacturing of pasta, bread and muffins [17]. Buckwheat flour is used widely in the manufacturing of gluten-free products because of higher amounts of fiber, minerals, and proteins [37]. The pseudocereals have high antioxidant properties that makes it of greater significance for the manufacturing of bakery products with higher antioxidant activity. The bakery products made from buckwheat and quinoa flours are softer and have good crumb texture.

6.7 MISCELLENOUS BAKERY PRODUCTS

Pseudocereals in combination with other grains are being used to produce various food products (Table 6.5). Gluten-free pasta can be prepared by mixing ideal amount of pseudocereal flours with enzymes, emulsifiers and albumen [24]. Protein-rich high energy gluten-free crackers and biscuits can be manufactured by using pseudocereal flours. Shelf-life of biscuits and crackers can be extended by adding 0.1% butylated hydroxyl toluene without affecting the flavor. Although incorporation of pseudocereal flours is increasing day-by-day in various gluten-free preparations, yet only types of such products are available in local vicinity due to their limited commercial production. For instance, amaranth bread and quinoa pasta are available only at a few numbers of stores in developed countries, that is, the United State [51].

Although pseudocereal based gluten-free products are commercially manufactured at a small scale, yet its availability in the market is very low. Thus, commercial production of pseudocereal based gluten-free products on large scale will ensure a nutritionally adequate diet, which is not only a tastier substitute but also a healthier choice for celiac patients.

TABLE 6.5 Formulations of Different Bakery Products Using Pseudocereals

Product	Ingredients	Method of preparation
Cookies	Pseudocereal flour (50 g)	a. Mix and sift the flour together.
	Rice flour (25 g)	b. Add baking powder and sugar into it.
	Oats flour (25 g)	c. Sift again and form smooth dough by adding
	Powdered sugar (50 g)	milk.
	Fat (58 g)	d. Roll it into thin sheets of 5 mm thickness and cut into desired shapes.
	Milk (11 mL)	e. Grease the pan and bake the cut cookies at 180 °C for 15–20 min.
	Baking powder (2 g)	
Cakes	Pseudocereal flour (50 g)	a. Mix and sift the flours along with baking powder twice.
	Rice flour (25 g)	
	Oats flour (25 g)	b. Mix fat and sugar to form a light cream.
	Sugar (100 g)	c. Add the beaten eggs along with essence into the cream by slow mixing.
	Fat (Butter) (50 g)	
	Eggs (80 g)	d. Gently, fold the flour into the mix.
	Baking powder (2 g)	e. Now, add milk to form a dropping consistency.
	Vanilla essence (4 mL)	f. Pour the batter into the greased cake-mould and bake at 180 °C for about 20 min.
Muffins	Pseudocereal flour (10 g)	a. Mix flour and baking powder and sift to remove lumps.
	Rice flour (45 g)	
	Oats flour (45 g)	b. Beat the eggs to form light foam.
	Sugar (100 g)	c. Then, add sugar and essence and mix them gently.
	Fat/oil (42 g)	
	Butter (33 g)	d. Add melted butter and fat and mix it well.
	Eggs (167 g)	e. Gently, fold the flour into the mix.
	Baking powder (2 g)	f. Pour the batter into the greased muffin tray and bake at 200 °C for 20 min.
	Vanilla essence (4 mL)	
Pies	Pseudocereal flour (50 g)	a. Mix the flour and sift twice.
	Rice flour (25 g)	b. Mix butter and sugar to form light cream.
	Oats flour (25 g)	c. Mix it with the flour and form dough.
	Butter (67 g)	d. Sheet the dough with rolling pin to 6 mm thickness.
	Powdered sugar (33 g)	

TABLE 6.5 *(Continued)*

Product	Ingredients	Method of preparation
	Baking powder (3 g)	e. Place the dough sheet in the pie-mould and shape the dough sheet along the mould.
	For filling:	
	Sponge cake (gluten-free) (20 g)	f. Then, trim the extra edges using knife and prick it with fork.
	Chopped apple (20 g)	g. Prepare the filling mixture by cooking the ingredients in small amount of water and fill it into the pie.
	Sugar (10 g)	
	Cinnamon powder (10 g)	h. Cover with dough strips on the top.
	Raisins (5 g)	i. Baking of pies is carried out at 200–220 °C for about 15–20 min.
Tarts	Pseudocereal flour (10 g)	a. Mix the flour and baking powder together and add creamed fat and sugar into it.
	Rice flour (45 g)	
	Oats flour (45 g)	b. Prepare dough ball and allow it to rest for 15–20 min.
	Butter (67 g)	
	Powdered Sugar (35 g)	c. Roll the balls into thin sheets and cut into desired shapes.
	Baking powder (2.5 g)	d. Place the sheets into tart mould and prick it with the fork.
	For filling:	
	Whipped cream (20 g)	e. Trim the extra edges and bake them at 180–190 °C for about 15–20 min.
	Chocolate sauce (20 g)	
	Mango (10 g)	
	Kiwi (10 g)	

6.8 SUMMARY

Pseudocereals (buckwheat, quinoa, and amaranth) are beneficial food ingredients in reducing various types of disorders in our body. These can be used to replace traditional cereals for producing gluten-free foods. These can be processed and manufactured without many specialized equipments. Pseudocereals being rich source of nutrients and bioactive components (antioxidants) play major role in prevention of cancers, cardiovascular diseases, celiac diseases, diabetes, and others. The future lies in the growing business and industrialization of using pseudocereals as raw or functional ingredients for a healthier lifestyle.

KEYWORDS

- bioactive compounds
- gluten free
- health benefits
- morphology
- nutrition
- pseudocereals

REFERENCES

1. Aguilar, E.C.; Leonel, A.J.; Teixeira, L.G.; Silva, A.R. Butyrate Impairs Atherogenesis by Reducing Plaque Inflammation and Vulnerability and Decreasing Nfkb Activation. *Nutrition Metabolism Cardiovascular Diseases*, **2014**, *24*, 606–613.
2. Alvarez-Jubete, L.; Wijngaarda, H.; Arendt, E.K.; Gallagher, E. Polyphenol Composition and *In Vitro* Antioxidant Activity of Amaranth, Quinoa Buckwheat and Wheat as Affected by Sprouting and Baking. *Food Chemistry*, **2010**, *119*, 770–778.
3. Blaise, P.; Alexender, M. Processing of Top Fermented Beer Brewed From 100% Buckwheat Malt with Sensory and Analytical Characterization. *Journal of Institute of Brewing*, **2010**, *116*, 265–274.
4. Bonafaccia, G.; Gambelli, L.; Fabjan, N.; Kreft, I. Trace Elements in Flour and Bran from Common and Tartary Buckwheat. *Food Chemistry*, **2003**, *83*, 1–5.
5. Bonafaccia, G.; Kreft, I. Technological and Qualitative Characteristics of Food Products Made with Buckwheat. *Fagopyrum*, **1994**, *14*, 35–42.
6. Bressani, R.; Caballero B. (Eds.), *Encyclopedia of Food Sciences and Nutrition*. Oxford, UK: Academic Press; **2003**; pp. 166–173.
7. Champ, M.; Langkilde, A.M.; Brouns, F.; Kettlitz, B.; Bail-Collet, Y.L. Advances in Dietary Fiber Characterization; Consumption, Chemistry, Physiology and Measurement of Resistant Starch; Implications for Health and Food Labeling. *Nutrition Research Reviews*, **2003**, *16*, 143–161.
8. Chaturvedi, N.; Shukla, K.; Vishnoi, D. Appraisal of Alkali Treated Malting on Proximate Composition and Antioxidant Activity of *Amaranthus Cruentus*. *Journal of Global Biosciences*, **2015**, *4*, 3291–3300.
9. Colin, W.; Harold, C.; Charles, E.W. *Encyclopedia of Grain Science*. New York: Elsevier Academic Press; **2004**; p. 1700.
10. Das, S. *Amaranthus: A Promising Crop of Future*. Singapore: Business Media; **2016**; p. 119.
11. De Carvalho, F.G.; Ovidio, P.P.; Padovan, G.J. Metabolic Parameters of Postmenopausal Women after Quinoa or Corn Flakes Intake: A Prospective and Double-Blind Study. *International Journal of Food Science and Nutrition*, **2014**, *65*, 380–385.

12. De Lumen, B.O. Lunasin: A Novel Cancer Preventive Seed Peptide that Modifies Chromatin. *Journal AOAC International,* **2008**, *91,* 932–935.
13. Durazzo, A.; Zaccaria, M.; Polito, A.; Maiani, G.; Carcea, M. Lignan Content in Cereals, Buckwheat and Derived Foods. *Foods*, **2013**, *2*, 53–63.
14. Esan, Y.O.; Omoba, O.; Enujiugha, V. Biochemical and Nutritional Compositions of Two Accessions of *Amaranthus Cruentus* Seed Flour. *American Journal of Food Science and Technology*, **2018**, *6,* 145–150.
15. Filho, A.M.; Pirozi, M.R.; Borges, J.T.; Pinheiro, H. M.; Chaves, J. B.; Coimbra, J. S. Quinoa: Nutritional, Functional and Antinutritional Aspects. *Critical Reviews in Food Science and Nutrition*, **2015**, *57,* 1618–1630.
16. Gallagher, E.; Gormley, T.R.; Arendt, E.K. Recent Advances in the Formulation of Gluten-Free Cereal-Based Products. *Trends Food Science and Technology,* **2004**, *15,* 143–152.
17. Gambus, H.; Gambus, F.; Sabat, R. Quality Improvement of Gluten-Free Bread by Amaranthus Flour. *Zywnosc-Nauka,* **2002**, *9,* 99–112.
18. *Glossary of Agricultural Production: Programs and Policy.* University of Arkansas Division of Agriculture; **2014**; p. 29.
19. Gorinstein, S.; Drzewiecki, J.; Delgado-Licon, E. Relationship between Dicotyledone-Amaranth, Quinoa, Fagopyrum, Soybean and Monocots (Sorghum and Rice) Based on Protein Analyses and Their Use as Substitution of Each Other. *European Food Research and Technology,* **2005**, *221*, 69–77.
20. Hernández-Ledesma, B.; Hsieh, C.C.; de Lumen, B.O. Chemo Preventive Properties of Peptide Lunasin: A Review. *Protein and Peptide Letter*, **2013**, *20,* 424–432.
21. Ikeda, S.; Yamashita, Y. Buckwheat as a Dietary Source of Zinc, Copper and Manganese. *Fagopyrum*, **1994**, *14,* 29 34.
22. Jacobsen, S. E. The Worldwide Potential for Quinoa (*Chenopodium quinoa* Wild.). *Food Reviews International,* **2003**, *19* (1–2), 167–177.
23. Kalinov, J.; Dadakova, E. Rutin and Total Quercetin Content in Amaranth (*Amaranthus* spp.). *Plant Foods for Human Nutrition,* **2009**, *64*, 68–72.
24. Kaur, S.; Kaur, N.; Grover, K. Development and Nutritional Evaluation of Gluten Free Bakery Products Using Pseudocereal quinoa (*Chenopodium quinoa*). *International Journal of Pure and Applied Bioscience,* **2018**, *6,* 810–820.
25. Kim, S.H.; Cui, C.B.; Kang, I.J.; Kim, S.Y.; Ham, S.S. Cytotoxic Effect of Buckwheat (*Fagopyrum Esculentum Moench*) Hull against Cancer Cells. *Journal of Medicinal Food,* **2007**, *10,* 232–238.
26. Kiprovski, B.; Mikulic-Petkovsek, M.; Slatnar, A. Comparison of Phenolic Profiles and Antioxidant Properties of European *Fagopyrum Esculentum* Cultivars. *Food Chemistry,* **2015**, *185,* 41–47.
27. Koziol, M.J. Chemical Composition and Nutritional Evaluation of Quinoa (*Chenopodium quinoa* Willd.). *Journal of Food Composition and Analysis (USA),* **1992**, *5,* 35–68.
28. Klvarez-Jubete, L.; Arendt, E.K.; Gallagher, E. Nutritive Value of Pseudocereals and Their Increasing Use as Functional Gluten-Free Ingredients. *Trends in Food Science & Technology,* **2010**, 21, 106–113.
29. Larson, A.J.; Symons, J.D.; Jalili, T. Therapeutic Potential of Quercetin to Decrease Blood Pressure: Review of Efficacy and Mechanisms. *Advances in Nutrition,* **2012**, *3,* 39–46.
30. Lazaridou, A.; Biliaderis, C.G. Gluten-free Doughs: Rheological Properties, Testing Procedures- Methods and Potential Problems in Gluten-free Food. In: *Gluten Free Food*

Science and Technology; Gallagher, E. (Ed.); Chichester, UK: Wiley Blackwell; **2009**; pp. 52–82.

31. Lee, A.R.; Ng, D.L.; Dave, E.; Ciaccio, E.J. The Effect of Substituting Alternative Grains in the Diet on the Nutritional Profile of the Gluten-Free Diet. *Journal of Human Nutrition and Dietetics*, **2009**, *22*, 359–363.

32. Li, F.; Zhang, X.; Li, Y.; Lu, K.; Yin, R.; Ming, J. Phenolics Extracted from Tartary (*Fagopyrum Tartaricum* L. Gaerth) Buckwheat Bran Exhibit Antioxidant Activity and an Antiproliferative Effect on Human Breast Cancer MDA-MB-231 Cells Through the P38/MAP Kinase Pathway. *Food and Functions*, **2017**, *8*, 177–188.

33. Li, G.; Zhu, F. Quinoa Starch: Structure, Properties, and Applications. *Carbohydrate Polymer*, **2018**, *1*, 851–861.

34. Majewska, M.; Skrzycki, M.; Podsiad, M.; Czeczot, H. Evaluation of Antioxidant Potential of Flavonoids: An *In Vitro* Study. *Acta Poloniae Pharmaceutica*, **2011**, *68*, 611–615.

35. Repo-Carrasco-Valencia, R.A.M.; Serna, L.A. Quinoa (*Chenopodium quinoa*, Willd.) as a Source of Dietary Fiber and other Functional Components. *Food Science and Technology*, **2011**, *31*, 225–230.

36. Mendonça, S.; Saldiva, P.H.; Cruz, R.J.; Areasd, J.A.G. Amaranth Protein Presents Cholesterol-Lowering Effect. *Food Chemistry*, **2009**, *116*, 738–742.

37. Moore, M.M.; Schober, T.J.; Dockery, P.; Arendt, E.K. Textural Comparisons of Gluten-Free and Wheat-Based Doughs, Batters and Breads. *Cereal Chemistry*, **2004**, *81*, 567–575.

38. Moses, T.; Papadopoulou, K.K.; Osbourn, A. Metabolic and Functional Diversity of Saponins, Biosynthetic Intermediates and Semi-Synthetic Derivatives by Tessa Moses. *Critical Reviews in Biochemistry Molecular Biology*, **2014**, *49*, 439–462.

39. Murakami, A.; Ashida, H.; Terao, J. Multitargeted Cancer Prevention by Quercetin. *Cancer Letters*, **2008**, *269*, 315–325.

40. Nancy, R.; Garlick, P.J. Introduction to Protein Summit 2007: Exploring the Impact of High-Quality Protein on Optimal Health. *The American Journal of Clinical Nutrition*, **2008**, *87*, 1551S–1553S.

41. Narwade, S.; Pinto, S. Amaranth: A Functional Food. *Concepts of Dairy & Veterinary Sciences*, **2018**, *2018*, 2637–4749.

42. National Nutrient Database for Standard Reference Release. Washington, DC: USDA (United States Department of Agriculture); April, **2018**; pp. 102.

43. Obendorf, R.L.; Steadman, K.J.; Fuller, D.J.; Horbowicz, M.; Lewis, B.A. Molecular Structure of Fagopyritol A1 (0-α-d-galactopyranosyl-(1→3)-d-chiro-inositol) by NMR. *Carbohydrate Research*, **2000**, *328*, 623–627.

44. Oomah, B.D.; Mazza, G. Flavonoids and Antioxidative Activities in Buckwheat. *Journal of Agricultural and Food Chemistry*, **1996**, *44*, 1746–1750.

45. Prakash, S.; Yadav K., Buckwheat (*Fagopyrum esculentum*) as a Functional Food: A Neutraceutical Pseudocereal. *International Journal of Current Trends in Pharmacobiology and Medical Sciences*, 2016, *1*, 1–15.

46. Ranilla, L.G.; Apostolidis, E.; Genovese, M.I.; Lajolo, F.M.; Shetty, K. Evaluation of Indigenous Grains from the Peruvian Andean Region for Anti-Diabetes and Antihypertension Potential using *in Vitro* Methods. *Journal of Medicinal Foods*, **2009**, *12*, 704–713.

47. Ren, G.; Zhu, Y.; Shi, Z.; Li, J. Detection of Lunasin in Quinoa (*Chenopodium quinoa* Willd.) and the *In Vitro* Evaluation of its Antioxidant and Anti-Inflammatory Activities. *Journal of Food Science and Technology,* **2017**, *97,* 4110–4116.

48. Renzetti, S.; Rosell, C.M. Role of Enzymes in Improving the Functionality of Proteins in Non-Wheat Dough Systems. *Journal of Cereal Sciences*, **2016**, *67*, 35–45.

49. Repo-Carrasco-Valencia, R.; Hellstrom, J.K.; Pihlava, J.M.; Mattila, P.H. Flavonoids and Other Phenolic Compounds in Andean Indigenous Grains: Quinoa (*Chenopodium quinoa*), Kaniwa (*Chenopodium pallidicaule*) and Kiwicha (*Amaranthus caudatus*). *Food Chemistry,* **2010**, *120*, 128–133.

50. Rodas, B.; Bressani, R. The Oil, Fatty Acid and Squalene Content of Varieties of Raw and Processed Amaranth Grain. *Archivos Latinoamericanos de Nutrition,* **2009**, *59*, 82–87.

51. Scalbert, A.; Manach, C.; Morand, C.; Remesy, C.; Jimenez, L. Dietary Polyphenols and the Prevention of Diseases. *Food Science and Nutrition*, **2005**, *45*, 287–306.

52. Schoenlechner, R.; Siebenhandl, S.; Berghofer, E. Pseudocereals. Chapter 7; In: *Gluten-free Cereal Products and Beverages*; Arendt, E. and Dal Bello, F. (Eds.); New York: Elsevier; **2008**; p. 464.

53. Schulze, M.B.; Schulz, M.; Heidemann, C.; Schienkiewitz, A. Fiber and Magnesium Intake and Incidence of Type 2 Diabetes: A Prospective Study and Meta-analysis. *Archives of Internal Medicine*, **2007**, *167*, 956–965.

54. Shukla, S.; Pandey, V.; Pachauri, G.; Dixit, B.S.; Banerji, R.; Singh, S.P. Nutritional Contents of Different Foliage Cuttings of Vegetable Amaranth. *Plant Foods for Human Nutrition*, **2003**, *58*, 1–8.

55. Singh, S.; Singh, R.; Singh, K.V. Quinoa (*Chenopodium Quinoa Willd*), Functional Superfood for Today's World: A Review. *World Scientific News*, **2016**, *58*, 84–96.

56. Skrabanja, V.; Kreft, I.; Golob, T.; Modic, M. Nutrient Content in Buckwheat Milling Fractions. *Cereal Chemistry*, **2004**, *81,* 172–176.

57. Steadman, K.J.; Burgoon, M.S.; Lewis, B.A. Minerals, Phytic Acid, Tannin and Rutin in Buckwheat Seed Milling Fractions. *Journal of the Science of Food and Agriculture*, **2000**, *81*, 1094–1100.

58. Tian, X.; Tang, H.; Lin, H.; Cheng, G.; Wang, S.; Zhang, X. Saponins: The Potential Chemotherapeutic Agents in Pursuing New Anti-Glioblastoma Drugs. *Mini-Reviews in Medicinal Chemistry*, **2013**, *13,* 1709–1724.

59. Zannini, E.; Jones, J.M.; Renzetti, S.; Arendt, E.K. Functional Replacements for Gluten. *Annual Review of Food Science Technology,* **2012**, *3*, 227–245.

CHAPTER 7

CHIA SEEDS: COMPOSITION, HEALTH BENEFITS, AND POTENTIAL APPLICATIONS

RESHU RAJPUT, HARINDERJEET KAUR BHULLAR, AMARJEET KAUR, and JASPREET KAUR

ABSTRACT

Chia (*Salvia hispanica*) is an oilseed plant, which contains natural polyphenols, omega-3 fatty acids mainly α-linolenic acid, high-quality proteins, dietary fibers, vitamins, and minerals. Because of its rich nutritional composition and health benefits, it has grabbed attention of many food industries and educators and has been used in the production of various food products (such as cakes, pasta, ice cream, chips, etc.). Chia can be considered a functional food because of health beneficial effects to prevent many diseases. This chapter focuses on nutritional composition and potential health benefits of chia.

7.1 INTRODUCTION

Chia (*Salvia hispanica*) seeds are edible seeds with about 900 species that are used as folk medicines and widely cultivated because of its wide applications. The name *Salvia* originated from the Latin word "*Salvare*," which means "the healer" [24]. It originated from the valleys of Central America, southern Mexico, and northern Guatemala. Thousands of years ago, chia seed was a staple food of the ancient Mayas and Aztecs. In the Columbian societies, it was the second major crop in prehistoric times after beans [3]. Chia is an integral part of the diet of the Columbian society, which has made their diets superior to today's diet and it has also been accepted by the modern consumers. The word chia is being obtained from Spanish word "chian," meaning oily. In Mayan language, chia means "strength" and the warriors of Aztec depended on chia seed to enhance their energy and stamina. Today this small seed has become favorite of athletes. Chia is an oilseed which has omega-3 fatty acids, appreciable content of dietary fiber, good quality protein, minerals, vitamins, and large amount of antioxidant compounds (such as tocopherols, carotenoids, sterols, phenolic compounds) including chlorogenic acid, quercetin, kaempferol, caffeic acid, and myricetin [16].

This chapter focuses on the nutritional and phytochemical composition and properties of chia and their potential health benefits.

7.2 HISTORY OF CHIA SEED

Chia belongs to central Mexico and northern Guatemala. Chia seeds came into knowledge for human consumption in 3500 BCE and attained attention as main grain crop around 1500 and 900 BCE in Central America [7, 16]. The seeds of chia were used by the people of Mayas and Aztecs for the purpose of food, folk medicines, and paintings. The Aztecs gets chia seeds as a yearly tribute and also used them in religious ceremonies as an offering to the gods [25]. Chia seeds were consumed as whole or grounded for food but were also pressed for the extraction of oil, which is used for making face paints. For thousands of years, chia seeds have been largely consumed in many countries because of its huge nutritional and therapeutic potential. The pre-Columbian society people used to prepare a popular drink from chia seeds known as "chia fresca," which means fresh chia and is still popular today. Modern science has accepted that the diets of the pre-Columbian people were superior to today's diet.

Current studies have facilitated to enlighten the fact that why the ancient civilizations have chia as their staple diet component. The nutritional and chemical composition of the seed makes it a potential crop for commercialization; and technological advancements have developed outstanding opportunities to establish an agricultural sector, which is able to contribute a new and an old crop to the world at the same time.

7.3 BOTANICAL DESCRIPTION OF CHIA

Chia (*Salvia hispanica*) is a flowering plant, which is cultivated biannually, belonging to the mint family *Lamiaceae*. It is commonly known as chia, Mexican chia, Spanish sage, and black chia [89]. The plant of chia can grow up to 1 m in height, bears flowers in summer, and has reverse petiolate and oppositely arranged leaves having length 4–8 cm and width 3–5 cm. Chia flowers are small in size and are white or purple in color (Figure 7.1). These are hermaphrodite flowers of 3–4 mm with corollas small in size and have high self-pollination system due to the presence of merged flower parts. The seeds are smooth, shiny, and oval and the size varies from 1 to 2 mm (Figure 7.2). The herb can be grown in an extensive range of sandy soils and drained clay and has a great resistance to acidic and salty soils [7, 54]. There is only little difference between the wild and domesticated chia varieties. Currently only *Salvia hispanica* species can be grown domestically among the genus *Salvia.* Under appropriate agronomic conditions, it can produce 1250–1500 kg of seeds/ha [16].

FIGURE 7.1 Chia flowers.

Source: Photo by Dick Culbert from Gibsons, B.C., Canada. https://en.wikipedia.org/wiki/Salvia_hispanica#/media/File:Salvia_hispanica_(10461546364).jpg https://creativecommons.org/licenses/by/2.0/deed.en

FIGURE 7.2   Chia seeds.

Source: Photo by Magister Mathematicae. https://en.wikipedia.org/wiki/Chia_seed#/media/
File:Semillas_de_Chia.jpg https://creativecommons.org/licenses/by-sa/3.0/

7.4   CHIA PRODUCTION

The chia is commercially produced in Paraguay and Bolivia. Also, Argentina, Mexico, and Australia have started growing chia plants with good results during the past several years. In 2014, an unusual increase in price prompted sudden production of chia in various other countries, with selling price of 8000–12,000 US$ per ton. The price of chia keeps on fluctuating depending on the demand, production volume, and quality. Bolivia ranks first in chia production and also the highest exporter of chia.

The area of cultivation has increased from 50,000 ha producing around 18,000 tons of chia seeds to a cultivation area of 80,000 ha producing around 30,000 tons per year [67]. The ideal yield per ha is dependent on climatic regions, harvest equipments and techniques of cultivation, crop varieties, soil texture, and type. Under some conditions, the yield of chia seeds is around 500–600 kg/ha with minimum inputs, though some cultivators can yield 1200 kg per ha [16, 19, 89]. Bolivia tried was successful in increasing chia seed yield from 350 kg per ha in 2013 to 650 kg per ha in 2014, which ranked after Paraguay and Argentina.

7.5 CONSUMPTION OF CHIA FOR HEALTHCARE

Chia seeds are greatly valued due to their high nutritional and medicinal values. Chia is supplemented and added as an ingredient in different food formulations due to presence of healthy omega-3 fatty acids. Beneficial health effects and nutritional composition of chia seeds are the supreme reasons, which have convinced the customers for buying chia [67, 75]. The various preparations of packaged chia seeds are available in the market and the recommended daily intake is around 15–25 g/day.

There are several beneficial effects of chia seeds that can be attained by consuming chia daily, such as it helps in reducing blood pressure and cholesterol, reduction in weight, reduced joint pain and antioxidant effects. Several research works have revealed beneficial effects of consumption of chia daily by consuming 35–37 g of chia powder of seeds [85, 97]. The chia seeds consumption offers omega-3 fatty acids for the athletes that require stamina [39]. Research has established that the dietary fibers and omega-3 fatty acids in chia seeds are indulged in the reduction of weight. However, it is regarded safe for short-term use, and only limited data is available for its safety in the long-term use [62].

Consumption of grinded chia seed, chia oil and proteins of biological importance has a positive effect on our health because of its antioxidant capacity and the safeguarding of cells from free radicals [49, 65]. Numerous food products in the market have been supplemented with chia seeds, such as seeds, seed flour, chia oil capsules as nutritional supplements, and different food formulations (such as ice cream, margarine, biscuits, yoghurt, etc.). Chia seeds and flour can be incorporated to crisps, fruit, nut, peanut butter, breakfast cereals, and other seed mixes, shakes, drinks, or bakery products (i.e., cake mixes and bread). Cosmetic products also contain chia, which is related to skin creams and liquid eye essence.

7.6 NUTRITIONAL COMPOSITION OF CHIA

This ancient grain is tremendously gaining popularity in many of the countries in the modern food regimen due to its massive nutritional and therapeutic perspectives. European parliament and the European Council approved chia as a novel food in 2009 [54]. Chia seed is composed of 15%–25% protein, 30%–33% fats, 18%–30% dietary fiber, 4%–5% ash, 26%–41% carbohydrates, minerals, vitamins and dry matter (Tables 7.1 and 7.2). Large amounts of antioxidants are also present in chia seeds [40].

TABLE 7.1   Amino Acid Content in Chia Seeds

Amino Acid	Content (g per 100 g Seeds)
Alanine	1.044
Arginine	2.143
Aspartic acid	1.689
Cysteine	0.407
Glutamic acid	3.500
Glycine	0.943
Histidine	0.531
Leucine	1.371
Lysine	0.970
Methionine	0.588
Phenylalanine	1.016
Proline	0.776
Threonine	0.709
Isoleucine	0.801
Tryptophan	0.436
Tyrosine	0.563
Valine	0.950

TABLE 7.2   Vitamin Content in Chia Seeds

Vitamin	Amount
Folate, µg	49
Niacin, mg	8.83
Riboflavin, mg	0.17
Thiamine, mg	0.62
Vitamin A, IU	54
Vitamin C, mg	1.6
Vitamin E, mg	0.5

Evaluation of heavy metals in chia revealed that they are present at safer level making them suitable for human consumption, because the seed is also free from mycotoxins [69]. Also. chia seeds are gluten-free so that chia food products can be safely consumed by people with celiac disease. Currently chia seed is largely used for the extraction of bioactive compounds for developing functional foods [13].

They are rich source of essential amino acids, particularly valine, lysine, leucine and isoleucine (Table 7.1). Therefore the seeds can be used in place of cereal proteins, which are deficient in essential amino acids. The seed has a higher content of α-linolenic acid, which makes it an excellent source of omega-3 fatty acid (about 65% of the total oil content). It is a great source of dietary fiber, which is characterized by high water absorption and has the ability to form aqueous solutions with higher viscosity. This helps in maintaining the desired moisture in the final product, so that it can stay fresh for longer time. Dietary fiber is beneficial for the digestive system and helps in controlling *diabetes mellitus*. Chia seed is also a potential source of antioxidants containing caffeic acid, kaempferol, myricetin, chlorogenic acid, and quercetin, which protect against various adverse conditions and have hepatic protective, antiageing and anticarcinogenic properties [89].

7.6.1 FIBER CONTENT

In the daily diet, dietary fiber plays an important role because of its beneficial effect on human health. Some of the beneficial effects of dietary fiber include modification of the glycemic and insulin responses, decrease in the risk for coronary heart disease, reduction of cholesterol, alteration in the intestinal function, controls type II *diabetes mellitus*, and several type of cancers [75]. In addition, the consumption of dietary fiber has been linked with the increase of postmeal satiety and reduces subsequent hunger.

The American Dietetic Association has revealed that dietary fibers possess beneficial health effects for maintaining good human health and prevents diseases [91, 93]. Chia seed has 34 to 40 g of dietary fiber (Table 7.3), which meets the per day requirement for the human consumption specifically for the adult population. The chia flour contains 40% of dietary fiber of which 5% to 10% is soluble dietary fiber, which forms the mucilage [54]. Chia seeds are hydrophilic in nature; when soaked in liquid, they become plump as a gelatinous mucilage capsule forms around the seed. The mucilage of chia has around 48% total sugar content, 1% fat, 4% protein, and 8% ash.

TABLE 7.3   Fiber Content of Selected Foods

Food	Fiber (g per 100 g)
Almonds	12.2
Amaranth	6.7
Chia	34.4
Dried apple	8.7
Dried banana	9.9
Dried fig	9.8
Dried peaches	8.2
Dried pears	7.5
Dried plums	7.1
Flax seed	27.3
Peanuts	8.5
Quinoa	7.0
Soybean	9.6

The chia seed fiber has large molecular weight and has tetrasaccharide as basic structure with 4-O-methyl-a-D-glucoronopyranosyl residues that get branched on the main chain with b-D-xylopyranosyl. Chia mucilage consists of mainly sugars, such as galactose, xylose, glucose arabinose, and galacturonic acids and very less information is available regarding the chemical structure of mucilage. Insoluble dietary fiber is abundant in chia seeds that accounts for approximately 87% of total fiber content. The main component in insoluble dietary fiber is Klason lignin that performs a vital function in protecting unsaturated fats [31]. The high fiber content and antioxidant properties of chia seeds are thought to protect the unsaturated fats present in them, where structural support is provided by fiber and the antioxidant capacity protects it from oxidative rancidity. The fiber content in chia seed is higher compared to amaranth, flaxseed and quinoa and even higher compared to different dried products (Table 7.3). Therefore, chia seed can be used in the prevention of many cardiovascular diseases [89].

7.6.2   PROTEINS

Chia is a nutrient-rich food, which has a great potential for being a functional food [77]. Though chia is not a globally recognized food, it has gained popularity commercially because of its outstanding content of biochemical compounds in Mexico and North and South America [7]. Due to proteins in chia seeds, it is a vital nutraceutical food that provides several health benefits.

The protein content in chia seeds ranges from 15% to 24% (Table 7.4), higher than that of corn, wheat, barley, oats, and rice, thus making chia an exceptional source of good quality proteins containing an appreciable amount of essential amino acids [7]. Therefore chia has proved to be a potential source for preventing health problems arising from protein deficiency. The chia protein content varies for different seeds grown in different habitats and it ranges from 18% to 22%, and the differences in protein content is possibly due to different climatic, soil and agronomic conditions [89].

TABLE 7.4 Protein Content of Some Cereals

Cereal	Protein Content (%)
Barley	12.48
Chia	20.70
Corn	9.42
Oats	16.89
Rice	6.50

Additional outstanding characteristic of chia is that it is gluten free, therefore the patients with celiac disease can digest it easily. Several studies have shown that there are total nine amino acids in chia in appreciable amounts. Also, various vital fractions of storage protein are present in chia seeds. Among these, globulins are the major protein fractions approximately 52%–54% as compared 17.3%–18.6% to albumins, 13.6% of glutelin, 13.6% and 17.9% of prolamin. Thermal analysis of protein fractions indicated that the albumins and globulins showed a better thermal stability; and the denaturation temperature of globulins, albumins, glutelin and prolamins was 105, 103, 91, and 85.6 °C, respectively.

Foods rich in protein content helps in weight loss because of the loss in fats from the body. Study on the effect of protein on weight loss showed that the consumption of 25% of the total protein results in significant loss in fats [82]. Consumption of high protein diet in daily routine might also help in maintaining the weight.

7.6.3   LIPIDS IN CHIA

The significant feature of chia seeds is high content of unsaturated fatty acids that are essential for healthy body. It consists of approximately 25%–40% of polyunsaturated fatty acids (PUFAs) that includes 55%–60% of omega-3 fatty acid (linolenic acid) and 18%–20% of linoleic acid. The highest percentage of α-linolenic acid is exhibited by chia seeds in comparison with any of other plant sources; and it is precursor of the long-chain unsaturated fatty acid. Since ancient times, chia oil has been used for treating infections of the eyes and disorders of the stomach [75]. Several research works have revealed that the oil of chia seeds contains phytocompounds, such as carotenoids, tocopherols, and phytosterols with antioxidant activities for the protection of lipids from deteriorating due to lipid oxidation [41, 51]. Research has shown that 100% of the recommended daily intake of omega-3 fatty acids can be achieved by consuming 7.3 g chia seed per day that is helpful in preventing several chronic diet-related diseases.

The omega-3 fatty acid in chia has a positive impact on human health due to its antiarrhythmic, antithrombotic, and anti-inflammatory activities [34, 35]. Saponification value, iodine value, and tocopherol content of chia oil is 193.3, 207, and 480 mg/kg, respectively and the amount of Cu and Fe is 0.1 and 0.3 ppm, respectively. Research has revealed that the chia seed oil has lower anisidine and peroxide value of 0.3 and 1.0 meqO$_2$/kg. The yield of the extracted oil from chia seeds varies with different habitats. Coates and Ayerza [20] revealed that the yield of chia seed oil was higher in Argentina compared to other locations in South America, due to agronomic, fertilization, irrigation practices, and so on. A natural antioxidant tocopherol in chia seed oil can be beneficial in extending its storage life.

An essential oil from chia seeds is produced in higher concentration that is being used to prepare supplements of omega-3. A metabolic syndrome dyslipidemia due to higher low-density lipoprotein (LDL) triglyceride and lower high-density lipoprotein (HDL cholesterol can be prevented by increased concentration of omega-3 fatty acids in oil.

7.6.4   COMPOSITION OF FATTY ACIDS IN OATS

The incidence of cardiovascular diseases, diabetes, obesity, hypertension, and various health-related problems has led the consumers to consume foods with unsaturated fatty oils than the saturated fats for maintaining good health for healthy lifestyle [36]. Saturated fats result in hypercholesterolemic effect with negative health impact [47]. The American Heart Association also guides people to shift from saturated fats to unsaturated oils [91].

Due to the presence of the unsaturated fatty acids chia seed is undoubtedly very correctly said to be as energy house of omega fatty acids. The chemical structure of fatty acids has been widely studied. There are three carbon atoms present in omega-3 fatty acids between the double bond methyl group at the end; and omega-6 fatty acids have 6 carbon atoms between the methyl group and the double bond at the end [68]. Omega-3 fatty acid has three essential fatty acids, such as eicosapentaenoic acid (EPA), α-linolenic, and docosahexaenoic acid (DHA) compared to arachidonic acid and linoleic acid in omega-6 fatty acid [68]. The popularity and cultivation of chia have increased because of the presence of huge concentration of PUFA. All these fatty acids have many health beneficial effects, such as cardioprotection [48].

7.6.5   MINERALS

Chia seed is a huge reservoir of nutrients and has a great potential to be used for the development of functional food. The chia seed contains 11 times higher phosphorous, 6 times higher calcium, and 4 times higher magnesium than those in rice, corn, wheat, and oats. The potassium, calcium, and phosphorous content of chia is 6 times higher than that in spinach [10]. Chia seeds contain an appreciable amount of macronutrients (mg/100 g), such as magnesium: 335, potassium: 407, calcium: 631, and phosphorous: 860. Chia seeds also contain micronutrients (μg/100 g), such as manganese: 2.72, molybdenum: 0.2, copper: 0.924, selenium: 55.2, iron: 7.72 and zinc: 4.58 [92].

7.6.6   VITAMINS

Chia seed contains appreciable amounts of vitamins (Table 7.2). It has higher niacin content compared to corn, rice, and soybean [93]. The vitamins of B group (thiamine and riboflavin) has the content similar that is present in rice and corn [15].

7.6.7   PHENOLIC COMPOUNDS IN CHIA SEEDS

Chia is a rich source of phenolic compounds with antioxidant activities. The most common polyphenols in chia are derivatives of cinnamic acid and flavonoids. The phenolic compounds content in chia comprises of 0.88–1.6 mg of gallic acid equivalent per g. The potential antioxidants (Table 7.5) in chia seeds are caffeic acid, chlorogenic acid, kaempferol, ferulic acid, quercetin, myricetin, and rosmarinic acids that exhibit antihypertensive, anticarcinogenic, neuroprotective, hepaticprotective and antiageing activities [81]. Some isoflavones are also present in chia seeds, such as, genistin, glycitin, daidzin, glycitein and genistein [66].

TABLE 7.5 Concentration of Phenolic Compounds in Chia Seeds

Bioactive Compound	Concentration (mg per g Seed)	Bioactive Compound	Concentration (mg per g Seed)
Phenolic acids		**Isoflavones**	
Caffeic acid	0.027–0.086	Daidzin	0.0066
Chlorogenic acid	0.013–0.074	Genistein	0.0051
Gallic acid	0.0115	Genistin	0.0034
Rosmarinic acid	0.9267	Glycitein	0.0005
Flavanols		Glycitin	0.0014
Kaempferol	0.0057–0.0435		
Myricetin	0.0095		
Quercetin	0.0181–0.209		

Antioxidants can fight against oxidative stresses and may reduce risks against diabetes, cancers, arteriosclerosis, thrombosis, inflammation, and others, thus keep our body healthy and stress free [43, 62, 93]. Reyes-Caudillo et al. [75] indicated that the amount of phenolics in chia seed was observed to be 8.8% on the basis of dry matter. The total amount of tocopherol in chia seed ranges from 238 to 427 mg/kg, which is comparable to the amount present in peanut oil but lower than that in soybean (1797.6 mg/kg), flaxseed (5588.5 mg/kg) and sunflower (634.4 mg/kg) [87]. Uribe et al. [90] indicated that chia seed can aid in preserving food lipid systems. Ayerza and Coates [8] isolated polyphenols (such as quercetin, myricetin, chlorogenic acid, kaempferol, and caffeic acid) from chia seeds. Moreover, chia seed antioxidant activity was higher than that of the extracts from *Moringa oleifera* and sesame according to Nadeem et al. [59]. Tepe et al. [83] concluded that the lipid peroxidation phenomenon can be significantly inhibited by the phenolics in chia.

7.6.8 CAFFEIC ACID DERIVATIVES

In chia seeds, caffeic acid has significant vital function from biological and chemical point of view. Caffeic acid is mainly composed of a phenyl group with two hydroxyl group, which is attached to acrylic acid and it represents a basic molecular structure of secondary metabolites in *Lamiaceae* family. Caffeic acid is also categorized as hydroxycinnamic acid, which is linked to quinic acid to form caffeoylquinic acids. Among these acids, chlorogenic acid is most dominant in chia [50]. Various monomers of caffeic acid are present in chia seeds and also polymers, which are products of condensation. By applying Ultra high-performance liquid chromatography (UHP-LC), monomers like caffeic acid and ferulic acid have been isolated from chia seeds. Dimers of caffeic acid are also found in chia and rosmarinic acid is more abundant, which is about 0.9267 mg/kg according to Martinez-Cruz et al. [50]. Rosmarinic acid possesses immunoregulatory functions, such as antibacterial, anti-inflammatory, antioxidant, antiviral, antimutagen, and antithrombotic [38]. Both caffeic and rosmarinic acids have an important role in the management and prevention of various neurological disorders, i.e., epilepsy [23]. Caffeic acid possesses properties, i.e., memory protective and boosting effect and also hypoglycemic activity.

7.6.9 FLAVONOIDS

Flavonoids are plant metabolites that provide health benefits through cell signaling pathways and antioxidant effects. These are polyphenolic compounds that contain 15 carbon atoms in their structure made up of two A- and B-benzene rings connected through a pyran ring C, which is heterocyclic and these are soluble in water. Flavonoids are mainly responsible for appearance, taste, and help in prevention of fat from oxidation [98]. Like caffeic acid, flavonoids are also extensively present in chia seeds and also possess bioactive activities, such as: anticancer, antiviral, antioxidant, antibacterial, hepatoprotective, and anti-inflammatory [31]. The synthesis of flavonoids is enhanced in chia seeds when there is microbial attack. Many researcher workers have extensively studied the flavonoid compounds in chia obtained from seed extract through methanolic hydrolysis and have evaluated their antioxidant activity. Myricetin is the major flavonol in the Iztac and Tzotzol chia seeds containing 0.121 and 0.115 mg/g, respectively [6].

7.7   HEALTH BENEFITS AND THERAPEUTIC PERSPECTIVES OF CHIA SEEDS

7.7.1   POTENTIAL EFFECTS ON IMMUNE SYSTEM

Fernandez et al. [33] conducted a study one month on 23-days old Weanling male Wistar rats to evaluate the effect of chia seed on the immune system. They used thymus and serum IgE (immunoglobulin E) concentrations as indicators of immunity. Authors found that the concentration of IgE was considerably higher with chia diet compared to the control. Also, authors observed no symptoms of diarrhea, abnormal behavior, dermatitis or any other allergy with chia seeds or oil, whereas fishy flavor, diarrhea, allergy, problems with gastrointestinal tract were observed with supplementation of diets with different sources of omega-3 fatty acids (such as marine products or flaxseeds) [7].

7.7.2   CARDIOPROTECTIVE EFFECTS

EPA and α-linolenic acid play important role in the formation of beneficial biochemical compounds (such as leukotriene, prostaglandin, and thromboxane) [68]. According to USDA [91–93], chia is an excellent choice for daily supplementation as it fulfils the daily dietary allowance for linoleic and α-linolenic acids. Omega-3 fatty acids aid in maintaining good health of the heart by improving heart rate variability, parasympathetic tone, and ventricular arrhythmia [46]. Increased ingestion of alpha-linolenic acid prominently reduces the risk of heart failures. An investigation conducted in St. Michael Hospital, Toronto, Canada revealed outstanding benefits of chia, such as [33]; (1) higher contents of iron, fiber content, magnesium and calcium than those in milk; (2) gluten-free; (3) blood glucose level in diabetic patients got stabilized by consuming 37 g seeds on daily basis; (4) excellent source of omega-3 fatty acids; (5) reduced systolic blood pressure up to 6 mm Hg; and (6) prevented myocardial infarction and strokes by suppressing aggregation of platelets. Some excellent medicinal properties have also been encountered in chia seeds including inhibition of blood clotting, prevent neurological disorders (such as epilepsy and stress), decreases blood cholesterol, and boost immune system. It has been established that during pregnancy consumption of chia seeds can be effective as it promotes development of retina and brain of fetus [33].

7.7.3 INFLUENCE OF CHIA OIL ON INSULIN SIGNALING

The research results indicated that mice fed with chia oil reduced fat mass accumulation and increased lean mass, increased levels of high-density lipoprotein cholesterol, and improved glucose levels and insulin tolerance. The dietary chia seeds improved the altered metabolic fate of glucose in rats fed on a sucrose-rich diet and also reduced the collagen deposition in the heart of dyslipidemia insulin-resistant rats.

7.7.4 ACE (ANGIOTENSIN I-CONVERTING ENZYME) INHIBITORY PEPTIDES

Due to the known negative effects of artificial angiotensin converting enzymes-I (ACE-I) on human body, researchers have been focusing on natural bioactive peptides with ACE-I inhibitory effect, for which one peptide of chia was examined for hydrolysate and ultra-filtered fractions. It was revealed that the inhibition of hydrolysate fraction was 58.46% compared to 69.3% of ultrafiltered fractions. The results recommended the possible isolation of bioactive peptides to reduce risk factors in a more natural way by controlling ACE-I. Segura-Campos et al. [80] conducted a study by the addition of hydrolyzed chia proteins into white bread and carrot cream and they observed that the hydrolysis generated peptides have potential ACE inhibitory activities without any significant effect on the quality of these food products.

7.7.5 OMEGA-3 FATTY ACIDS AS NUTRACEUTICAL FROM CHIA

Chia seed is a significant source of omega-3, especially α-linolenic acid. The fish oil contains higher levels of omega-3 fatty acids compared with other available oils. However, chia oil has a higher percentage of omega-3 fatty acids per 100 g compared with salmon, herring, cod liver, and sardine oils. Due the contents of α-linolenic acid in chia, these omega-3 fatty acids in chia have potential in reducing the cardiovascular diseases and some types of cancer, serum LDL triglyceride, and in increasing HDL cholesterol.

The chia seeds have good ratio of omega-3 and omega-6 fatty acids and has a high content of ALA (56.98 g/100 g oil) with health benefits of reduction in diabetes and metabolic syndrome. Significant benefits of chia oil can be found in its skin applications because of omega-3 and omega-6 in chia seeds, which show good epidermal barrier ability for prevention of melanin hyperpigmentation [2, 73]. Ayerza [5] studied the effect of extract of chia oil rich in omega-3 and omega-6 against melanogenesis and it was proposed that the possible mechanism was the expression of genes responsible for melanogenesis encoding main melanogenic enzymes that were suppressed by chia seeds [30]. These results are beneficial to cosmetic and pharmaceutical industries.

In a Wistar rat experiment, where standard basal diet (soybean oil and cellulose) was replaced by heat-treated or untreated chia seeds and chia flour, studies revealed a reduction in weight compared with the group of test animals on a control diet [27].

7.7.6   ANTICHOLESTEROLEMIC

Chia seeds consumption has revealed its ability to reduce the levels of serum cholesterol, due to high contents of dietary fiber and omega-3 fatty acids [26, 72]. Recently, it has been determined that bioactive peptides of chia and chia proteins can block key markers responsible for cholesterol synthesis, such as 3-hydroxy-3-methylglutaryl coenzyme-A (HMG-CoA) reductase [22]. A clinical study was conducted in which 10 menopausal women consumed 25 g/day milled chia seeds for 7 weeks and it was observed that there was an increase in the plasma levels of DHA and alpha-linolenic acid (ALA) by 30% and 138%, respectively compared with the baseline levels [44]. Toscano et al. [85] reported that consuming 35 g/day of chia flour by obese and overweight subjects aided in reduction in low-density lipoprotein and total cholesterol and increases HDL cholesterol. Moreover, the 62 post-menopausal women on 25 g/day of milled chia for 10 weeks showed increased plasma levels of EPA and ALA [63].

7.7.7   HYPOGLYCEMIC EFFECTS

Several research works have proved that chia seeds possess hypoglycaemic effect. The consumption of chia seeds and chia flour helped in reducing the level of blood glucose compared with control [27]. Ho et al. [37] reported that people consuming chia seeds fortified bread has shown lower postprandial glycaemia compared to those consuming bread without chia seeds. With the addition of 24 g of chia seeds, lowest level of glycaemia was observed, whereas with 7 g addition it was highest, therefore this effect was dose-dependent. Research workers established that the ability of chia seeds to exhibit hypoglycaemic effect was due to the presence of high dietary fiber content [37].

7.7.8   ANTIOXIDANT ACTIVITY

Several research works have documented higher antioxidant activity of chia seeds [21, 78] compared to flaxseed. The ability to scavenge DPPH radicals and to reduce ions of iron is also shown by chia seeds [21]. Coelho and Salas-Mellado [21] confirmed that enzymatic oxidation of guaiacol was inhibited by the extracts of chia seeds. Segura-Campos et al. [80] confirmed that protein hydrolysates of chia seeds have also the potential to reduce ABTS cation radicals. Antioxidant activity of chia seeds compounds has been confirmed in the fat emulsion system. Marineli et al. [49] conducted an experiment with obese rats on diet with chia seeds at 133 g/kg for 6 weeks and chia oil at 40 g/kg for 12 weeks. Few days after consumption, activity of antioxidant enzymes (catalase, glutathione reductase, glutathione, and glutathione peroxidase) was increased significantly in animals fed on seeds or oil compared to animals consuming a high-fructose diet without chia supplementation. Also, when chia seeds were consumed for a period of 12 weeks, there was an increase in activity of hepatic antioxidant enzymes (such as glutathione reductase and glutathione).

7.8   APPLICATIONS OF CHIA

7.8.1   FOOD USES

Nowadays, chia is being consumed for several purposes in New Zealand, Mexico, Argentina, Japan, Chile, Australia, Canada, and the United States. The European Parliament and Council of Europe in 2009 approved chia seed as Novel food. According to the several research works, chia does not impose any antinutritional or toxic effects. Chia seed can be consumed as such or it can be hydrated. Consumption of soaked chia seed provides cooling effect to body. On the other hand, these seeds can also be added in fruit juices or sugar solutions. Mexicans have prepared a drink with chia seeds naming *"agua de chia"* by soaking seeds in the mixture of water and lemon juice [55]. The important functions of chia seeds in food systems include maintaining structure of gel, texture, appearance, and others. The important applications of the seeds are nutritional supplementation. The chia seeds are deoiled and milled to 150 μm size for manufacturing commercial chia flour that can be incorporated in numerous food products, such as cakes, biscuits, pasta, cereal bars, snacks and yoghurt for enhancing their nutritional value.

As already discussed earlier, oil is extracted from chia seeds by pressing or Soxhlet. The essential omega-3 fatty acids are abundant in oil. The chia seed oil is exploited for the production of capsules supplemented with omega-3. The oil of chia is beneficial for commercial applications as it can be directly added to food products without flavor-alteration, especially in products containing fish oil [1].

Another application of chia is the use of mucilage in food industry. Mucilage is majorly a polysaccharide, which is a soluble fiber. Several recent works have established that mucilage can generate various blends of polymers to form coatings and films with enhanced characteristics [56]. In ice creams and other frozen desserts, fresh mucilage can be added as stabilizer. Chia seed mucilage can be used as a thickener (i.e., guar gum and gelatin) in different stages of food preparation. Also, chia mucilage can be dried by freeze drying, spray drying, and oven drying for preparing mucilage powder that can be used in preparation of various health and energy drinks [42].

7.8.2   CHIA SEEDS

Due to hydrophilic properties of chia, its seeds are used as substitutes for eggs and fats. The water can be absorbed by the seeds as much as 12-folds greater than their own mass [57]. This property of chia seeds helps in providing food with specific consistency. Chia seeds currently are used as whole or grinded to flour, gel, and oil [45]. The chia seeds aids in calorie and fat reduction in bakery products. Moreover, the content of omega-3 fatty acids is slightly higher in case of baked products. Borneo et al. [12] revealed that the gel of chia in cakes can replace as much as 25% of eggs or oils for improving the taste, color, texture, and overall acceptance. However, some negative change has been observed in the overall quality of baked products with 50%–75% replacement of oil in dough [12].

Oliveira et al. [64] showed that chia seed flour can used to prepare pasta, which can act as a substitute for wheat flour. With incorporation of 7.5% of chia flour in wheat flour exhibited higher nutritive value than the control pasta. This pasta showed good technological properties with greater acceptability of taste. Menga et al. [52] prepared gluten-free fresh pasta by the addition of chia seeds and mucilage to the rice flour. They proposed that healthy and nutritious gluten-free pasta can be made with addition of 10% of mucilage or chia seeds, which is attributed to good cooking characteristics and its firmness was equivalent to the commercial product.

Coelho and Salas-Mellado [21] reported that breads prepared with the addition of chia flour or chia seeds resulted in good quality bread with greater acceptance. They revealed that the incorporation of chia flour into breads in two trials with two different concentrations of chia flour (7.8 and 11.0 g/100 g) resulted in a significant beneficial ratio of PUFAs to saturated fatty acids in the final product compared to the control bread. Coorey et al. [24] studied the effects of incorporating flour of chia on the nutritional and consumer acceptability of chips and a positive effect was observed on nutritional and sensory qualities of chips. They showed that substitution of 5% potato and rice flour with chia flour was beneficial for texture, color, appearance, aroma, taste and overall acceptability of the final product.

Pintado et al. [70] examined different strategies to incorporate 10% chia flour and olive oil to modulate the fat content in frankfurters. Incorporation of chia increased the total dietary fiber and mineral contents, also PUFAs in frankfurters irrespective of incorporation strategy. Reduction in energy and fat contents were achieved in frankfurters formulated with olive oil and chia.

7.8.3   CHIA GUM

Dietary fibers present in foods, apart from having physiological functions, have positive impact on health, performs different functions that mainly depends on water holding properties [11] (such as selling, solubility, water absorption and holding capacity, viscosity, and gelling). Chia gum is a product obtained from chia seed through extraction from dietary fiber fraction by treating seeds with water and this is used as an additive in food systems to provide stability, texture, control viscosity, and consistency [18]. Segura-Campos et al. [79] evaluated the chemical and functional characteristics of chia seed gum and suggested that the capacity to hold water is a major physicochemical property exploited in the food industry.

Campos et al. [17] reported that chia seed gruel is used as a substitute of stabilizers and emulsifiers in the production of ice cream. However, a negative effect was observed in the color of ice cream, which was attributed to the dark color of chia gel. The chia gum can withstand high temperature up to 244 °C, hence making it a promising and preferred agent in value-added food preparations [84]. Fernandez and Salas-Mellado [32] conducted a research to examine the technological quality of breads and pound cakes by incorporating chia mucilage; and observed reduction in amount of fat and proposed its effectiveness to be used as a fat substitute.

Currently various products are prepared, which are either based on chia seeds or fortified with them. These products include breakfast cereals, cakes, fruit juices, yoghurts, jams, and preserves [94, 98]. Chia is used in industrial food production as whole seeds, ground or mucilage to increase the nutritional value of the product (Table 7.6).

7.8.4   INSECT CONTROL IN CHIA CULTIVATION

Most chia plants (dark colored or black grains) are not attacked by insects, as they contain some phytocompounds that provide protection. From the leaves of chia, an essential oil can be extracted and this oil consists of β-pinene, globulol, widdrol, β-caryophyllene, γ-muroleno, germacren-B, and α-humoleno. These compounds are believed to have effective insect repellent characteristics.

7.8.5   MEDICAL EVIDENCES OF CHIA SEEDS

Till today, no evidence has been found of adverse or allergenic effects caused by consuming whole or ground chia seeds. Chia is rich in soluble fiber and >5 g of dietary fiber can be provided by only 12 g. The insoluble fiber has the ability to absorb large amount of water, which helps to provide a feeling of fullness and slows digestion that leads to a stable rise in blood sugar levels and a more stable release of insulin.

Due to its high omega-3 fatty acids content, several works have attributed various medicinal properties of chia seeds, such as reduction in cholesterol, inhibition of blood clotting, reduction in time for digestion of carbohydrates, decreasing risk of neurological disorders, promoting tissue regeneration, assisting in controlling blood sugar levels, improving immune system, and helps in development of brain and retina in fetus. Numerous epidemiological and clinical studies have suggested that higher consumption of α-linolenic acid is associated with reduction of cardiovascular diseases [14].

7.9   SUMMARY

This chapter provides information on functional, nutritional, medical, and physicochemical properties of chia seeds. *Salvia hispanica* L. has been used

TABLE 7.6 Food Applications of Chia

Raw Material Used	Food Product	Ref.	Raw Material Used	Food Product	Ref.
Chia seed	Cake;	[12]	Chia seed	Restructured ham-like product;	[29]
	Ice cream supplemented with olein fraction of chia oil;	[88]		Chia seed fortified pineapple jam;	[61]
oil	Margarine;	[58]		Corn tortillas supplemented with chia seeds.	[74]
Chia	Pasta;	[64]		Fruit punch, smoothie and kheer;	[9]
flour	Chips;	[24]		Bread.	[76]
	Frankfurters;	[70]			
	Wheat-based biscuits.	[53]			
Chia seeds and mucilage	Gluten-free pasta.	[52]	Whole chia flour	Pound cake.	[71]
Chia	Ice-cream;	[17]			
mucilage	Low fat yoghurt.	[28]			

since ancient times as a basic food in the diet of Aztec and Mayan populations. Chia seeds have tremendous potential as a source of nutrients and nutraceuticals of importance to technology, science, medicine, and engineering due to the presence of bioactive compounds with health benefits. Chia seeds are also rich in phytochemicals (caffeic acid, quercetin, myricetin, kaempferol and chlorogenic acid, etc.), PUFAs, dietary fibers (soluble and insoluble), proteins with a high level of essential amino acids, vitamins and minerals.

KEYWORDS

- antioxidant activity
- cardiovascular diseases
- chia gum
- chia seeds
- mucilage
- neurological disorders
- omega-3 fatty acids
- phytochemicals
- *Salvia hispanica*

REFERENCES

1. Ahmed, M.; Hamed, R.; Ali, M.; Hassan, A.; Babiker, E.; Proximate Composition, Anti-nutritional Factors and Protein Fractions of Guar Gum Seeds as Influenced by Processing Treatments. Pakistan Journal of Nutrition, **2006**, 5 (5), 340–345.
2. Ando, H.; Ryu, A.; Hashimoto, A.; Oka, M.; Ichihashi, M. Linoleic acid and α-Linolenic Acid Lightens Ultraviolet-Induced Hyperpigmentation of the Skin. Archives of Dermatological Research, **1998**, 290, 375–381.
3. Armstrong, D. Application for Approval of Whole Chia (Salvia hispanica L.) Seed and Ground Whole Seed as Novel Food ingredient. **2011**; https://acnfp.food.gov.uk/sites/default/files/mnt/drupal_data/sources/files/multimedia/pdfs/applicdosschiacompany.pdf; Accessed on June 25, 2019.
4. Ayaz, A.; Akyol, A.; Inan-Eroglu, E.; Cetin, A. K.; Samur, G.; Akbiyik, F. Chia Seed (Salvia hispanica L.) Added Yogurt Reduces Short-Term Food Intake and Increases Satiety: Randomized Controlled Trial. Nutrition Research and Practice, **2018**, 11 (5), 412–418.

5. Ayerza, R. Oil Content and Fatty acid Composition of Chia (Salvia hispanica L.) from Five North Western Locations in Argentina. Journal of the American Chemical Society, **1995**, 72, 1079–1081.

6. Ayerza, R. Seed Composition of Two Chia (Salvia hispanica L.) Genotypes, which Differ in Seed Color. Emirates Journal of Food Agriculture, **2013**, 25, 495–500.

7. Ayerza, R.; Coates, W. Ground Chia Seed and Chia Oil Effects on Plasma Lipids and Fatty Acids in the Rat. Nutrition Research, **2005**, 25, 995–1003.

8. Ayerza, R.; Coates, W. Omega-3 Enriched Eggs: The Influence of Dietary α-Linolenic Fatty Acid Source on Egg Production and Composition. Canadian Journal of Animal Science, **2001**, 81, 355–362.

9. Battalwar, R.; Shah, V. Incorporation of Chia Seeds in Fruit Punch, Kheer, Smoothie and its Sensory Evaluation. *International Journal of Food and Nutrition Science*, **2015**, *4*, 84–90.

10. Beltran-Orozco, M. C.; Romero, M. R. *La Chia, Alimento Milenario* (The Chia: Millenium Food). Departamento de Graduados e Investigación en Alimentos, ENCB, IPN, México; **2003**; p. 25.

11. Borderias, A. J.; Sanchez-Alonso, I.; Perez-Mateos, M. New Applications of Fibers in Food: Addition to Fishery Products. *Trends in Food Science and Technology*, **2005**, *16*, 458–465.

12. Borneo, R.; Aguirre, A.; Leon, A. E. Chia (*Salvia hispanica* L.) Gel can be used as Egg or Oil Replacer in Cake Formulations. *Journal of American Dietetic Association*, **2010**, *110*, 946–949.

13. Bresson, J. L.; Flynn, A.; Heinonen, M. Opinion on the Safety of Chia Seeds (*Salvia hispanica* L.) and Ground Whole Chia Seeds, as a Food Ingredient. *Journal of European Food Safety Authority*, **2009**, *99*, 1–26.

14. Brouwer, I. A.; Katan, M. B.; Zock, P. L. Dietary α-Linolenic Acid is Associated with Reduced Risk of Fatal Coronary Heart Disease, but Increased Prostate Cancer Risk: A Meta-Analysis. *Journal of Nutrition*, **2004**, *134*, 919–922.

15. Bushway, A. A.; Belyea, P. R.; Bushway, R. J. Chia Seed as a Source of Oil, Polysaccharide and Protein. *Journal of Food Science*, **1981**, *46*, 1349–1350.

16. Cahill, J. Ethnobotany of Chia (*Salvia hispanica L.*). *Economic Botany*, **2003**, *57*, 604–618.

17. Campos, B. E.; Ruivo, T. D.; Scapin, M.; Madrona, G. S.; Bergamasco, R. C. Optimization of the Mucilage Extraction Process from Chia Seeds and Application in Ice Cream as a Stabilizer and Emulsifier. *LWT Food Science and Technology*, **2016**, *65*, 874–883.

18. Capitani, M. I.; Corzo-Rios, L. J.; Chel-Guerrero, L. A. Rheological Properties of Aqueous Dispersions of Chia (*Salvia hispanica L.*) Mucilage. *Journal of Food Engineering*, **2015**, *149*, 70–77.

19. Coates, W. Whole and ground chia (*Salvia hispanica* L.) Seeds, Chia Oil: Effects on Plasma Lipids and Fatty Acids. In: *Nuts and Seeds in Health and Disease Prevention*; V. Patel, R. Preedy, & V. Watson (Eds.); Academic Publisher, San Diego, CA; **2011**; pp. 309–314.

20. Coates, W.; Ayerza, R. Production Potential of Chia in North Western Argentina. *Industrial Crops and Products*, **1996**, *5* (3), 229–233.

21. Coelho, M. S.; Salas-Mellado, M. M. Effects of Substituting Chia (*Salvia hispanica* L.) Flour or Seeds for Wheat Flour on the Quality of the Bread. *LWT Food Science and Technology*, **2015**, *60*, 729–736.

22. Coelho, M. S.; Soares-Freitas, R. A.; Areas, J. A. Peptides from Chia Present Antibacterial Activity and Inhibit Cholesterol Synthesis. *Plant Foods for Human Nutrition*, **2018**, *73*, 101–107.

23. Coelho, V. R.; Vieira, C. G.; Souza, L. P.; Moyses, F. Antiepileptogenic, Antioxidant and Genotoxic Evaluation of Rosmarinus Acid and its Metabolite Caffeic Acid in Mice. *Life Sciences*, **2015**, *122*, 65–71.

24. Coorey, R.; Grant, A.; Jayasena, V. Effects of Chia Flour Incorporation on the Nutritive Quality and Consumer Acceptance of Chips. *Journal of Food Research*, **2012**, *1* (4), 85–95.

25. Craig, R. *Application for Approval of Whole Chia (Salvia hispanica L.) Seed and Ground Whole Seed as Novel Food Ingredient.* Craig & Sons: Northern Ireland; *2004*; https://acnfp.food.gov.uk/sites/default/files/mnt/drupal_data/sources/files/multi-media/pdfs/chiaapplication.pdf; Accessed on June 25, 2019.

26. da Silva, B. P.; Anunciacao, C.; da Silva Matyelka, J. C. Chemical Composition of Brazilian Chia Seeds Grown in Different Places. *Food Chemistry*, **2017**, *221*, 1709–1716.

27. da Silva, B. P.; Dias, D. M.; de Castro Moreira, M. E. Chia Seed Shows Good Protein Quality, Hypoglycemic Effect and Improves the Lipid Profile and Liver and Intestinal Morphology of Wistar Rats. *Plant Foods for Human Nutrition*, **2016**, *71* (3), 225–230.

28. Darwish, A.; El-Sohaimy, S. A. Functional Properties of Chia Seed Mucilage Supplemented in Low-Fat Yoghurt. *Alexandria Medical Journal*, **2018**, *39* (3), 450–458.

29. Ding, Y.; Lin, H. W.; Lin, Y. L.; Yang, D. J. Nutritional Composition in the Chia Seed and Its Processing Properties on Restructured Ham-Like Products. *Journal of Food and Drug Analysis*, **2018**, *26*, 124–134.

30. Diwakar, G.; Rana, J.; Saito, L.; Vredeveld, D.; Zemaitis, D.; Scholten, J. Inhibitory Effect of a Novel Combination of Chia Seed and Pomegranate Fruit Extracts on Melanin Production. *Fitoterapia*, **2014**, *97*, 164–171.

31. Falco, B. D.; Amato, M.; Lanzotti, V. Chia Seeds Products: An Overview. *Phytochemistry Reviews*, **2017**, *16*, 745–760.

32. Fernandes, S. S.; Salas-Mellado, M. M. Addition of Chia Seed Mucilage for Reduction of Fat Content in Bread and Cakes. *Food Chemistry*, **2017**, *227*, 237–244.

33. Fernandez, I.; Vidueiros, S. M.; Ayerza, R.; Coates, W.; Pallaro, A. Impact of Chia (*Salvia hispanica* L.) on the Immune System: Preliminary Study. *Proceedings of the Nutrition Society*, **2008**, *67*, E12.

34. Garg, M. L.; Wood, L. G.; Singh, H.; Moughan, P. J. Means of Delivering Recommended Levels of Long Chain Polyunsaturated Fatty Acids in Human Diets. *Journal of Food Science*, **2006**, *71*, 66–71.

35. Geelan, A.; Brouwer, I. A.; Zock, P. L.; Katan, M. B. Antiarrhythmic Effects of n-3 Fatty Acids: Evidence from Human Studies. *Current Opinion in Lipidology*, **2004**, *15*, 25–30.

36. Hansel, B.; Nicolle, C.; Lalanne, F.; Tondu, F.; Lassel, T. Effect of Low-fat, Fermented Milk Enriched with Plant Sterols on Serum Lipid Profile and Oxidative Stress in moderate Hypercholesterolemia. *American Journal of Clinical Nutrition*, **2007**, *86* (3), 790–796.

37. Ho, H.; Lee, A. S.; Jovanovski, E.; Jenkins, A. L.; Desouza, R.; Vuksan, V. Effect of Whole and Ground Chia seeds (*Salvia hispanica* L.) on Postprandial Glycemia in Healthy Volunteers: A Randomized Controlled, Dose-Response Trial. *European Journal of Clinical Nutrition*, **2013**, *67*, 786–788.

38. Huang, Y. S.; Zhang, J. T. Antioxidative Effect of Three Water Soluble Components Isolated from *Salvia miltiorrhiza* In Vitro. *Acta Pharmceutica Sinica B*, **1991**, *27*, 96–100.

39. Illian, T. G.; Casey, J. C.; Bishop, P. A. Omega-3 Chia Seed Loading as a Means of Carbohydrate Loading. *Journal of Strength and Conditioning Research*, **2011**, *25*, 61–65.

40. Ixtaina, V. Y.; Nolasco, S. M.; Tomas, M. C. Physical Properties of Chia (*Salvia hispanica L.*) Seeds. *Industrial Crops and Products*, **2008**, *28* (3), 286–293.

41. Ixtaina, V.Y.; Martinez, M. L.; Spotorno, V.; Mateo, C. M. Characterization of Chia Seed Oil Obtained by Pressing and Solvent Extraction. *Journal of Food Composition Analysis*, **2011**, *24*, 166–174.

42. Jaddu, S.; Yedida, H. V. Chia Seed: A Magical Medicine. *Journal of Pharmacognosy and Phytochemistry*, **2018**, *7* (2), 1320–1322.

43. Jeong, S. M.; Kim, S. Y.; Kim, D. R.; Man, K. C. Effect of Heat Treatment on the Antioxidant Activity of Extracts from Citrus Peels. *Journal of Agricultural and Food Chemistry*, **2014**, *52*, 3389–3393.

44. Jin, F.; Nieman, D.C.; Sha, W.; Xie, G.; Qiu, Y.; Jia, W. Supplementation of Milled Chia Seeds Increases Plasma ALA and EPA in Postmenopausal Women. *Plant Foods for Human Nutr*ition, **2012**, *67*, 105–110.

45. Kulczynski, B.; Kobus-Cisowska, J.; Taczanowski, M. Chemical Composition and Nutritional Value of Chia Seeds: Current State of Knowledge. *Nutrients*, **2019**, *11*, 1–16.

46. Leaf, A.; Kang, J. X. Omega-3 Fatty Acids and Cardiovascular Disease. In: *The Return of T-3 Fatty Acids into the Food Supply, I: Land-Based Animal Food Products and Their Health Effects*; Simopoulos, A. P. (Ed.); Karger Publishers: Basel, Switzerland; **1998**; pp. 24–37.

47. Lokuruka, M. N. Role of Fatty Acids of Milk and Dairy Products in Cardiovascular Diseases: A Review. *African Journal of Food Agricultural Nutrition and Development*, **2007**, *7* (1), online; https://www.ajol.info/index.php/ajfand/article/view/136148; Accessed on December 31, 2019.

48. Manzella, D.; Paolisso, G. Cardiac Autonomic Activity and Type II *Diabetes Mellitus*. *Clinical Science*, **2005**, *108*, 93–97.

49. Marineli, R. S.; Lenquiste, S. A.; Moraes, E. A.; Marostica, M. R. Antioxidant Potential of Dietary Chia (*Salvia hispanica* L.) Seed and Oil in Diet-Induced Obese Rats. *Food Research International*, **2015**, *76*, 666–674.

50. Martinez-Cruz, O.; Paredes-Lopez, O. Phytochemical Profile and Nutraceutical Potential of Chia (*Salvia Hispanic L.*) Seeds by Ultra High Performance Liquid Chromatography (UHPLC). *Journal of Chromatography A*, **2014**, *1346*, 43–48.

51. Matthaus, B. Antioxidant Activity of Extracts Obtained from Residues of Different Oilseeds. *Journal of Agricultural and Food Chemistry*, **2002**, *50*, 3444–3452.

52. Menga, V.; Menga, V.; Amato, M.; Phillips, T. D. Gluten-free Pasta Incorporating Chia (*Salvia hispanica* L.) as Thickening Agent: An Approach to Naturally Improve the Nutritional Profile and the *In Vitro* Carbohydrate Digestibility. *Food Chemistry*, **2017**, *221*, 1954–1961.

53. Mesias, M.; Holgado, F.; Marquez-Ruiz, G.; Morales, F. J. Risk-benefit Considerations of a New Formulation of Wheat-Based Biscuit Supplemented with Different Amounts of Chia Flour. *LWT Food Science and Technology*, **2016**, *73*, 528–535.

54. Mohd, A. N.; Yeap, S. K.; Ho, W. Y.; Beh, B. K. The Promising Future of Chia (*Salvia hispanica* L). *Journal of Biomedical Biotechnology*, **2012**, *2012*, 1–9.

55. Monroy, T. R.; Mancilla Escobar, M. L. Protein Digestibility of Chia (*Salvia hispanica* L.) Seeds. *Revista Salud Publica y Nutrición*, **2008**, *9* (1), 5–19.

56. Munoz, L. A.; Aguilera, J. M.; Rodriguez-Turienzo, L. Characterization and Microstructure of Films Made from Mucilage of *Salvia hispanica* and Whey Protein Concentrate. *Journal of Food Engineering*, **2012**, *111*, 511–518.

57. Munoz, L. A.; Cobos, A.; Diaz, O.; Aguilera, J. M. Chia Seed (*Salvia hispanica* L.): An Ancient Grain and a New Functional Food. *Food Reviews International*, **2013**, *29* (4), 394–408.

58. Nadeem, M.; Abdullah, M.; Mahumd, A.; Hussain, I.; Inayat, S. Stabilization of Butter Oil with Modified Fatty Acid Profile by Using *Moringa oleifera* Extract as Antioxidant. *Journal of Agricultural Science and Technology*, **2013**, *15*, 919–928.

59. Nadeem, M.; Imran, M.; Taj, I.; Ajaml, M.; Junaid, M. Omega-3 Fatty Acids, Phenolic Compounds and Antioxidant Characteristics of Chia Oil Supplemented Margarine. *Lipids in Health and Diseases*, **2017**, *16* (102), 2–12.

60. Nadeem, M.; Situ, C.; Mahmud, A.; Khalique, A. Antioxidant Activity of Sesame (*Sesamum indicum*) Cake Extract for the Stabilization of Olein Based Butter. *Journal of American Oil Chemist Society*, **2014**, *91* (6), 967–977.

61. Nduko, J. M.; Maina, R. W.; Muchina R. K.; Kibitok S. K. Application of Chia (*Salvia hispanica*) Seeds as a Functional Component in the Fortification of Pineapple Jam. *Food Science and Nutrition*, **2018**, *6*, 2344–2349.

62. Nieman, D. C.; Cayea, E. J.; Austin, M. D. Chia Seeds do not Promote Weight Loss or Alter Disease Risk Factors in Overweight Adults. *Nutrition Research* **2009**, *29*, 414–418.

63. Nieman, D. C.; Gillitt, N.; Jin, F.; Henson, D. A. Chia Seed Supplementation and Disease Risk Factors in Overweight Women: A Metabolomics Investigation. *Journal of Alternative and Complementary Medicine*, **2012**, *18*, 700–708.

64. Oliveira, M. R.; Novack, M. E.; Santos, C. P.; Kubota, E.; Rosa, C. S. Evaluation of Replacing Wheat Flour with Chia Flour (*Salvia hispanica* L.) in Pasta. *Semina Ciencias Agrarias*, **2015**, *36*, 25–45.

65. Orona-Tamayo, D.; Valverde, M. E. Inhibitory Activity of Chia (*Salvia hispanica* L.) Protein Fractions Against Angiotensin I-Converting Enzyme and Antioxidant Capacity. *LWT—Food Science and Technology*, **2015**, *64*, 236–242.

66. Orona-Tamayo, D.; Valverde, M. E.; Paredes-Lopez, O. Chia: New Golden Seed for the 21st Century: Nutraceutical Properties and Technological Uses. *Plant Derived Proteins*, **2016**, *2016*, 265–281.

67. Paperkamp, M. *Chia from Bolivia: A Modern Super Seed in a Classic Pork Cycle*. Netherland: CBI Marked Intelligence, Ministry of Foreign Affairs; **2014**; pp. 1–15; https://www.cbi.ca/market-information/grains-pulses/chia; Accessed on June 25, 2019.

68. Pawlosky, R.; Hibbeln, J.; Lin, Y.; Salem, N. Omega-3 Fatty Acid Metabolism in Women. *British Journal of Nutrition*, **2003**, *90*, 993–994.

69. Peiretti, G. O.; Gai, F. Fatty Acid and Nutritive Quality of Chia (*Salvia hispanica* L.) Seeds and Plant during Growth. *Animal Feed Science and Technology*, **2009**, *148*, 267–275.

70. Pintado, T.; Herrero, A. M.; Jimenez-Colmenero, J.; Ruiz-Capillas, C. Strategies for Incorporation of Chia (*Salvia hispanica* L.) in Frankfurters as a Health-Promoting Ingredient. *Meat Science*, **2016**, *114*, 75–84.

71. Pizarro, P. L.; Almeida, E. L.; Samman, N. C.; Chang, Y. K. Evaluation of Whole Chia (*Salvia hispanica* L.) Flour and Hydrogenated Vegetable Fat in Pound Cake. *LWT Food Science and Technology*, **2013**, *54*, 73–79.

72. Rasheed, A.; Cummins, C. Beyond the Foam Cell: The role of LXRS in Preventing Atherogenesis. *International Journal of Molecular Sciences*, **2018**, *19* (8), 2307–2311.

73. Rawlings, A. V. Trends in Stratum Corneum Research and the Management of Dry Skin Conditions. *International Journal of Cosmetic Science*, **2003**, *25*, 63–95.

74. Rendon-Villalobos, R.; Ortiz-Sánchez, A. Formulation, Physicochemical, Nutritional and Sensorial Evaluation of Corn Tortillas Supplemented with Chia Seed (*Salvia hispanica* L.). *Czech. Journal of Food Science*, **2012**, *30*, 118–125.

75. Reyes-Caudillo, E.; Tecante, A.; Valdivia-Lopez, M. A. Dietary Fiber Content and Antioxidant Activity of Phenolic Compounds Present in Mexican Chia (*Salvia hispanica* L.) Seeds. *Food Chemistry* **2008**, *107* (2), 656–663.

76. Romankiewicz, D.; Hassoon, W. H.; Cacak-Pietrzal, G. The Effect of Chia Seeds (*Salvia hispanica* L.) Addition on Quality and Nutritional Value of Wheat Bread. *Journal of Food Quality*, **2017**, *2017*, 1–7.

77. Sandoval-Oliveros, M. R.; Paredes-Lopez, O. Isolation and Characterization of Proteins from Chia Seeds (*Salvia hispanica* L.). *Journal of Agricultural and Food Chemistry*, **2013**, *61*, 193–201.

78. Sargi, S. C.; Silva, B. C.; Santos, H. M. C. Antioxidant Capacity and Chemical Composition in Seeds Rich in Omega-3: Chia, Flax and Perilla. *Food Science and Technology*, **2013**, *33*, 541–548.

79. Segura-Campos, M. R.; Salazar-Vega, I. M. Biological Potential of Chia (*Salvia hispanica* L.) Protein Hydrolysates and Their Incorporation into Functional Foods. *LWT Food Science and Technology*, **2013**, *50*, 723–731.

80. Segura-Campos, M. R.; Solis, N. C.; Rubio, G. R. Chemical and Functional Properties of Chia Seed (*Salvia hispanica* L.) Gum. Mexico. *International Journal of Food Science*, **2014**, *2014*, 1–5.

81. Shahidi, F.; Naczk, M. Phenolic Compounds in Grains. In: *Food Phenolics. Source, Chemistry, Effects, Applications*; Technomic Publishing Company: Lancaster, PA; **1995**; pp. 36–45.

82. Skov, A. R.; Toubro, S.; Ronn, B.; Holm, L.; Astrup, A. Randomized Trial on Protein vs Carbohydrate in *Ad Libitum* Fat Reduced Diet for the Treatment of Obesity. *International Journal of Obesity and Related Metabolic Disorders*, **1999**, *23*, 528–536.

83. Tepe, B.; Sokmen, M.; Akpulat, A. H.; Sokmen, A. Screening of the Antioxidant Activity of Six Salvia Species from Turkey. *Food Chemistry*, **2006**, *95*, 200–204.

84. Timilsena, Y. P.; Adhikari, R.; Kasapis, S.; Adhikari, B. Molecular and Functional Characteristics of Purified Gum from Australian Chia Seeds. *Carbohydrate Polymer*, **2016**, *136*, 128–136.

85. Toscano, L. T.; da Silva, C.S.O.; Toscano, L. T. Chia Flour Supplementation Reduces Blood Pressure in Hypertensive Subjects. *Plant Foods for Human Nutrition*, **2014**, *69*, 392–398.

86. Toscano, L. T.; Toscano, L. T.; Tavares, R. L.; Oliveira, C. S.; Silva, A. S. Chia Induces Clinically Discrete Weight Loss and Improves Lipid Profile Only in Altered Previous Values. *Nutrition Hospitalaria*, **2015**, *31* (3), 1176–1182.

87. Tuberoso, C. I. G.; Kowalczyk, A.; Sarritzu, E.; Cabras, P. Determination of Antioxidant Compounds and Antioxidant Activity in Commercial Oilseeds for Food Use. *Food Chemistry*, **2007**, *103*, 1494–1501.

88. Ullah, R.; Nadeem, M.; Imran, M. Omega-3 Fatty Acids and Oxidative Stability of Ice Cream Supplemented with Olein Fraction of Chia (*Salvia hispanica* L.) Oil. *Lipids in Health and Disease*, **2017**, *16* (34), 1–8.

89. Ullah, R.; Nadeem, M.; Khalique, A.; Mehmood, S.; Javid, A.; Hussain, J. Nutritional and Therapeutic Perspectives of Chia (*Salvia hispanica L.*): A Review. *Journal of Food Science and Technology*, **2016**, *53* (4), 1750–1758.

90. Uribe, J. A. R.; Perez, J. I. N.; Kauil, H. C.; Rubio, G. R.; Alcocer, C. G. Extraction of Oil from Chia Seeds with Supercritical CO_2. *Journal of Supercritical Fluids*, **2011**, *56* (2), 174–178.

91. USDA. *Dietary Guidelines for Americans*. 5th edition; Home and Garden Bulletin 232; **2000**; https://health.gov/dietaryguidelines/2015/guidelines/; Accessed on July 13, 2019.

92. USDA. *National Nutrient Database for Standard Reference*. Release 28: Nutrient Data; Laboratory Home Page—US Department of Agriculture, Agricultural Research Service; **2015**; https://www.ars.usda.gov/northeast-area/beltsville-md-bhnrc/beltsville-human-nutrition-research-center/methods-and-application-of-food-composition-laboratory/mafcl-site-pages/sr17-sr28/; Accessed on July 13, 2019.

93. USDA. *National Nutrient Database for Standard Reference*. Release 24: Nutrient Data; Laboratory Home Page—US Department of Agriculture, Agricultural Research Service; **2011**; http://www.irondisorders.org/Websites/idi/files/Content/854266/USDA_National_Nutrient_Database_iron.pdf; Accessed on July 13, 2019.

94. Valdivia-Lopez, M. A.; Tecante, A. Chia (*Salvia hispanica*): A Review of Native Mexican Seed and its Nutritional and Functional Properties. *Advances in Food Nutrition Research*, **2015**, *75*, 53–75.

95. Vuksan, V.; Jenkins, A. L.; Brissette, C. Chia (*Salvia hispanica L.*) Seeds in the Treatment of Overweight and Obese Patients with Type-2 Diabetes: A Double-Blind Randomized Controlled Trial. *Nutrition Metabolism and Cardiovascular Diseases*, **2017**, *27* (2), 138–146.

96. Vuksan, V.; Whitman, D; Sievenpiper, J. L. Supplementation of Conventional Therapy with the Novel Grain Chia (*Salvia hispanica L.*) Improves Major and Emerging Cardiovascular Risk Factors in Type-2 Diabetes: Results of a Randomized Controlled Trial. *Diabetes Care*, **2007**, *30*, 2804–2810.

97. Yao, L. H.; Jiang, Y. M.; Shi, j.; Tomas-Barberan, F. A. Flavonoids in Food and Their Health Benefits. *Plant Foods for Human Nutrition*, **2004**, *59*, 113–122.

98. Zettel, V.; Hitzmann, B. Applications of Chia (*Salvia hispanica L.*) in Food Products. *Trends in Food Science and Technology*, **2018**, *80*, 43–50.

CHAPTER 8

NOVEL WHOLE-GRAIN FOODS: NUTRITIONAL AND PHYTOCHEMICAL PROPERTIES FOR HEALTHCARE

MANREET SINGH BHULLAR, MANDEEP TAYAL,
SAMNEET KASHYAP, and RAVNEET SANDHU

ABSTRACT

This chapter reviews novel whole-grain foods with primary focus on Healthcare and nutrition. Increase in incidences of gluten intolerance and rise in the awareness of the health-conscious consumers are the factors that attracted food manufacturers to develop food product formulations that meet the definitions of whole-grain foods and dietary requirements of consumers. The chapter focuses on whole grains, nutrient composition, alternatives to cereal grains, dietary fibers as the driving element for enhanced consumption of cereals, and few examples of novel trends in the food industry. The chapter lays down critical information for opportunities to explore and literature studies that promise the potential use of noncereal ingredients that enhance the overall performance of whole-grain foods.

8.1 INTRODUCTION

Being a living entity on earth, human body demands ample amount of energy at every moment. To satisfy this requirement, humans rely on consumption of different energy sources, for example, food carbohydrates which include cereals, vegetables, fruit, and many other products. Among these, cereals have been known to make major proportion of human diet during the last 3000–4000 years [39]. Most widely grown cereals are wheat, rice, and corn, and other cereal grains are grown in specific regions of the world. Scientifically, a cereal is defined as a grain or edible seed that belongs to the Gramineae family [13].

With the evolution of human race on planet earth and significant changes occurring on earth including climate, water scantiness, ever increasing world population, and unaffordability of food have brought in severe threat to food security. The whole grains have been reported to lower the risk of chronic diseases including cardiovascular diseases, diabetes, cancers, and others [35]. Food researchers are looking forward to develop high-yielding varieties. Food nutritionists are focusing on the dietary recommendations on the whole grains with health benefits. Consequently, cereal whole grains are attracting consumers and food industries (such as General Mills) to develop new products with increasing interest from food scientists, technologists, and nutritionists [28].

A whole-grain ingredient is defined as "the caryopsis (intact, ground, cracked or flaked), with starchy endosperm, germ and bran, present in significantly similar proportions as in the intact caryopsis." There are three major components of the whole-grain flour including germ, bran, and endosperm that are present in similar amounts as in its original grain form. The health benefits from all these three components together is better than any single fraction of the grain [33]. There are several standards used internationally to validate the whole-grain food ingredient and its product. However, 51% whole-grain flour and 1.7 g of dietary fiber are the set requirements for a food to meet whole-grain health standard in the United States. Many small factors lead to nonpopularity of whole grains with other food products, such as less knowledge or confusion about whole grains, limited availability of whole grain products, less variation, and taste, and more preference toward refined products.

The consumption of cereal grains around the world has been significant to meet the food demand. The types of cereal grains include rice, wheat, corn, rice, and oats. Other cereals that are consumed in low quantities include triticale, barley, sorghum, and millets. Wheat ranks the top position with leading consumption among cereal grains around the world. Novel manufactured foods can help in increasing whole-grain consumption in daily diet. Presently, the whole-grain products consumed are whole-grain breads and cereals for breakfast. Though there are possibilities to introduce new whole-grain noodles, biscuits, and pasta. Innovation and technology can add up to this aim, but the taste and texture of the product are among the most significant factors to consider before launching a product in the market [34].

This chapter reviewed the latest trends in whole-grain components, consumption trends, nutritional composition, and products prepared from whole grains.

8.2   WHOLE GRAINS: CONSUMPTION, TRENDS, AND DIETARY RECOMMENDATIONS

Human beings have been consuming cereal grains since early times and these have been an integral part of the human diet. With the evolution of food over the time, product development and the consumer demand has driven whole-grain industry into developing several products. This shift in production and consumer demands for cereal and cereal products results in the decrease of dietary fiber and other nutrients in the cereal products. Several studies have reported low consumption of the whole grains in the American diets [7, 20]. MyPyramid (also called US food guidance system) recommends whole grains daily intake of three servings [9, 36]. Knowing the health benefits of consumption of whole grains and high demand in the health and nutrition industry, a much-needed consensus is required among the global scientists, health researchers and government agencies to design policies that focus on increasing the intake of whole grains and reducing the gaps in consumption knowing significant health benefits of whole grains.

8.3   CHALLENGES FOR CONSUMPTION OF WHOLE GRAIN

Dietary recommendations by the health officials or other government agencies do not reach out to all users and consumers are ignorant to health benefits of key food ingredients including whole grain in the diet. Little attention has been paid toward reading and understanding the label on the food product. The demand for refined cereal products due to bitter or astringent taste of bran in whole grain foods marks the ignorance and lack of knowledge among consumers. Food companies have been trying hard to incorporate whole-grain ingredient in other food products that mask the stringent flavor of the bran and develop healthy foods. With taste being the biggest challenge, price, texture, and softness of the whole-grain products are other challenges that add a hindrance to product development and consumer acceptance.

Additionally, there lies a lot of variation in defining the standards for whole-grain products with respect to the origin of food. Efforts have been made by various organizations (Whole Grain Council, Association of American Cereal Chemists) to define whole-grain intake standards; but very little consideration has been shown by the food industry. The current definition of whole-grain food product does tell any information about the minimum amount of whole-grain content to be present in the reconstituted/processed product to claim it as whole-grain food product. The need for a new definition is required to develop a clear framework of comparison among different whole-grain products [32]. The change would contribute to innovations in food manufacturing and food product development, changing consumer behavior, diet-based recommendations, and public health policy [5].

Furthermore, the food industry still lacks the availability of rapid analytical methods to validate the whole-grain composition in processed foods [34]. In addition, the use of whole grains poses significant safety issues in many regions with extensive use of pesticides and environmental pollution. The outer layer of grains can carry significant amounts of heavy metals and pesticides residues [17]. Stringent quality control and monitoring would be required to achieve the desired goal of developing novel whole-grain products with enhanced health benefits.

8.4   BASIC WHOLE GRAINS

The major whole grains consumed worldwide include wheat, rice, oats, corn, and barley with wheat heading the list. However, several other grain products have recently been added to the market. High-fiber content in whole grain foods has prominent benefits to consumer health. Due to increased health diseases including diabetes and obesity in the western countries, world health agencies are focusing to incorporate more whole grain into the diets [10]. The regular intake of whole cereal grains has been shown to protect consumers from several chronic diseases including diabetes (*Diabetes mellitus*) and colon cancer [14, 38]. Other than common cereal grains commonly consumed in the world, buckwheat and wild rice are two other noncereals (pseudograins) commonly consumed in the United States [19]. The cereals are classified as follows:

- *True Cereals*: Barley, brown rice, maize (corn), millets, oat, rye, sorghum, teff, and wheat.
- *Pseudocereals*: Amaranth, buckwheat, and quinoa.

Additionally, pseudograins have gained significant attraction of consumers and commercial foods with development of new foods. Pseudograins belong to dicot grasses with broad leaves unlike cereals, which are all monocots. In food market, three major pseudograins include quinoa, buckwheat, and amaranth. Also, alternate cereals in the new term used in North America that comprises of pseudocereals and other cereals that are not commonly consumed in North America including sorghum, kamut, spelt, teff, and pearl millet. These cereals are generally grown in Asia and Africa but with evolution in nutritional research and higher incidence of chronic disease, these have become of great interest to researchers and food producers in North America [16]. The nutritional composition and health benefits of pseudograins and alternate grains are very similar to those of cereal grains [1]. Knowing that the bran, germ, and endosperm make up the whole grain and if any one of these is removed through processing, it is called refined grains (e.g., white flour). Refining process takes away the bran, the brown pigmentation and significantly the nutritional component, and thus the whole-grain foods including fiber and proteins [2].

Whole grains have been well known for enhancing carbohydrate metabolism and preventing several diseases, such as cardiovascular diseases, diabetes, weight management, obesity, hypertension, strokes, coronary heart diseases, colon cancer, and others [25, 27]. With the increasing number of people affected from these chronic diseases, consumers tend to adapt to new dietary recommendations rich in fiber and whole grains grab the top spot in fulfilling the consumer needs.

8.5   NUTRITIONAL COMPONENTS IN WHOLE GRAINS

Whole grains comprise of three major components, that is, bran, germ, and endosperm. The important biologically active elements come from the bran and germ proportions of the grain. Table 8.1 shows the nutritional content of the cereal grains that are marked by good presence of B-vitamins (including riboflavin and thiamin), minerals (including Ca and Mg), and high concentration of basic amino acids (lysine and arginine) [40]. The whole grains possess good antioxidant quantity due to presence of folates, phenolic compounds, avenanthramides, avenalumic acid [23], dietary fiber, antinutrients (phytic acid and tannins), lignans, and tocotrienols. The dietary fiber content of major whole grains is presented in Figure 8.1. USDA Nutrition Data Laboratory—US Government [https://www.nal.usda.gov/fnic/usda-nutrient-data-laboratory] has provided the nutrient database that can be accessed and used by anyone free of cost: "The current version, Survey-SR 2013-2014, is mainly based on the USDA National Nutrient Database for Standard Reference (SR)-28-(2) and contains sixty-six nutrient search for 3,404 foods. These nutrient data will be used for assessing intake data from WWEIA, NHANES 2013-2014."

Knowing whole grain are rich in dietary content, the milling and refining process removes the bran and thus the dietary fiber and other nutrient components (e.g., minerals, vitamins) and phenolic compounds (Table 8.2). The refining processing step ends up with product with a high proportion of the starch compared to raw cereal grain composition.

TABLE 8.1 Nutritional Composition of Selected Whole Grain Cereals (USDA National Nutrient Database)

Grain Cereal	Energy kJ/100 g	Total Carbohydrate	Protein	Total Fat	Total Dietary Fiber
		%			
Barley (hulled)	1480	73.5	12.48	2.3	17.3
Corn (yellow)	1526	74.3	9.4	4.7	7.3
Millets	1580	72.8	11	4.2	8.5
Oat	1626	66.3	16.89	6.9	10.6
Rice (brown, long grain)	1547	77.2	7.9	2.9	3.5
Rye	1413	75.9	10.3	1.6	15.1
Sorghum	1413	74.6	11.3	3.3	6.3
Teff	1534	73.1	13.3	2.4	8
Triticale	1404	72.1	13	2.1	NA
Wheat (soft white)	1421	75.4	10.7	1.99	12.7

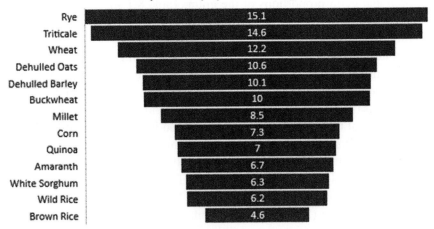

FIGURE 8.1 Dietary fiber (%) of whole grains (USDA National Nutrient Database).

TABLE 8.2 Nutrient Composition of Refined, Whole and Enriched Wheat Flour (Adapted from USDA National Nutrient Database)

Nutrients[a]	Refined Wheat Flour (%)	Whole Wheat Flour (%)	Enriched Wheat Flour (%)
Fiber	25	100	NA
Folate	59	100	661
Iron	33	100	129
Magnesium	16	100	NA
Niacin (B3)	25	100	119
Potassium	29	100	NA
Riboflavin (B2)	24	100	299
Thiamin (B1)	24	100	156
Vitamin B6	11	100	NA
Vitamin E	8	100	NA
Protein	78	100	NA

[a]Assuming composition of nutrients in whole wheat flour as 100%.

8.6 DIETARY FIBER: THE MAJOR COMPONENT OF WHOLE GRAINS

Dietary fiber constitutes the proportion of whole grains that cannot be broken down by the digestion enzymes. Dietary fibers are classified as:

- Water-soluble fraction mainly contains nonstarchy polysaccharides that aid in lowering cholesterol and insulin levels in humans.
- water-insoluble dietary fibers are cellulose, hemicellulose, and lignin.

The percentage of dietary fiber in different cereals varies with variety, agroclimatic regions, cultivation practices, and processing methods. The reported values for dietary content in barley, wheat, and oats are 10%, 12%, and 14% (dry weight basis) [36]. Among cereal grains, wheat is one of the major sources of dietary fiber when compared to other commonly consumed whole-grain cereals (rice, maize, oats). The research has shown that lower intake of dietary fiber may lead to the incidence of constipation, cancer of large bowl, Crohn disease, ulcers, gallstones, obesity, coronary heart disease, and so many others [10].

8.7   OTHER NUTRIENT COMPONENTS

The β-glucans are polysaccharides present in the cell wall of cereal grains and are largely found in barley (3%–11%) and oats (3%–7%). These have been promoted to reduce plasma cholesterol and blood sugar levels in consumers. Cereal grains are also good sources of oligosaccharides, which aid in growth of *Bifidobacteria* in human large intestine [39], which helps the body perform essential functions such as digestion and staving off harmful bacteria. Other nutrient components are phenolic compounds, carotenoids, phytic acid, phytosterols and tocols [14].

8.8   INDUSTRIES VERSUS WHOLE GRAINS

Scientific studies on whole grains have greatly influenced health-conscious consumers over the past decade. A greater proportion of population has started following public health recommendations that greatly endorse incorporating whole grains in daily diet for better health and prevention of chronic diseases. The development of new product is driven by the consumer demand while maintaining the nutritional needs and food market sustainability. Authors of this chapter have discussed some novel functional foods that claim health benefits. Some of the products include baked products made with oats and Psyllium husk, which is a high-quality source of soluble fiber and helps to lower total cholesterol.

8.9   NOVEL WHOLE-GRAIN FOOD PRODUCTS

The consumer demand for novel food products with greater proportions of whole grains have driven food companies to manufacture foods that meet the expectation of their health-conscious consumers. With a well-known list of health benefits, this chapter discusses some of the novel food products and their health benefits.

8.9.1   CHIA SEEDS

Chia seeds are the small edible seeds of a flowering plant, greatly grown in Mexico, Argentina and the Northern United States. The use of chia seeds in making bread and reducing the additional fat substituting wheat with chia seeds is reported in several studies [8, 43]. The substitution with chia seeds resulted in a softer bread with increased yeast activity. However, using 25% chia seeds provided promising results with reduced retrogradation and a further increase in the proportion of chia seeds resulted in lipid oxidation and lowered shelf-life [12, 26]. Chia seeds have proven to be a potential substitute for common cereal grains in use while adding significant nutritional value to the new foods.

With the rising incidence of gluten intolerance and celiac disease among consumers, the needs for developing gluten-free products are the necessity of time. Food companies are spending good chunk of money into research and development to find novel ingredients to manufacture food items that do not pose risk to gluten-intolerant consumers. Since chia seeds are gluten free, therefore they pose significant potential ingredient for gluten-free breads. In rice and soya-based breads, chia seeds have been successfully used at 2.5% in composition with minimal differences compared to the control samples. Also, pseudocereals in gluten-free foods and rendering similar nutritional benefits as that of cereal grains are promising ingredients for the development of new whole-grain foods. The use of chia seeds has been reported in the literature in several food products and provides potential use for better health benefits. Examples of such products are ice cream [6, 41], yogurt [3], and several meat products [11].

8.9.2 PASTA AND NOODLE PRODUCTS WITH GLUTEN-FREE INGREDIENTS

From 2013 to 2015, the gluten-free food industry has grown by 136%, providing considerable prevention of celiac disease [31]. Pseudocereals (buckwheat, quinoa, amaranth), legume flours (soybean), and fruit/vegetable powders are potential substitutes for wheat flour in making gluten-free pasta products. Pseudocereals have become increasingly used in the pasta formulations to develop gluten-free nutritious pasta product. Rice and corn are other new cereal ingredients that have substituted wheat in many new gluten-free foods. Several studies have reported better texture and cooking ability of gluten-free pasta when brown rice, corn starch, corn flour, and rice flour are used [37]. Gluten-free egg-free pasta and gluten-free spaghetti are other new foods developed using nongluten cereal grains [15].

8.9.3   HIGH-AMYLOSE WHEAT

The increase in the amylose content as resistant starch is considered as the dietary fiber content and poses similar benefits to the gut health. High-amylose content in foods has significant potential to add nutrient value and health benefits. The examples are as follows:

- Bread products resulted in increase of fiber while maintaining the visual and sensory properties of the white bread and promise higher acceptance by the consumers;
- Noodles resulted in enhancing the firmness. It also resulted in higher content of total dietary fiber (resistant starch); and
- Popped wheat grains with enhanced fiber content.

Thus a minor manipulation with the composition of whole-grain foods resulting in enhanced nutritional quality with better health benefits to the consumers would mark the potential future of whole-grain foods [29].

8.9.4   WHOLE-GRAIN-BASED INFANT CEREALS

These are processed cereal food products that can be simple cereals or cereals with added high-protein foods that may have to be reconstituted in milk or water. Infant cereals are foods that are given to kids during complementary feeding period to enhance nutritional intake, compensate iron deficiency, foster an adult-like microbiota, and feed adequate food for transition to solids. Knowing the low intake of cereal grains in adult human diet, targeting infants and developing taste, and bodily needs for a healthier lifestyle can pose a potential opportunity for food manufacturers and health organizations to meet the daily intake rate of whole-grain cereal grains. A consensus is required to implement the consumption of cereal grains to infants as it poses critical challenges including gluten intolerances, heavy metal contamination, and pesticide residues [21].

8.9.5   CANARY SEED AS NOVEL WHOLE-GRAIN CEREAL

The hairy canary seed (*Phalaris canariensis* L.) is an ancient grain used as bird feed and was considered harmful for human consumption. Research scientists in Canada have developed a hairless variety of canary seed that is regarded as "generally regarded as safe" for human consumption by Food and Drug Administration and Health Canada. Canary seed is a true cereal grain and contains significant amounts of starch (61%), protein (20%), and dietary fiber (7%). The protein content of canary seed is higher than other common whole grain cereals. As gluten-free canary seed, it marks a potential novel whole grain food-product for gluten-allergic consumers. More research is needed to identify health benefits of canary seeds and its anti-nutritional constituents to develop healthier foods [24].

8.9.6 SALBA AND DERIVED FOOD PRODUCTS

Salba (*Salvia hispanica* L.) is an heirloom variety of chia seeds grown in South America. The basic difference between two seeds is the color, chia seeds are black and salba seeds are white. The pleasant taste of salba seeds provides potential opportunities to food manufacturers to incorporate seeds in baked and other food products [42]. Salba seeds are good sources of dietary fiber and polyunsaturated fatty acids, which can lower the incidence of chronic diseases. Further research is required to find potential uses of salba seeds in food manu-facturing and fulfill the health-promoting demands of the consumer [18].

8.9.7 SPROUTED WHOLE GRAINS

Sprouting of grain seeds is an effective way of enriching nutritional composition of grains, reducing the anti-nutritional content and enhancing the digestibility. With increased consumer awareness of healthy foods, sprouts find an easier way to hit retail market shelves. Although there is no standard definition of sprouted whole grains, yet the sprouted grain can be considered whole grains if the sprout length is less than the kernel and all the three components of whole grain (i.e., bran, endosperm, and germ) are intact [22]. Significant research has been conducted to identify the effects of consuming sprouted grains on the prevention of chronic diseases. The sprouted whole grains has the advantage of being consumed directly or milled into flour to use as a raw ingredient for manufacturing other food products, such as germinated brown rice [30].

8.9.8 USE OF MICROPELLETS

Micropellet refers to fine particle size flour or powders that may constitute protein, vitamin, fiber, whole grain and can be used to manufacture a variety of snack foods enriched with desired product characteristics. It have the advantage of being small in size (<250 μm) and can be easily used in an extruder formulation. The ability to use micropellets in a wide variety of foods and enhancing the overall nutrition of the food as per the needs would enable whole-grain production companies to enrich food products with required key ingredients [4].

8.9.9 SPORTS BARS

Sports bars have marked the food shelves long time ago. With the health and nutrition needs of the sportsmen, bars rich in nutrients are the key demand. Also, the demand for lower sugar intake with more of protein and dietary fiber is attracting the food industries to manufacture novel sport bar meeting the consumer demands. These bars go by the name of cereal bars or meal replacement bars; and they are commonly known as power bars. Understanding the short-time availability, the consumption of high fiber bars provides steady fuel and natural sugars to meet the energy requirements of the body.

8.10 SUMMARY

Modern times are equipped with significant research data, artificial intelligence, and machine learning to develop novel whole-grain food product that meets the challenge of demands of taste and nutrition of consumers. The consumer acceptability plays a pivotal role for the food manufacturers before moving a new product on the food shelf. However, with the consumer exposure to health recommendations and accommodating the lifestyle of the modern population, developing new whole-grain food products will require strategic decision-making for food product developers to ensure convenience, affordability, and better health. More and more whole-grain cereals have been identified as potential ingredients for manufacturing whole-grain fiber-rich foods. Further research is underway at different institutes around the world to understand the mechanisms, through which whole-grain cereals impart health benefits and can prevent the incidence of chronic diseases. A better understanding of the fundamentals of consuming whole grains would assist the food industry to develop new products.

KEYWORDS

- cereal
- dietary fiber
- pseudocereals
- whole-grain product
- whole grains

REFERENCES

1. Alvarez-Jubete, L.; Arendt, E. K.; Gallagher, E. 2010. Nutritive Value of Pseudocereals and their Increasing Use as Functional Gluten-free Ingredients. *Trends in Food Science and Technology*, **2010**, *21*, 106–113.

2. Ayaz, A.; Akyol, A.; Inan-Eroglu, E. Chia Seed (*Salvia hispanica* L.) Added Yogurt Reduces Short-term Food Intake and Increases Satiety: Randomized Controlled Trial. *Nutrition Research and Practice*, **2017**, *11* (5), 412–418.

3. Baier, S. K.; Bhaskar, A. R.; Bortone, E.; Faa, P. *Methods of Incorporating Micropellets of Fine Particle Nutrients into Snack Food Products.* US20170006910A1; **2018**; https://patents.google.com; Accessed on June 10, 2019.

4. Călinoiu, L. F.; Vodnar, D. C. Whole Grains and Phenolic Acids: A Review on Bioactivity, Functionality, Health Benefits and Bioavailability. *Nutrients*, **2018**, *10* (11), 1615.

5. Chavan, V. R.; Gadhe, K. S.; Dipak, S.; Hingade, S. T. Studies on Extraction and Utilization of Chia Seed Gel in Ice Cream as a Stabilizer. *Journal of Pharmacognosy and Phytochemistry*, **2017**, *6* (5), 1367–1370.

6. Cleveland, L. E.; Moshfegh, A. J.; Albertson, A. M.; Goldman, J. D. Dietary Intake of Whole Grains. *Journal of the American College of Nutrition*, **2000**, *19* (3), 331–338.

7. Coelho, M. S.; de las Mercedes Salas-Mellado, M. Effects of Substituting Chia (*Salvia hispanica* L.) Flour or Seeds for Wheat Flour on the Quality of the Bread. *LWT Food Science and Technology*, **2015**, *60* (2), 729–736.

8. Collar, C. *Novel High-Fiber and Whole Grain Breads.* Woodhead Publishing Limited: Cambridge, England; **2007**; p. 496.

9. Ding, Y.; Lin, H.-W.; Lin, Y. L.; Yang, D. J. Nutritional Composition in the Chia Seed and its Processing Properties on Restructured Ham-like Products. *Journal of Food and Drug Analysis*, **2018**, *26* (1), 124–134.

10. Fernandes, S.S.; de las Mercedes Salas-Mellado, M. Addition of Chia Seed Mucilage for Reduction of Fat Content in Bread and Cakes. *Food Chemistry*, **2017**, *227*, 237–244.

11. Gani, A.; Sm, W.; Fa, M. Whole Grain Cereal Bioactive Compounds and their Health Benefits: A Review. *Journal of Food Processing and Technology*, **2012**, *3* (3), 146–150; DOI :10.4172/2157-7110.1000146.

12. Gao, Y.; Janes, M. E.; Chaiya, B.; Brennan, M. A. Gluten☐free Bakery and Pasta Products: Prevalence and Quality Improvement. *International Journal of Food Science and Technology,* **2018**, *53* (1), 19–32.

13. Gasiorowski, K.; Szyba, K.; Antimutagenic Activity of Alkylresorcinols from Cereal Grains. *Cancer Letters*, 1996, 106 (1), 109–115.

14. Gosine, L.; McSweeney, M. B. Consumers' Attitudes towards Alternative Grains: A Conjoint Analysis Study. *International Journal of Food Science and Technology,* **2019**, *54* (5), 1588–1596.

15. Herrera-Herrera, A. V.; González-Sálamo, J. Organophosphorus Pesticides (OPPs) in Bread and Flours. In: *Flour and Breads and Their Fortification in Health and Disease Prevention;* Elsevier: New York; **2019**; pp. 53–70.

16. Ho, H.; Lee, A. S.; Jovanovski, E.; Jenkins, A. L.; Desouza, R.; Vuksan, V. Effect of Whole and Ground Salba Seeds (*Salvia hispanica* L.) on Postprandial Glycemia in Healthy Volunteers: A Randomized Controlled, Dose-Response Trial. *European Journal of Clinical Nutrition,* **2013**, *67* (7), 786.

17. Jonnalagadda, S. S.; Harnack, L.; Hai Liu, R. Putting the Whole Grain Puzzle Together: Health Benefits Associated with Whole Grains—Summary of American Society for Nutrition 2010 Satellite Symposium. *Journal of Nutrition*, **2011**, *141* (5), 1011–1022.

18. Kantor, L. S.; Variyam, J. N.; Allshouse, J. E.; Putnam, J. J.; Lin, B. Dietary Intake of Whole Grains: A Challenge for Consumers. *Whole-Grain Foods in Health and Disease,* **2002**, *2002*, 301–325.

19. Klerks, M.; Bernal, M. J.; Roman, S.; Bodenstab, S. Infant Cereals: Current Status, Challenges and Future Opportunities for Whole Grains. *Nutrients,* **2019**, *11* (2), 1–25.

20. Liu, T.; Hou, G. G. Trends in Whole Grain Processing Technology and Product Development. In: *Whole Grains: Processing, Product Development, and Nutritional Aspects;* CRC Press, Taylor and Francis Group: Boca Raton, FL; **2019**; p. 257.

21. Marquart, L.; Slavin, J. L.; Fulcher, R. G. *Whole-grain Foods in Health and Disease.* American Association of Cereal Chemists: St. Paul, MN; **2002**; p. 382.

22. Mason, E.; L'hocine, L.; Achouri, A; Karboune, S. Hairless Canary Seed: A Novel Cereal with Health Promoting Potential. *Nutrients*, **2018**, *10* (9), 1–16.

23. McMackin, E.; Dean, M.; Woodside, J. V.; McKinley, M. C. Whole Grains and Health: Attitudes to Whole Grains against a Prevailing Background of Increased Marketing and Promotion. *Public Health Nutrition*, **2013**, *16* (4), 743–751.

24. Mesías, M.; Holgado, F.; Márquez-Ruiz, G.; Morales, F. J. Risk/Benefit Considerations of New Formulation of Wheat-based Biscuit Supplemented with Different Amounts of Chia Flour. *LWT Food Science and Technology,* **2016**, *73*, 528–535.

25. Miller Jones, J. Fiber, Whole Grains and Disease Prevention. In: *Technology of Functional Cereal Products;* Hamaker, B. R. (Ed.); Woodhead Publishing Limited: Cambridge, England; **2008**; pp. 46–58.

26. Mir, S. A.; Manickavasagan, A.; Shah, M. A. *Whole Grains: Processing, Product Development, and Nutritional Aspects.* CRC Press, Taylor & Francis Group: Boca Raton, FL; **2019**; p. 309.

27. Newberry, M.; Berbezy, P.; Belobrajdic, D. High-amylose Wheat Foods: A New Opportunity to Meet Dietary Fiber Targets for Health. *Cereal Foods World,* **2018**, *63* (5), 188–193.

28. Patil, S. B.; Khan, M. K. Germinated Brown Rice as a Value-Added Rice Product: A Review. *Journal of Food Science and Technology*, **2011**, *48* (6), 661–667.

29. Reilly, N. R. The Gluten-free Diet: Recognizing Fact, Fiction, and Fad. *Journal of Pediatrics*, **2016**, *175*, 206–210.

30. Ross, A. B.; van der Kamp, J. W.; King, R. A Definition for Whole-Grain Food Products: Recommendations from the Health Grain Forum. *Advances in Nutrition,* **2017**, *8* (4), 525–531.

31. Schaffer-Lequart, C.; Lehmann, U.; Ross, A. B. Whole Grain in Manufactured Foods: Current Use, Challenges and the Way Forward. *Critical Reviews in Food Science and Nutrition,* **2017**, *57* (8), 1562–1568.

32. Seal, C. J.; Brownlee, I. A. Whole Grains and Health, Evidence from Observational and Intervention Studies. *Cereal Chemistry,* **2010**, *87* (2), 167–174.

33. Sidhu, J. S.; Kabir, Y.; Huffman, F. G. Functional Foods from Cereal Grains. *International Journal of Food Properties*, **2007**, *10* (2), 231–244.

34. da Silva, E. M. M.; Ascheri, J. L. R.; Ascheri, D. P. R. Quality Assessment of Gluten-Free Pasta Prepared with a Brown Rice and Corn Meal Blend via Thermoplastic Extrusion. *LWT Food Science and Technology*, **2016**, *68*, 698–706.

35. Singhal, P.; Kaushik, G. Therapeutic Effect of Cereal Grains: A Review. *Critical Reviews in Food Science and Nutrition,* **2016**, *56* (5), 748–759.

36. Slavin, J. Whole Grains and Human Health. *Nutrition Research Reviews*, **2004**, *17* (1), 99–110.

37. Temelli, F.; Stobbe, K.; Rezaei, K.; Vasanthan, T. Tocol Composition and Supercritical Carbon Dioxide Extraction of Lipids from Barley Pearling Flour. *Journal of Food Science,* **2013**, *78* (11), 1643–1650.

38. Ullah, R.; Nadeem, M.; Khalique, A.; Imran, M. Nutritional and Therapeutic Perspectives of Chia (*Salvia hispanica* L.): A Review. *Journal of Food Science and Technology*, **2016**, *53* (4), 1750–1758.

39. Vuksan, V.; Whitham, D.; Sievenpiper, J. L. Supplementation of Conventional Therapy with the Novel Grain Salba (*Salvia hispanica* L.) Improves Major and Emerging Cardiovascular Risk Factors in Type 2 Diabetes: Results of a Randomized Controlled Trial. *Diabetes Care,* **2007** *30* (11), 2804–2810.

40. Zettel, V.; Hitzmann, B. Applications of Chia (*Salvia hispanica* L.) in Food Products. *Trends in Food Science and Technology,* **2018**, *80*, 43–50.

PART III
Novel Strategies to Enhance Bioactive Compounds in Cereals

CHAPTER 9

BIOFORTIFICATION STRATEGIES FOR WHEAT AND WHEAT-BASED PRODUCTS

DIKSHA BASSI, KAMALJIT KAUR, TARVINDER PAL SINGH, and JASPREET KAUR

ABSTRACT

Biofortification of wheat with micronutrients can be one of the sustainable and cost-effective strategies to tackle zinc, iron, iodine, and selenium deficiencies to prevent malnutrition among target population. Anthocyanin biofortified colored wheat has also created interest in industry due to its antioxidant and anti-inflammatory activity. HarvestPlus has been developed as a worldwide leader in generating biofortified staple foods to diminish micronutrient malnutrition. In this chapter, methods of biofortification and strategies for wheat biofortification with micronutrients are detailed.

9.1 INTRODUCTION

Wheat is one of the most commonly consumed cereal grains all over the world. The worldwide production of wheat is 729 metric tons (MT) on a land area of 220 million ha (Mha). China ranks first in wheat production with 157 MT on 24.5 Mha; India ranks first in the area (30.5 Mha) and second in production (95.9 MT) [1]. Worldwide, around >3 billion people suffer from micronutrient deficiencies, mostly in regions where cereal-based foods are consumed. Wheat contains very low amounts of important micronutrients (i.e., zinc and iron) and most of these are removed during milling. Fortifying wheat flour during processing is the common method to increase the micronutrient level but fortified food often has less public acceptability due to the changes occurring in color or flavor while adding micronutrients. A new approach to increase micronutrient concentration during developing wheat varieties is by biofortification to increase the bioavailability and mineral contents in the grains [25]. Biotechnology and plant breeding practices are for micronutrient enhancement of crops to prevent malnutrition [34].

Hidden hunger poses severe hazard to the health of people [57]. Around 2 billion people suffer from iron deficiency out of which 500 million persons have iron deficiency anemia (IDA). IDA decreases the ability of a person to do physical activities, whereas anemic workers exhibit reduced efficiency, lesser working capability, and weakness. Iron plays a vital role in electron transport system and in plants it exists as a transition metal. The deficiency of zinc leads to reduced resistance to diseases, reduced growth rate, neuro-sensory defects (such as taste abnormalities), and delayed wound healing.

The most commonly used medium for the fortification of zinc, iron, and calcium are cereal products. Micronutrient deficiency is mainly due to insufficient dietary intake because of poverty, vegetarian diets rich in phytates and fiber, food taboos, poor hygiene that increases infections and infestations, and adverse iron zinc interactions. Women at the reproductive stage are more vulnerable to nutritional deficiencies. Zinc deficiency causes reduced immune function, neurobehavioral abnormalities, poor growth, high susceptibility to infection, and harmful consequences of pregnancy. Zinc deficiency in many developing countries is due to less intake of foods rich in zinc, that is, animal source and high consumption of legumes and cereals having considerable amount of phytate (myo-inositol hexaphosphate, which hinders absorption of zinc) [24]. Zinc acts as cofactor in >300 enzymes. A total of 17.3% of the global population is under the threat of zinc deficiency [52].

Zinc and iron deficiencies mainly occur in the developing countries due to large intake of cereal foods [such as wheat (*Triticum aestivum* L.), rice (*Oryza sativa* L.), and maize (*Zea mays* L.)] [42]. Zinc and iron are deficient in endosperm, which is an edible part of modern cultivated wheat. They are present in whole-grain wheat in the concentration of 7–85 mg/Kg zinc and 29–73 mg/Kg iron [18]. About 75% of zinc and iron (which is present in the seedy parts and not in endosperm) is lost during milling [40].

Cereals cultivated on zinc-deficient soils are one of the main causes for their incapability to accumulate high zinc and iron [3], due to large-scale irrigation, use of macronutrient fertilizers (i.e., P, N, K), and high yielding varieties [21]. In cereal grains, a reduction in the accumulation of zinc and iron occurred due to negative correlation between uptake of Zn and Fe and irrigation [48] and a comparable negative correlation between uptake of Zn and Fe, and P [46]. Zn and Fe malnutrition occurred because people from developing countries largely depend upon cereal grains, whose edible portions contain a low level of Zn and Fe; therefore there is a need to increase concentration of Zn to 30 mg/kg and Fe to 8 mg/kg in the endosperm. Dietary diversification, pharmaceutical supplementation, biofortification, and industrial fortification have been gaining attention recently to combat micronutrient malnutrition [38, 53].

Rice, wheat, and maize are the staple crops in developed countries to provide major calories to the diet. Therefore micronutrient deficiency can be overcome by biofortification, grain flour fortification, agronomic forti-fication, and diversification of food habits. Implementation of HarvestPlus Challenge Program has been implemented by CGIAR (the Consultative Group for International Agricultural Research) for biofortification of maize, cassava, rice, wheat, and others, using plant breeding [9].

This chapter focuses on the biofortification techniques, strategies for zinc, iron, iodine, and selenium biofortification, anthocyanin fortified colored wheat and HarvestPlus Challenge Program.

9.2 BIOFORTIFICATION METHODS

Biofortification is generally performed by (1) agronomic biofortification by fertilizer management and it is further of two types: soil and foliar application; and (2) genetic biofortification by genetic engineering or cross-breeding strategies.

9.2.1 AGRONOMIC BIOFORTIFICATION

Agronomic biofortification is attained through fertilizer application on foliar application on leaves or to the soil. This technique is inexpensive and simple, though it is applicable only in limited geographical locations due to limitation of soil chemistry and fertilizers together with problems of nutrient mobility and storage within plant.

Foliar application of fertilizers is more effective than soil application for durum wheat [16]. Biofortification via agronomic methods has been an effective approach to increase micronutrients in staple food grains. Zn levels can be tripled or quadrupled contingent on the kind of fertilizers used. There can be a direct application of Zn in soils by organic and inorganic mixtures. Oxides, nitrates, and sulfates are available as inorganic fertilizers, but $ZnSO_4$ is mostly used due to its low cost and great solubility.

Zinc application by $ZnSO_4$ presents better results than ZnO or ZnEDTA on its accumulation in grains [13]. Use of $ZnSO_4$ can increase zinc level up to 0.47-folds. By foliar application, mineral level in the kernels was improved by 28%–68% [59]. Timing of foliar application plays an important role in increasing the concentration of micronutrient in grain. For example, foliar application in wheat grain during milk stage of grain development is more effective than the application during reproductive stage [40].

Foliar application of is generally done by zinc sulfate and EDTA-chelated zinc. Foliar application of Zn is effective in enhancing the in both rice and wheat grains, when applied at advanced developmental stage (i.e., grain filling). Fertilizer strategy enhances the micronutrient concentration in whole grain but particularly in the endosperm [14].

A study was conducted to evaluate the effect of Zn treatments on growth, quality, and produce of generally planted cultivars of wheat. Field trials were done via two methods of Zn application: soil and foliage + soil. Pure Zn doses of 0, 5, 10, 20, 30, and 40 kg/ha were applied to the soil in both treatments. Results indicated that there was a significant impact of Zn treatments on grain produce, concentration of Zn, concentration of P, and weight of thousand kernels of wheat, but the impact was nonsignificant for protein concentrations. Foliar + soil treatment improved the Zn concentration in grains. It was concluded that zinc treatment of 10–20 kg/ha was effective on grain Zn concentration [5].

9.2.2 GENETIC BIOFORTIFICATION

Genetic biofortification methods involve characterization and exploitation of genetic variation for mineral percentage and new approaches that involve marker-aided breeding and discovery of gene. It helps in increasing the level of substances that promote nutrient absorption and reducing level of antinutrients, thus producing food crops with improved micronutrients [8]. These techniques are more labor-intensive and complex than the agronomic methods but are cost-effective and based on long-term strategies.

Germplasm screening of wheat is done with hexaploid, tetraploid, and diploid sources from the CIMMYT gene-bank for iron and zinc variations [39]. The sources used for high zinc and iron grain concentration are *Triticum dicoccoides, Aegilops tauschii, Triticum monococcum,* and *Triticum boeticum* [15]. Wheat germplasm showed two- to threefolds differences in Zn and Fe concentrations [4]: Zn by 2.03-folds, from 36.4 to 73.8 mg/kg; and Fe by 1.64-folds, from 41.4 to 67.7 mg/kg.

Biofortification was done by conventional breeding by selecting wild varieties with high micronutrient levels and then crossed with genetically superior plant lines. This method needs a lot of workers and molecular markers used in close relation to the traits of interest [51]. On the contrary, genetic engineering is used for constructing trait of interest using molecular marker by mixing promoters and genes. This trait of interest was then transformed in plants and its expression confirmed their integration due to which multiplication of transgenic plants was done [47].

9.3 STRATEGIES FOR Zn BIOFORTIFICATION

9.3.1 ZINC BIOAVAILABILITY AND ITS LOCALIZATION IN WHEAT KERNEL AND ITS OUTCOMES AFTER GRINDING

Zn is generally present in the embryo and aleurone parts of the wheat grain. Endosperm is poor in zinc. The concentration of zinc in aleurone layer and embryo is above 100 mg of Zn/kg compared to 10 mg of zinc/kg in the endosperm [40]. As the consumption of wheat is mostly after grinding, it separates the parts rich in zinc and the endosperm poor in Zn is left behind. The amount of zinc in white wheat flour is ranges from 5 to 10 mg of Zn/kg depending upon the rate of extraction [16], and these levels fail to match the nutritional zinc requirement. The methods and the information on nutritional zinc requirement have persisted since the 1970s along with consideration of Zn bioavailability and its absorption [23]. According to International Zinc Nutrition Consultative Group, recommended daily allowance (RDA) for adults lies between 9 and 19 mg of Zn/day depending upon sex and specific conditions of gestation and lactation [12]. For the daily intake of about 400 g wheat flour by the target persons residing in the main wheat consuming countries, the zinc concentration in flour will be about 8 mg Zn/kg and regular zinc intake will be close to 3.2 mg only. With no involvement of supplementation, food fortification or dietary modification, this population may undergo severe deficiencies of zinc and its effects. Brown et al. [11] suggested an intervention strategy for target countries, where Zn fortification was done at the concentration of 15–30 mg Zn/kg due to the presence of a very low amount of Zn in flour.

Wheat grains contain components, such as phenolic acids and phytates that reduce the intake of zinc and decrease its bioavailability. Evidence is available, in which dephytinization of wheat is done to reduce the phytate content, which in turn will increase the bioavailability of Zn and hence the Zn intake [23].

9.3.2 AGRONOMIC BIOFORTIFICATION WITH ZINC VIA APPLICATION OF FERTILIZER

Genetic modification, agronomic interventions, and conventional and molecular plant breeding are principal methods implemented for biofortification with zinc in food crops. Availability of zinc in plant is reduced due to various physical and chemical issues arising in the cultivation of wheat and other cereals. This might occur in achieving a marked biological effect in target subjects under the conditions of insufficient availability of zinc when newly created biofortified zinc genotypes by genetic engineering or breeding might become ineffective to accumulate the quantity of Zn in the wheat grain [13]. Therefore it is important to understand the sources of zinc in the wheat kernel (Figure 9.1), such as [35] (1) Zn is translocated into the wheat grain by absorption by the roots from the soil; and (2) Zn is translocated into the wheat grain by remobilization of Zn deposited in stems and leaves (vegetative tissues) during the reproductive stage. These two sources differ with respect to the Zn accumulation in wheat grain depending on various soil and plant factors, such as timing of senescence, nitrogen, dietary status of plant, water, and micronutrient availability during grain filling and span of the grain-filling period. Maintenance of plant-obtainable zinc via vegetative tissues or soil is a critical phenomenon required for Zn biofortification of cereals.

Zhou et al. [60] conducted a field experiment to examine the effect of Zn fertilization in wheat biofortification with four treatments: no zinc (control), foliar zinc, soil zinc, and foliar + soil zinc. The results exhibited that (soil + foliar) zinc and foliar zinc treatments considerably improved the grain zinc concentration by 27 mg/kg at zero zinc to 49 and 48 mg/kg resulting in rise in kernel zinc by 90% and 84%, consecutively. They concluded that the application of foliar Zn led to successful wheat grain biofortification with Zn with no effect on the yield. Application of foliar zinc to wheat included EDTA☐chelated Zn and zinc sulfate ($ZnSO_4$). Both Zn☐EDTA and zinc sulfate were effective for increasing Zn concentrations in tissues and correcting the Zn deficiency. Therefore zinc sulfate is considered as the most cost-effective option due to high price of Zn-EDTA. In relation to biofortification, the effectiveness of foliar zinc fertilizer treatment is determined by the time taken by its application [55].

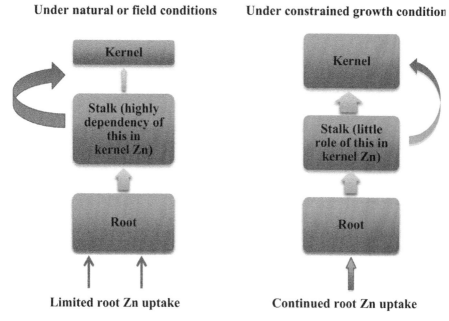

Under natural or field conditions   **Under constrained growth condition**

Limited root Zn uptake   Continued root Zn uptake

FIGURE 9.1 Uptake of zinc by roots and Zn accumulation in grains under natural and controlled growth conditions.

Abdoli et al. [2] conducted field experiments at different developmental stages on grain Zn content and agronomy traits to examine the effects of foliar application of zinc sulfate. Foliar applications of Zn (@ 0.44 g of Zn/liter) and zero Zn (control) were used during the elongation stage of stem and stages of stemming and grain filling. The results exhibited that foliar zinc treatment at the stem elongation phase was less effective compared to foliar zinc treatment at the stemming and grain filling phases on grain yield and its components of wheat. Persson et al. [41] studied that an increase in the concentration of N and Zn in durum wheat had a major effect on concentration of Zn in endosperm, attaining higher concentration compared to current target of breeding, because most of the population consumes endosperm part of wheat, which is low in Zn concentration.

When zinc fertilizers applied in soil to the regions coming under production of wheat in Sindh and Punjab, an increase in dietary Zn supply could be seen (i.e., around 12.6–14.6 mg/capita/day), and the occurrence of zinc deficiency was lessened by almost 50% under the assumption that no other changes take place during food consumption. An increase in gross wheat yield could be by 0.6 and 2.0 MT grain/year in Sindh and Punjab, respectively, representing a yearly rise in kernel supply of 19 kg/capita [32].

9.4 STRATEGIES FOR IRON BIOFORTIFICATION

IDA impacts 2 billion people around the world. Mostly children, expecting and nonexpecting mothers are more prone to IDA. It decreases physical and mental developments and reduces resistance to diseases. It increases the death rate and work performance is also retarded among all generations. The RDA of iron is 30 mg/day for expecting mother and varies from 8–18 mg/day on the basis of body weight, age, and sex. In this section, following three strategies for biofortification of iron are presented.

9.4.1 ENRICHMENT OF FERRITIN CONTENT

The main location for storage of iron in the plant is ferritin. As compared to its ferric state, ferritin has more holding capacity for iron. It carries up to 4000 iron molecules in the center. In this manner, ferritin provides iron at the developmental phase of plants and certain species of plant seeds have large amounts of iron and ferritin. The high bioavailability of iron in plant foods is attributed to its ferritin-form and oxidative stress is also lowered by it [10].

In biofortification strategies, ferritin genes belonging to the *Leguminosae* family of soybean were inserted in other plants and their overexpression in plants gave rise to the ferritin proteins. However, in cereal grains and seeds, overexpression of ferritin resulted into increased iron content in plant's edible parts [36]; but the overexpression of ferritin did not show the same effect in vegetative tissues rather deficiency symptoms were found. Therefore a detailed study on ferritin overexpression is needed and simultaneously it could be considered necessary to improve iron absorption for full effectiveness [43].

9.4.2 REDUCTION OF PHYTIC ACID CONTENT

Reliability on biofortification of iron is due to the reduction of antinutrients metabolites (such as phytic acid, tannins, and phenolic polymers), which results in the formation of iron complex that tends to exhibit lower solubility in the gut [54]. Phytic acid comprises 80% of the total amount of phosphorus in the seed. In seeds of legumes (such as soybean cereal germ and peripheral endosperm cells), it is present in the form of mineral storage compounds and phosphorus [7]. Major reason for anemia and deficiency of iron in developing countries is the presence of phytic acid in plant-based diets. Despite this, negativity was seen in reducing phytic acid as it prevents kidney stones and enhances immune system in balanced diet [49]. There will be formation of low phytic acid (lpa) phenotype on breakdown of phytic acid biosynthesis chain [44, 45].

An alternate method was the addition of phytase enzyme, which transformed the plants by disruption of phytic acid as it was produced. Various microorganisms were used in the separation of the phytase enzyme. While considering the procedure of food processing, enzyme activity, and heat stability are important standards [7]. As a result, several experiments have to be done to eradicate negative effects of antinutrient, that is, phytic acid in addition to retaining its positive effects on plant growth.

9.4.3 IMPROVEMENT IN NICOTIANAMINE CONTENT

Nicotianamine present in plants is a vital compound of metal homeostasis. In nicotianamine synthase, an enzyme acts on nicotianamine leading to the formation of S-adenosyl methionine. It can bind to various types of metals comprising of ferrous and ferric according to the pH of the surroundings. In case of iron, it confirms the solubility of iron in the cell. Moreover, it also plays a part in numerous important submechanisms of metal homeostasis in plant, for example, inter- and intracellular transport, sequestration, detoxification, and uptake and mobilization. Various studies indicated that nicotianamine had positive effects on the accumulation of iron in seeds and its uptake. Hence, nicotianamine is a vital compound for iron biofortification in seeds and kernels [20, 33].

Various research works also revealed that the expression of a gene nicotianamine synthase in increased quantity can accelerate the amount of nicotianamine but not essentially used by plants. Therefore availability of iron in apoplast may be restricted due to the increased amount of nicotianamine [19]. Moreover, a rise in the amount of iron in the leaves (but not in the seeds) was observed due to the increased expression of nicotianamine synthase.

9.5   STRATEGIES FOR IODINE BIOFORTIFICATION

Iodine is an essential micronutrient required for synthesis of thyroid hormones. RDA for iodine is 90 to 250 mg. Inadequate intake results in insufficient secretion of thyroid hormones that result in goiter. Iodization of salt is universally practiced in several countries; however, many people are diverting toward lowering the salt intake to prevent cardiovascular diseases and hypertension. Iodine is generally not studied as a trace element for higher plants; however, studies reported that it is associated with the plant biochemical and physiological activities [56]. Iodine is a nonessential nutrient for plant but can be absorbed via roots and leaves. Plant growth is supported at low dose, whereas phytotoxin is produced at high dose.

The application of iodine as agrochemical is the simplest approach for solving major issues of lack of iodine in soil and further in the human diet. Recent works have optimized the agreements regarding iodine treatment to the soil by foliar spray, by irrigation or in hydroponic solution. Transfer factor of iodine is determined by the ratio of concentration in edible plant part and its concentration in soil.

Foliar spray methods reported that uptake of iodine via xylem is faster in contrast to phloem and accumulation of iodine in tubers, fruits, and seeds is lower compared to uptake by leaves [26]. Leafy vegetables (such as Chinese cabbage, celery, and lettuce) are ideal crops for biofortification of iodine. In a greenhouse experiment on wheat, KI and KIO_3 were applied as fertilizer to the soil. These iodine compounds raised the iodine concentration in the shoots. Results suggested that iodine got translocated through phloem from shoots to the grains [17].

9.6   STRATEGIES FOR SELENIUM BIOFORTIFICATION

Selenium (Se) is important for animal and human health but in excess it may be poisonous. Hesketh [28] studied the significance of Se in metabolic process of cells through its important part in the role of specific enzymes or seleno-proteins. Glutathione peroxidases, deiodinases, and the thioredoxin reductases and seleno-protein P are the best characterized seleno-proteins.

Johnson et al. [31] studied the significance of geological structure of a region for deciding the amount. in which selenium becomes available to the soil and then to food sources of agricultural importance. For instance, due to these variations in geology, US wheat showed ten times higher selenium compared to UK wheat; therefore soils from the regions of America growing wheat are richer in selenium to a same order of magnitude. Therefore Se-deficiency is widespread in Keshan, where this deficiency was first stated with incidence of myocardiopathy [30]. Se deficiency due to geological differences can be countered with effective biofortification of crops with Se-rich fertilizers.

Wheat is generally a major dietary Se-source in countries, where wheat is a staple food crop. The amount of Se in wheat can be improved by agronomic biofortification (Figure 9.2). During fertilization of developing crop by the suitable inorganic form of trace mineral, the plant transformed it to a number of organic forms of selenium, particularly seleno-methionine that is more appropriate for consumption by humans [37]. The usefulness of foliar or soil treatment of selenium is based on soil characteristics, time of foliar treatment, Se-form and method of basal application. Ylaranta [58] discovered that foliar selenate and basal application had similar effectiveness at low level of 10 g/ha, foliar improved at 50 g/ha, and both similar at the high level of 500 g/ha. Further tests showed that basal application to the clayey soil having pH 6.3 was less effective than foliar selenate applied at the three-leaf stage and four-leaf stage; equally effective on alluvial soil having pH 4.6 and high humus; and somewhat more beneficial compared to foliar fertilizer on alluvial soil having pH 5.0. Foliar selenite when applied at the rate of 10 g/ha with the help of a wetting agent, elevated the level of selenium in wheat grain from 16–168 mg/kg onto the clay soil, whereas basal application at the rate of 9 g raised it to just 77 mg/kg. In general, the more effective method for barley and wheat was foliar application of selenium, excluding the areas, where there is reduced growth due to low rainfall. It was concluded that biofortified wheat improved status of selenium in humans. Further research is yet required to confirm its bioefficacy in populations deficient in selenium. On the basis of intervention study, it was shown that Se-biofortified wheat is fit for consumption by humans if consumed daily up to 300 mg of Se, as organic selenium over a period of 24 weeks. The level of organic selenium more than 430 mg of Se/L will cause cytotoxic and cytostatic effects in cells of an individual.

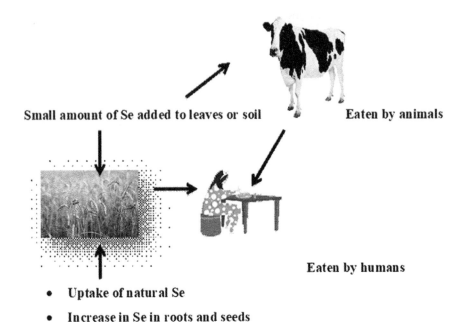

Small amount of Se added to leaves or soil   **Eaten by animals**

Eaten by humans

- **Uptake of natural Se**
- **Increase in Se in roots and seeds**

FIGURE 9.2   Selenium biofortification process.

9.7   ANTHOCYANIN BIOFORTIFIED COLORED WHEAT

Anthocyanins are the pigments present in colored plants. It has antioxidant effects and helps in preventive treatment of cancer, heart failure, inflammation, ageing, and others. Interest has been created for anthocyanin rich-colored wheat among the agronomists due to its antioxidant and anti-inflammatory activities and it differed according to the sequence of black > blue > purple > white wheat. Wheat commonly consumed around the world is amber in color. Blue color is localized to the aleurone layer and purple color to the pericarp. By combining genes of both blue and purple colors, black wheat was developed [22, 29].

Breeding was done by selecting three advanced colored breeding lines capable of producing abundant crops (blue, purple, black), one white wheat (PBW 621) and three-colored donor wheat lines from BC_1F_8 generations. They were grown in the farms of National Agri-Food Biotechnology Institute, Mohali, Punjab, India in 2015–2016.

Extraction of crude anthocyanin was done with acidified methanol for development of anthocyanin biofortified colored wheat. Advanced colored breeding lines were capable of producing abundant crops that showed high anti-inflammatory and antioxidant activities. These wheat lines had all the characteristics for their able utilization in the industry [6].

Sharma et al. [50] studied the product making and nutritional potential of colored wheat lines. Results revealed that colored wheat lines exhibited high antioxidant activity and high anthocyanin content, nutritional; and processing parameters were similar to white wheat lines capable of producing abundant crops. Wheat lines, which are colored, exhibited large amount of zinc and iron and were utilized commercially due to its desirable characteristics.

9.8 HARVESTPLUS PROGRAM

CGIAR started a HarvestPlus Challenge Program, which is an interdisciplinary association of implementing agencies and research institutes that is producing emerging varieties of pearl millet, wheat, rice, maize, sweet potato, and others through biofortification. HarvestPlus has become a world leader in generating biofortified staple crops and it is located at the International Food Policy Research Institute (IFPRI). HarvestPlus has aimed to reduce micronutrient deficiency and supply micronutrients to the population through staple crops. It is focused on three main micronutrients, that is, iron, zinc, and vitamin A, recognized by the World Health Organization (Table 9.1).

TABLE 9.1 Crops Released by HarvestPlus in Developing Countries

Cereals	Mineral	Countries Obtained First	Agronomic Feature	Release Year
Wheat	Zn and Fe	India and Pakistan	Resistance to infection and lodging resistance;	2013
Pearl millet	Fe and Zn	India	Resistance to mold growth; drought resistant.	2012
Rice	Zn and Fe	India and Bangladesh	Resistance to infection and pest resistance; cold and submergence tolerant.	2013

Note: Fe, iron; Zn, zinc.

Activities of HarvestPlus are shown by a pathway of impact and are categorized into three phases of discovery, development, and dissemination as depicted below:

1. *Discovery:* Identification of target populations and profile of staple food consumption; setting up of nutrient level target; screening and applied biotechnology.
2. *Development*: Improvement of crop; G×E interactions on concentration of nutrient; retention of nutrient and its bioavailability; studies regarding dietary efficiency in human cases.
3. *Delivery*: Crops released through biofortification; facilitation of marketing, dissemination and customer acceptance; nutritional status of target populations improved.

Dissemination depends on the success of the development and discovery phases, along with setting up partnerships among HarvestPlus and country organizations that will result in the supply of biofortified seeds to farmers and introducing biofortified crops to customers [27].

9.9 SUMMARY

Genetic techniques are more labor-intensive and complex than the agronomic methods, but they are cost-effective and involve long-term strategies. Biofortified wheat also improved the status of selenium in humans. Anthocyanin biofortified wheat produced through plant breeding possesses high anti-inflammatory activity, antioxidant activity, and anthocyanin content. HarvestPlus Challenge Program is developing biofortified varieties of staple crops and delivering biofortified seeds to farmers and introducing biofortified crops to customers. Rising atmospheric CO_2, rising temperature, draught, and unpredictable precipitation are forecasted in variation in climate framework for 21st century that may influence the soil fertility and in-turn intake of micronutrients by humans.

KEYWORDS

- biofortification
- colored wheat
- fertilizers
- foliar spray
- HarvestPlus

REFERENCES

1. *FAOSTAT On-Line Database*. Rome, Italy: Food and Agriculture Organization of the United Nations; **2016**; online; http://www.fao.org/faostat/en/; accessed on December 31, 2019.
2. Abdoli, M.; Esfandiari, E. Effects of Foliar Application of Zinc Sulfate at Different Phenological Stages on Yield Formation and Grain Zinc Content of Bread Wheat (cv. *Kohdasht*). *Azarian Journal of Agriculture*, **2014**, *1*, 11–16.
3. Alloway, B. J. Soil Factors Associated with Zinc Deficiency in Crops and Humans. *Environmental Geochemistry and Health*, **2009**, *31*, 537–548.
4. Badakhshan, H.; Namdar, M.; Mohammadzadeh, H.; Mohammad, R. Z. Genetic Variability Analysis of Grains Fe, Zn and Beta-Carotene Concentration of Prevalent Wheat Varieties in Iran. *International Journal of Agriculture and Crop Sciences*, **2013**, *6*, 57–62.
5. Barut, H.; Simsek, T.; Irmak, S.; Sevilmis, U.; Aykanat, S. The effect of Different Zinc Application of Bread Wheat Varieties. *Turkish Journal of Agriculture-Food Science and Technology*, **2017**, *5*, 898–907.
6. Boeing, J. S.; Barizao, E. O. Evaluation of Solvent Effect on the Extraction of Phenolic Compounds and Antioxidant Capacities from the Berries: Application of Principal Component Analysis. *Chemistry Central Journal*, **2014**, *8*, 48–52.
7. Bohn, L.; Meyer, A. S.; Rasmussen, S. K. Phytate: Impact on Environment and Human Nutrition: A Challenge for Molecular Breeding. *Journal of Zhejiang University Science B*, **2008**, *9*, 165–191.
8. Bouis, H. E. Micronutrient Fortification of Plants Through Plant Breeding: Can It Improve Nutrition in Man at Low Cost? *Proceedings of the Nutrition Society*, **2003**, *62*, 403–411.
9. Bouis, H. E.; Welch, R. M. Biofortification: A Sustainable Agricultural Strategy for Reducing Micronutrient Malnutrition in the Global South. *Crop Science*, **2010**, *50*, 20–32.
10. Briat, J. F.; Ravet, K.; Arnaud, N.; Gaymard, F. New Insights into Ferritin Synthesis and Function Highlight a Link Between Iron Homeostasis and Oxidative Stress in Plants. *Annals of Botany*, **2010**, *105*, 811–822.
11. Brown, K. H.; Hambridge, K. M.; Ranum, P. Zinc Fortification of Cereal Flours: Current Recommendations and Research Needs. *Food and Nutrition Bulletin*, **2010**, *31*, 62–74.
12. Brown, K. H.; Rivera, J. A.; Bhutta, Z. International Zinc Nutrition Consultative Group (IZiNCG) Technical Document #1: Assessment of the Risk of Zinc Deficiency in Populations and Options for its Control. *Food and Nutrition Bulletin*, **2004**, *25*, 99–203.
13. Cakmak, I. Enrichment of Cereal Grains with Zinc: Agronomic or Genetic Biofortification? *Plant and Soil*, **2008**, *302*, 1–17.
14. Cakmak, I.; Kutman, U. B. Agronomic Biofortification of Cereals with Zinc: A Review. *European Journal of Soil Science*, **2018**, *69*, 172–180.
15. Cakmak, I.; Ozkan, H.; Braun, H. J.; Welch, R. M.; Romheld, V. Zinc and Iron Concentrations in Seeds of Wild, Primitive and Modern Wheats. *Food and Nutrition Bulletin*, **2000**, *21*, 401–403.
16. Cakmak, I.; Pfeiffer, W. H.; McClafferty, B. Biofortification of Durum Wheat with Zinc and Iron. *Cereal Chemistry*, **2010**, *87*, 10–20.
17. Cakmak, I.; Prom-u-thai, C.; Guilherme, L. R. G. Iodine Biofortification of Wheat, Rice and Maize through Fertilizer Strategy. *Plant Soil*, **2017**, *418*, 319–335.

18. Cakmak, I.; Tourn, A.; Millet, E.; Feldman, M. Triticumdicoccoide: An important Genetic Resource for Increasing Zinc and Iron Concentration in Modern Cultivated Wheat. *Soil Science and Plant Nutrition*, **2004**, *50*, 1047–1054.

19. Cassin, G.; Mari, S.; Curie, C.; Briat, J. F.; Czernic, P. Increased Sensitivity to Iron Deficiency in Arabidopsis Thaliana over Accumulating Nicotianamine. *Journal of Experimental Botany,* **2009**, *60*, 1249–1259.

20. Cheng, L. J.; Wang, F.; Shou, H. X.; Huang, F. L. Mutation in Nicotianamine Amino-transferase Stimulated the Fe(II) Acquisition System and Lead to Iron Accumulation in Rice. *Plant Physiology,* **2007**, *145*, 1647–1657.

21. Dar, W. D. *Macro-Benefits from Micronutrients for Grey to Green Revolution in Agriculture*. Proceedings of the IFA International Symposium on Micronutrients; New Delhi, India: International Fertilizer Industry Association; February 23–25, **2004**; p. 13.

22. Garg, M.; Chawla, M.; Chunduri, V.; Kumar, R.; Sharma, S.; Sharma, N. K. Transfer of Grain Colors to Elite Wheat 532 Cultivars and their Characterization. *Journal of Cereal Science,* **2016**, *71*, 138–144.

23. Gibson, R. S.; Bailey, K. B.; Gibbs, M.; Ferguson, E. L. Review of Phytate, Iron, Zinc and Calcium Concentrations in Plant-Based Complementary Foods Used in Low-Income Countries and Implications for Bioavailability. *Food and Nutrition Bulletin,* **2010**, *31*, 134–146.

24. Gibson, R. Zinc Nutrition in Developing Countries. *Nutrition Research Reviews*, **1994**, *7*, 151–173.

25. Gomez-Galera, S.; Rojas, E.; Sudhakar, D. Critical Evaluation of Strategies for Mineral Fortification of Staple Food Crops. *Transgenic Research,* **2010**, *19*, 165–180.

26. Gonzali, S.; Kiferle, C.; Perata, P. Iodine Biofortification of Crops: Agronomic Biofortification, Metabolic Engineering and Iodine Bioavailability. *Current Opinion in Biotechnology*, **2017**, *44*, 16–26.

27. Grusak, M. A.; Cakmak, I. Methods to Improve the Crop-Delivery of Minerals to Humans and Livestock. In: *Plant Nutritional Genomics*; Broadley, M. R. and White, P. J. (Eds.); Oxford, UK: Blackwell Publishers; **2005**; pp. 265–286.

28. Hesketh, J. Nutrigenomics and Selenium: Gene Expression Patterns, Physiological Targets and Genetics. *Annual Review of Nutrition,* **2008**, *28*, 157–177.

29. Hosseinian, F. S.; Li, W.; Beta, T. Measurement of Anthocyanins and Other Phytochemicals in Purple Wheat. *Food Chemistry,* **2008**, *109*, 916–924.

30. Hou, J.; Wang, T.; Liu, M.; Li, S.; Chen, J.; Liu, C. Suboptimal Selenium Supply: Continuing Problem in Keshan Disease Areas in Heilongjiang Province. *Biological Trace Element Research,* **2011**, *143*, 1255–1263.

31. Johnson, C. C.; Fordyce, F. M.; Rayman, M. P. Symposium on Geographical and Geological Influences on Nutrition: Factors Controlling the Distribution of Selenium in the Environment and Their Impact on Health and Nutrition. *Proceedings of the Nutrition Society*, **2010**, *69*, 119–132.

32. Joy, E. J. M.; Ahmad, W.; Zia, M. H. Valuing Increased Zinc Fertilizer-Use in Pakistan. *Plant and Soil,* **2016**, *411*, 139–150.

33. Klatte, M.; Schuler, M.; Wirtz, M. The Analysis of Arabidopsis Nicotianamine Synthase Mutants Reveals Functions for Nicotianamine in Seed Iron Loading and Iron Deficiency Responses. *Plant Physiology,* **2009**, *150*, 257–271.

34. Kumar, S.; Rao, M.; Gupta, N. C. Breeding for High Iron and Zinc Content in Cultivated Wheat. *Journal of Crop Science and Technology,* **2014**, *3*, 7–9.

35. Kutman, U. B.; Kutman, B. Y.; Ceylan, Y.; Ova, E. A.; Cakmak, I. Contributions of Root Uptake and Remobilization to Grain Zinc Accumulation in Wheat Depending on Post-Anthesis Zinc Availability and Nitrogen Nutrition. *Plant and Soil,* **2012,** *361,* 177–187.

36. Lucca, P.; Hurrell, R.; Potrykus, I. Fighting Iron Deficiency Anemia with Iron-Rich Rice. *Journal of the American College of Nutrition,* **2002,** *21,* 184–190.

37. Lyons, G.; Ortiz-Monasterio, I.; Stangoulis, J.; Graham, G. Selenium Concentration in Wheat Grain: Is There Sufficient Genotypic Variation to Use in Breeding? *Plant Soil,* **2005,** *269,* 369–380.

38. Meenakshi, J. V.; Johnson, N.; Manyong, V. M. How Cost Effective is Biofortification in Combating Micronutrient Malnutrition? An Ex-Ante Assessment. *Harvest Plus,* **2007,** *2007,* 188–209.

39. Monasterio, I.; Graham, R. D. Breeding for Trace Minerals in Wheat. *Food and Nutrition Bulletin,* **2000,** *21,* 393–396.

40. Ozturk, L.; Yazici, M. A.; Yucel, C.; Torun, A. Concentration and Localization of Zinc During Seed Development and Germination in Wheat. *Plant Physiology,* **2006,** *128,* 144–152.

41. Persson, D. P.; Bang, T. C.; Pedas, P. R. Molecular Speciation and Tissue Compart-mentation of Zinc in Durum Wheat Grains with Contrasting Nutritional Status. *New Phytologist,* **2016,** *211,* 1255–1265.

42. Pfeiffer, W. H.; McClafferty, B. Biofortification: Breeding Micronutrient-Dense Crops. *Breeding Major Food Staples,* **2007,** *2007,* 61–91.

43. Qule, Q.; Yoshihara, T.; Ooyama, A.; Goto, F.; Takaiwa, F. Iron Accumulation Does Not Parallel the High Expression Level of Ferritin in Transgenic Rice Seeds. *Planta,* **2005,** *222,* 225–233.

44. Raboy, V. The ABCs of Low-Phytate Crops. *Nature Biotechnology,* **2007,** *25,* 874–875.

45. Rasmussen, S. K.; Ingvardsen, C. R.; Torp, A. M. Mutations in Genes Controlling the Biosynthesis and Accumulation of Inositol Phosphates in Seeds. *Biochemical Society Transactions,* **2010,** *38,* 689–694.

46. Saha, B.; Saha, S.; Poddar, P.; Murmu, S.; Singh, A. K. Uptake of Nutrients by Wheat as Influenced by Long-Term Phosphorus Fertilization. *The Bioscan,* **2013,** *8,* 1331–1335.

47. Sayre, R.; Beeching, J.; Cahoon, E. B.; Egesi, C. The BioCassava Plus Program: Biofortification of Cassava for Sub-Saharan Africa. *Annual Review of Plant Biology,* **2011,** *62,* 251–272.

48. Scagel, C.; Bi, G.; Fuchigami, L.; Regan, R. Irrigation Frequency Alters Nutrient Uptake in Container-Grown Rhododendron Plants Grown with Different Rates of Nitrogen. *Horticultural Science,* **2012,** *47,* 189–197.

49. Shamsuddin, A. M. Demonizing Phytate. *Nature Biotechnology,* **2008,** *26,* 496–497.

50. Sharma, S.; Chunduri, V.; Kumar, A. Anthocyanin Bio-fortified Colored Wheat: Nutritional and Functional Characterization. *PLoS One,* **2018,** *13,* E-article: 0194367; online; https://doi.org/10.1371/journal.pone.0194367; accessed on December 31, 2019.

51. Tester, M.; Langridge, P. Breeding Technologies to Increase Crop Production in a Changing World. *Science,* **2010,** *327,* 818–822.

52. The World Health Organization. Iron Deficiency Anemia: Assessment, Prevention and Control. Geneva: WHO; **2011**; p. 132; online; https://www.who.int/nutrition/publications/en/ida_assessment_prevention_control.pdf;

53. Tontisirin, K.; Nantel, G.; Bhattacharjee, L. Food-Based Strategies to Meet the Challenges of Micronutrient Malnutrition in the Developing World. *Proceedings of the Nutrition Society,* **2002,** *61,* 243–250.

54. Welch, R. M.; Graham, R. D. Breeding for Micronutrients in Staple Food Crops from a Human Nutrition Perspective. *Journal of Experimental Botany*, **2004**, *55*, 353–364.
55. Welch, R. M.; Graham, R. D.; Cakmak, I. *Linking Agricultural Production Practices to Improving Human Nutrition and Health*. ICN 2nd Conference on Nutrition: Better Nutrition for Better Lives; Rome, Italy: FAO/WHO; **2013**; p. 39.
56. White, P. J.; Broadley, M. R. Biofortification of Crops with Seven Mineral Elements Often Lacking in Human Diets: Iron, Zinc, Copper, Calcium, Magnesium, Selenium and Iodine. *New Phytologist*, **2009**, *182*, 49–84.
57. WHO Nutrition. *Highlights of Recent Activities in the Context of the World Declaration and Plan of Action for Nutrition*. Nutrition Programme: International Conference on Nutrition; Geneva: WHO; **1995**; p. 50; online; https://apps.who.int/iris/bitstream/handle/10665/61051/a34303.pdf;jsessionid=9DB3F67699F37F05473952013D19CD6A?sequence=1;
58. Ylaranta, T. Raising the Selenium Content of Spring Wheat and Barley using Selenite and Selenate. *Annales Agriculturae Fenniae*, **1984**, *23*, 75–84.
59. Zhang, Y.; Shi, R.; Rezaul, K. M.; Zhang, F.; Zou, C. Iron and Zinc Concentrations in Grain and Flour of Winter Wheat as Affected by Foliar Application. *Journal of Agricultural and Food Chemistry*, **2010**, *58*, 12268–12274.
60. Zhou, C. Q.; Zhang, Y. Q.; Rashid, A.; Ram, H.; Savasli, E. Biofortification of Wheat with Zinc through Zinc Fertilization in Seven Countries. *Plant and Soil*, **2012**, *361*, 119–130.

CHAPTER 10

INNOVATIVE PACKAGING FOR CEREALS AND CEREAL-BASED PRODUCTS

RAMANDEEP KAUR and KAMALJIT KAUR

ABSTRACT

Innovations in cereal and bakery industry and increased concern of consumers for food safety have enhanced the development of active, intelligent, bioactive, and nanocomposite packaging. This provides a ravishing future with greater opportunities for researchers and entrepreneurs to develop cereal-based novel products.

10.1 INTRODUCTION

The basic functions of packaging material are to elevate the shelf stability of the food by averting adverse changes triggered by chemical changes (such as moisture, oxygen, rancidity, and others), physical changes (such as temperature fluctuation, light, external force), and microbial spoilage. Nowadays innovative food packaging technologies (Figure 10.1) are getting more attention due to changes in consumer lifestyle or industrial development of novel products, such as ready-to-eat products with better shelf-life and quality. Innovative packaging performs additional roles in addition to providing an inert barrier against the external environment [11].

Active packaging is a system in which certain additives (antimicrobials, ethanol emitters and carbon dioxide emitters, and others) have been deliberately added in the headspace of the package to improve its performance [70]. In contrast to this, various devices [such as time–temperature indicators (TTIs), sensors, physical shock and microbial growth indicators, and others] in the intelligent packaging are used to record specific properties of the food in the packaging material and able to inform the manufacturer and consumers about the condition of these properties and to check the integrity and effectiveness of active packaging systems [38].

Bioactive packaging materials are developed by incorporation of plant extracts (such as grape seed, turmeric, green tea, cranberry, and blueberry) with antimicrobial and antioxidant activities, which are further responsible for required physical and chemical changes in the product resulting in extended shelf-life while maintaining the quality [85]. Various studies have reported the increase of utilization of bioactive compounds in order to diminish the threat of various diseases (such as carcinogenic, cardiovascular, coronary, and chronic diseases) due to their capacity to remove free radicals from the body while maintaining our body functions [22]. Hence, the addition of bioactive compounds (antioxidants, prebiotics, and probiotics) to the production of active packaging helps in developing the nutritional and functional films [51].

Also, nanocomposite-based technologies have been developed to provide new opportunities and challenges for the development of various products in the food industry. These packaging materials are eco-friendly, safe to use; and nanopackaging materials are light in weight than the conventional packaging materials [72].

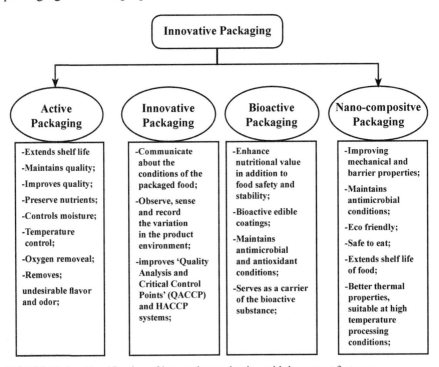

FIGURE 10.1 Classification of innovative packaging with important features.

Cereals are utilized for the production of various bakery products, such as bread, flour, biscuit, pasta, cookies and bars. Among all these products, bread is a more perishable product due to its undesirable physical, chemical, and microbial changes during the production and storage [57], because such changes can cause enormous food-waste and loss to the manufacturer and consumer [56]. Moreover, the growth of fungus leads to food intoxication, which is associated with the several acute and chronic diseases [63]. Flour, cookies, and biscuits also absorb moisture during storage, which may lead to physical (texture), chemical (rancidity), and microbial changes resulting in inferior shelf-life. To tackle these issues, innovative packaging is the best option to enhance the shelf-life of cereal and cereal products.

Numerous types of other packaging materials have been used for the packaging of cereal and cereal-based products. Traditionally, jute bags and woven sacks have been used for storage of grains. However, large amount of grains is deteriorated in such kind of packaging materials due to mold growth. Nowadays, varieties of packaging materials have been used for packaging of cereal and cereal-based products, which include polypropylene, polyethylene [high-density polyethylene (HDPE) and low-density polyethylene (LDPE)], aluminum laminates, outer duplex board carton, polyester BOPP-metalized for packaging of breakfast cereals.

LDPE and linear low-density polyethylene are the most commonly used packaging material for milled products (such as flour, semolina) due to its low cost and easy availability. Both films have good tensile, tear and bursting strength. It also possesses superior water and water vapor barrier properties. There are other two popular packaging materials for grounded cereals, such as cartons and pouches. The utilization of stand-up bags and plastic pouches is relatively new and is a preferred method of ground cereal packaging. There are numerous reasons for the increasing attractiveness of these plastic barrier bags, such as easy and economical manufacturing, easy to transfer from one place to another compared to the cartons.

This chapter focuses on the research trends in innovative packaging techniques applied in packaging of cereals and cereal-based products.

10.2 ACTIVE PACKAGING CEREALS AND CEREAL-BASED PRODUCTS

Active packaging plays a dynamic role in food as it allows package to interact with food as well as with the environment. Active packaging can provide several functions, such as scavenging of oxygen, moisture, or ethylene; emission of ethanol and flavors; and antimicrobial activity [50]. It is based on principle that either the intrinsic properties of the polymer are used itself as preservation material or on the inclusion and entrapment of specific substances inside the polymer [73]. The packaging material offers maintained shelf-life of product with the enhancement of value by regulating the migration of water vapor, air, CO2, oxygen, and odor of the food compounds between packaging material and its environment [8]. Active packaging can be divided into absorbers (active scavenging systems) and emitters (active releasing systems).

According to the European Commission, the active packaging (No. 450/2009) is a packaging material that absorbs undesirable components from the food or its surroundings (such as moisture, air, oxygen, carbon dioxide, ethylene gas, or odors). On the other hand, emitters are active releasing systems that add compounds to the packaged food, such as carbon dioxide, antimicrobial compounds, antioxidants, flavors, or ethylene to prevent the occurrence of undesired changes [20].

10.2.1   GAS SCAVENGERS AND EMITTERS

The most commonly used oxygen scavengers, carbon dioxide, ethylene scavengers, and emitters in cereals and bakery products are summarized in Table 10.1.

10.2.2   OXYGEN SCAVENGERS

The most commonly used active packaging technology in the bakery industry and other food industries are oxygen scavengers. Baked goods (such as bread, biscuit, cakes, pastries, and cookies) can have prolonged shelf-life with use of active packaging. Oxygen-scavenging packaging can be used as sachets or oxygen-scavenging compounds that are directly added to the film. These sachets or compounds absorb oxygen from the food product and its surroundings and keep the baked foods fresh. On the other hand, addition of oxygen absorber in the packaging material eradicates the risk of unintentional damage of the sachets and use of their contents. These compounds also blended or dispersed with high permeable films are polyethylene (PE), which causes quick transmission of H_2O and O_2 from the packaged food product to the components with reactive properties. Moreover, the oxygen-absorbing ability of other packaging materials, such as laminates, is significantly lower than the Fe-based oxygen absorber package [18].

TABLE 10.1 Applications of Innovative Packaging in Cereal and Cereal-Based Products

Packaging/Scavengers	Food Product	Ref.
Active packaging	Bakery goods	[30]
Ag/TiO_2 (PE)	Bread	[58]
Ag/TiO_2 -SiO_2; Ag/ N-TiO_2 and Au/TiO_2	Bread	[83]
Antimicrobial packaging	Gluten-free bread	[30]
Cellulose acetate films containing 0%, 2% and 4% sodium propionate	Bread	[75]
Combination of aqueous ozone treatment and MAP	Noodles	[6]
Ethanol emitters	Bread loaf	[45]
Ethanol emitters	Buns	[21]
Gliadin films incorporating cinnamaldehyde	Bread and cheese spread	[7]
Modified atmosphere packaging (MAP) and potassium sorbate	Bread	[19]
Nanoemulsions of essential oils from clove buds (*Syzygium aromaticum*) and oregano (*Origanum vulgare*).	Bread	[64]
Oxygen absorbers	Crackers	[10]
Oxygen scavengers	Bread	[36]
Poly acrylic acid and sodium salt reduce aflatoxin contamination	Maize	[54]

Oxygen-scavenging packagings are highly effective and can reduce oxygen levels in a package to as low as 100 ppm. Various compounds are used for oxygen scavengers packaging, however, when using for food these must be selected carefully. The ascorbic acid and iron powder are most commonly used in sachets [16]. Other various types of oxygen scavengers include photosensitive dye, enzymes (glucose oxidase or ethanol oxidase), saturated fatty acids, immobilized yeast on solid material, and others, which are commercially accessible [81].

The low-oxygen status in packaging materials prevents oxidation of lipids, growth of molds, and other spoilage bacteria. Modified atmosphere packaging (MAP) is also used in packaging material to control the amount of oxygen. However, it has one disadvantage that the oxygen can be leaked through the packaging material during storage [43].

Cruz et al. [17] reported the increase in the shelf-life of fresh lasagna pasta during storage (10 ± 2 °C) by using active packaging. Pasta was prepared with and without use of potassium sorbate and packed in active packaging containing oxygen scavengers in the headspace of the packaging material. It was concluded that oxygen absorbers were effective in retarding the growth of fungi, yeasts, Staphylococcus, *Escherichia coli*, and coliforms in fresh pasta.

Various studies have reported the effectiveness of oxygen scavengers in bakery goods. MAP with scavengers packaging of two dissimilar absorption capacities (i.e., 100 and 210 mL) was used for packaging of sponge cakes having water activity of 0.8–0.9. The cakes were stored for 28 days (at 25 °C) and observed for mold growth. It was found that MAP combination with oxygen absorbers could retard mold growth for 28 days [31].

In another study [10], oxygen scavengers were used in packaging of wheat crackers. Wheat crackers were prepared with high levels of oil and were stored at 15, 25, and 35 °C and with and without oxygen sachets. Sensory scores were observed at constant intervals using a hedonic scale. It was observed that with the increase in storage temperature (15, 25, and 35 °C), oxygen was decreased in the headspace of the metal can. The cans without oxygen absorbers had high level of rancidity after 24 weeks (at 25 and 35 °C), whereas cans with oxygen scavengers were shelf-stable even after 44 weeks of storage, irrespective of temperature of storage. Hence, shelf stability of canned crackers was prolonged to 20 weeks with oxygen scavengers.

Upasen and Wattanachai [80] reported the shelf-life stability of white bread without the use of any preservative. In this study, three kinds of packaging materials were used, such as LDPE layer containing an oxygen absorber, a one LDPE-layer incorporated with an oxygen absorber sachet, and three-layers of LDPE laminated with O-nylon. To check the stability and quality of bread, various parameters were evaluated, such as microbial count, oxygen transmission rate, headspace gases, and physical appearance. It was observed that sachet containing oxygen absorbers showed no microbial growth for 4 days and after that growth was very low compared to the packaging material (without the sachet).

The multilayer packaging film was prepared by using gallic acid and sodium carbonate in three steps—compounding, extrusion, and lamination. Film was brownish-red in color and changed to greenish-black after oxygen absorption under humid conditions. Gallic acid acts as a barrier to oxygen and water vapor. It can be used in products, which contain high water activity (aw >0.86) [65].

10.2.3 CARBON DIOXIDE ABSORBERS AND EMITTERS

Carbon dioxide absorbers and emitters can be incorporated to the headspace of packaging material to retard the progress of microbes in baked and other products (e.g., fresh meat, fish, poultry, and cheese) [50]. Oxygen scavengers generate a partial vacuum in the packaging material, which further leads to the collapse of the package. Also, when mixture of carbon dioxide and oxygen is flushed in the packaging material, it forms a partial vacuum. Carbon dioxide creates vacuum due to its solubility at low temperatures. Soluble gas stabilization process is also used in various products. In this process, sufficient amount of pure carbon dioxide is dissolved in the package for 1–2 h [42].

On the basis of characteristics of the food product, type of carbon dioxide emitter or absorber should be selected based on its carbon dioxide production quantity, desired level of CO_2, and the package type. The speed and capacity of the carbon dioxide scavenger absorption should be optimized [66].

Several carbon dioxide emitters and absorbers in the packaging material are based on the principle of chemical reaction, membrane separation, physical adsorption, and cryogenic condensation. Carbon dioxide adsorbents are classified into two categories: chemical and physical adsorbents [46]. Chemical adsorbent (such as calcium hydroxide) is most commonly used in food packaging [70]. In case of physical adsorbents, most widely activated carbon, zeolite, and activated charcoal are used [46].

In some food products, dual-action active packaging is used, such as labels and sachets that contain iron powder and CaOH for scavenging oxygen and CO_2 under high humid conditions. Various companies have made dual-action active packaging, such as Toppan Printing. Corporation Limited (Freshilizer™ type.CV), and "Mitsubishi Gas Chemical Corporation Limited" (Ageless™ type E Fresh Lock™) [42].

10.2.4 ETHYLENE ABSORBERS AND EMITTERS

Ethylene is a growth-stimulating hormone and it acts as a preservative in bakery products to prevent growth of microbes during storage. Various kinds of ethylene absorbers and emitters (such as potassium permanganate (KMnO4) are used, and these can be immobilized by inert minerals found in packaging materials and blankets, activated carbon base with various metal catalysts, activated charcoal impregnated with palladium catalyst, and others [71].

Latou et al. [45] used active packaging containing ethanol emitter alone and ethanol emitter combined with an oxygen absorber for the stability of wheat bread at 20 °C. Bread with preservatives and without preservatives was taken as controls. With treatment and storage, microbiological, physical, chemical, and organoleptic changes occurring in the product were determined for 30 days. It was observed that yeast and mold count, mainly growth of *Bacillus cereus*, was lower in samples containing ethanol emitters. Lipid oxidation and formation of off-flavors were also lower in samples with active packaging.

10.2.5 *MOISTURE ABSORBERS*

Another important class of active packaging is moisture absorbers. Moisture and water activity are considered as a major cause of food spoilage, generally in hygroscopic products. In case of dry bakery goods (such as biscuits, cookies, flour), moisture causes softening and caking of products during storage. Various kinds (such as pads, sheets, sachets, and blankets of moisture regulators) are used in food industry. In Active packaging, moisture absorbers can be classified into two main classes: (1) moisture removers that remove moisture from dry food products by absorption; and (2) humidity controllers that control humidity in the package headspace (such as desiccants) [86]. In the low moisture and water activity products, desiccants: calcium oxide, silica gel, activated clays, and minerals are used in the form of sheet and pads. These are generally tear-resistant plastic sachets. Numerous companies also used humidifier (controls humidity) in addition to moisture absorbers in the packaging material. Long sheets and blankets are used for high water activity products. Sheets generally comprise of two layers of a micro porous polymer (such as: polyethylene or polypropylene), which is sandwiched with superabsorbent polymers (carboxy methyl cellulose (CMC), polyacrylate salts, and starch copolymers) with higher ability for moisture. It has the ability to absorb up to 500 times more water than its own weight [71]. Commercially several kinds of moisture absorbent are available, such as Thermarite® Pvt. Ltd. and Peaksorb® in Australia, Toppan™ in Japan, Fresh-RPax™ in Atlanta, and Luquasorb™ in Germany [13].

In some foods, dual-action active packaging system is also used that consists of oxygen scavenging (iron powder) and odor adsorption (activated carbon). In modern days, various researchers focus on the development of organic moisture absorbers [25], which are used for organic-based moisture absorbers packaging materials, such as fructose, sorbitol, cellulose, xylitol, fructose, cellulose, and their derivatives (sodium carboxymethylcellulose, potassium carboxymethylcellulose, and ammonium carboxymethyl cellulose) [26].

Moisture absorbers based on polymers are the new era of research studies for the active packaging material. In this process, polymers appropriate for active packaging are combined with moisture absorbers [polyacrylic acid (PAA) and polyvinyl alcohol]. Among these compounds, polyvinyl alcohol is highly preferable due to its hydrophilic, highly film-forming, oxygen barrier, and biodegradability properties. Polyvinyl alcohol is combined with LDPE for developing moisture-absorbing packaging film. It was observed that polyvinyl alcohol combined LDPE film had highly moisture absorption capacity for the food package. However, it had inferior mechanical properties. Another film with the combination of polyvinyl alcohol (PVA) and green tea extract have been developed [15]. This film has antioxidant properties in addition to moisture absorption properties. Starch, cellulose, and its derivatives (CMC) were used to formulate active film to absorb moisture [29].

In selected studies, PAA and sodium salt powder were used to reduce the aflatoxin contamination in the maize and suppress the mold growth. Crosslinked PAA and sodium salt powder act as desiccant to absorb moisture. This super-absorbent polymer was kept in the form of tea bag in the maize, which had water content about 32%. Then, maize was dried at 40 °C up to 13% water content. It was observed that aflatoxin reduced to 3 ng/ g with the application of 1:5 and 1:1 ratio (super absorbent polymer to maize ratio). Therefore the application of PAA and sodium salt illustrates the potential for falling aflatoxin contamination, principally in developing countries due to its low economic value and reusable solution [54].

10.3 INTELLIGENT PACKAGING OR SMART PACKAGING

Intelligent packaging is a new emerging technology, which communicates on the conditions and properties of the food. It is based on the integrity and effectiveness of the packaging material [39]. In the intelligent packaging, indicators are incorporated for sensing and giving knowledge on the food condition, food function and food behavior in packaging material, safety and quality, pack integrity, tamper indication, product authenticity, product traceability, and antitheft [77]. Intelligent packaging devices include physical shock indicators, TTI, sensors, gas sensing dyes, and several types of antitheft, anticounterfeiting, and tamperproof techniques [42].

10.3.1 INDICATORS

Indicators refer to compounds that provide the information on the existence, not present or amount of any other component or the degree of change in other component, such as food [37]. Indicators (such as freshness indicator, TTI, and gas indicators) are used in food products to provide information on the quality parameters of the food within the packaging material.

10.3.1.1 FRESHNESS INDICATORS

Freshness indicator provides the information on the microbial growth, any undesirable odor, and flavor in the food product. In baked goods, it sheds light on the freshness of product. Food freshness indicators are usually prepared by using various organic acids, glucose, volatile nitrogen compounds, ethanol, biogenic amines, CO_2, and sulfur containing components to access the quality of the packaging for a food product [35]. Commercially available freshness indicators are Sensor QTM by Food Quality Sensor International in Massachusetts, United States, Toxin Guard' by Toxin Alert in Ontario, Canada, Food Sentinel System by SIRA Technologies in California, United States, and Fresh Tag by COX Technologies (Belmont, North Carolina, United States); and such devices are used for observing quality and progress of microorganisms in various packed food items [69].

10.3.1.2 GAS INDICATORS

Gas indicators are important for respiring commodities, such as fresh fruits and vegetables. It is very difficult to preserve freshness of raw fruits and vegetables due to exchange of gases within package. The best way to detect gas in a package is by ultraviolet activated oxygen concentration indicator, which was developed by the use of glycerol, thionine, encapsulating polymer (zein), and P25–TiO_2. The leakage of dye was about 80%, when the maize protein layered film was immersed in liquid for a 1 day. The leaching of dye into water can be halted by the ion binding ability and the inside the indicator system use of alginate can decrease the dye leakage to about 6% [82].

Smiddy et al. [74] used gas indicator in pizza and beef products for identifying inappropriate sealing and excellence deterioration of the package. Various gas indicators are available commercially for use in food products are Vitalon by Toagosei Chemical, Ageless EyeTM developed by Mitsubishi Gas Chemical Co. in Tokyo, Japan and Shelf-life Guard' by UPM-Kymmene Corporation in Helsinki, Finland.

Visual oxygen indicators are also used in MAP packaging of a food. It utilized redox dye, which alters its color with any change in oxygen concentration. However, this indicator has some disadvantages due to its high sensitivity and residual oxygen affects the efficiency of indicator [44]. Ageless Eye® is another oxygen-display tablet, which changes color with the existence or lack of "oxygen." It turns pink when the package lacks oxygen. At 0.5% oxygen level or more, it changes blue. However, if the existence of oxygen is shown in 5 min or less, then it takes 2–3 h from blue to pink change [59].

10.3.1.3 TIME TEMPERATURE INDICATOR (TTI)

Temperature is the main parameter to access the storage stability of various products. Abrupt fluctuations in the temperature can affect the quality of a packaged or stored food product. In recent years for the focus on maintained food quality, food technologist and researchers have focused on the innovations in food packaging [26]. TTI is based on the change in color, principle of enzymatic, chemical, electrochemical, mechanical, and microbiological reactions [55].

Recently, natural ingredients based TTI have been developed by the use of anthocyanin (red cabbage) and chitosan that are combined with poly-vinyl alcohol (PVA). It is very economical and safe to use. Anthocyanin alters its color with change in pH, hence it acts as a natural indicator of food quality. With the change in pH, it indirectly indicates the change in temperature [66].

10.3.2 SENSORS

The sensor is a device to perceive, spot, and measure changes in packaged food and then transfer indicators to check its physical and chemical properties. A sensor has the capability to regularly spot any alterations in nearby environment of the food [62]. Generally, sensors have two parts: a receptor and a transducer. The receptor converts physical or chemical data into energy, and a transducer converts the energy to a quantifiable signal. In the food industry, different gases are hydrogen, nitrogen dioxide, carbon dioxide, and oxygen; and gas-based sensors are numerous. The best alternative to time-consuming sensors is gas chromatography-mass spectrometer to detect gaseous molecule after the destruction of a package [48].

Recently an optochemical carbon dioxide sensor was developed by using Pt-TFPP dye, α-naphtholphthalein, and a tetraoctyl trimethyl ammonium hydroxide that presents as a phase transfer compound. It is used in food products, which are preserved by MAP packaging to sense any change in the composition of gases [12].

High-sensitive optical dual sensor was developed to the quantification of oxygen level at varied temperature conditions. This high-sensitive optical sensor comprises of two compounds, which emit light for the detection of temperature and oxygen respectively. In this, 10-phenanthroline luminescent, ruthenium tris-1, and temperature-sensitive dye were utilized. Fullerene C-70 probe was used for detection of oxygen. Results showed that high-sensitive optical dual sensor had the ability to identify temperature varying from 0 to 120 °C and 50 ppm least detection limit for oxygen [9].

10.4   BIOACTIVE PACKAGING

In recent years, utilization of bioactive substances in innovative packaging system and edible bioactive coating is gaining more attention due to rising consumer demands for healthy foods. The development of bioactive films and coatings can add worth to the packaged products by enhancing their stability and to develop eco-friendly and economical products. Since synthetic antioxidants are linked with possible harmful effects, therefore their use is not being preferred in the foods by consumers [47].

The bioactive substances (such as phytoestrogens, phenolic compounds, carotenoids, plant sterols, monoterpenes, plant extracts, bacteriocins, soluble dietary fibers, prebiotics, essential oils, organo-sulfur compounds, and enzymes) are most commonly added into the packaging materials. The bioactive coating containing antioxidant can enhance the shelf-stability, prevent browning of the food product, and maintain nutritional quality [41]. The various kinds of bioactive packaging films have been used for preservation of cereal products (Table 10.2).

Muratore et al. [61] developed the bioactive packaging for grain-based food products. Eugenol was attached to cellulose by using a polycarboxylic acid, which acts as a linking agent. The bioactive, sensory, and mechanical properties (such as water absorption, tear, tensile, and puncture strength) were evaluated. Further, insecticide activity and growth of microbes were evaluated using packaging prototypes. It was observed that bioactive material maintains good mechanical properties, with preservation of shelf-life. In addition, no migration of flavor and odor to the packaged product was observed.

TABLE 10.2 Utilization of Bioactive Packaging in Cereal Products.

Bioactive Films	Uses	Ref.
Bioactive	Quinoa and rice bread.	[5]
Cellulose and eugenol	Grain-based food products.	[61]
Cinnamaldehyde antimicrobial films	Enhance shelf-life of bakery products.	[49]
Edible films from methylcellulose; and nano emulsions of clove bud and oregano	Extends shelf-life of bread.	[64]
Essential oil	Prevents aflatoxigenic fungi and aflatoxin production in maize.	[53]
Poly-L-lysine based bioactive film	Controls mycotoxigenic fungi in bread.	[52]
Protamine salmine hydrochloride and ε-poly-L-lysine	Extends shelf-life of rice- or wheat-based confectioneries.	[34]
Rice starch	Films act as scavenging in food simulants tests.	[3]
Zein and zein wax	Effective in the prevention of oxidative changes.	[78]

10.4.1 ANTIOXIDANT PACKAGING

Antioxidant packaging is an important category of bioactive packaging to reduce, prevent, or retard the growth of microorganisms in the packaged food. Bioactive compounds can be added in or coated onto the active package. Various natural antioxidant compounds (such as α-tocopherol, polyphenols, aromatic plant extracts, green tea extract, and polyphenols) have been utilized to develop the bioactive food packaging. They are generally recognized as safe with "generally recognized as safe" status [68]. To prevent oxidation of packaged food, antioxidants and vitamins have also been incorporated into films. Prebiotics and probiotics have also been added in films in some dairy products and these have played an important role to reduce pathogenic bacteria, strengthen the immune system, increase the bioavailability of vitamins and nutrients, and regulate the intestinal microflora [60]. Therefore the addition of prebiotics and probiotics in the manufacturing of edible and biodegradable films provides nutrition in addition to the shelf-stability [78].

Plant-based antioxidant food packaging films have been developed with the utilization of ascorbic acid and butylated hydroxyl toluene (BHA) into a rice starch-glycerol matrix. Antioxidant active packaging, emits, or absorb component from or into food product or from the surrounding environment to enhance the shelf-life. Antioxidant capacity of the packaging material was evaluated by determining (2,2-diphenyl 1 picrylhydrazyl) (DPPH) radical into food simulants, such as water, 10% ethanol (as an alcoholic food simulant) and 95% ethanol (as fatty food simulant). The maximum amount of DPPH radical was observed with water and 10% ethanol. However, in case of 95% ethanol, no release of DPPH was observed. Hence, it was found that the antioxidant compound release was dependent on the type of food stimulant and antioxidant [4].

Butylated hydroxyl toluene (BHT) and BHA may be more suitable for dried products due to their high volatile nature than the natural antioxidant, such as α-tocopherol. However, α-tocopherol also provides the same antioxidant protection for the dried powder when exposed to oxygen and light. In another study, multilayer films (HDPE/ethylene-vinyl alcohol copolymer/LDPE) were joined with an inner layer of LDPE, which comprised of 4% of α-tocopherol to hinder the lipid oxidation of powder even at 30 and 40 °C [26]. Similarly, sealable LDPE films comprising of α-tocopherol (1.9% and 3%) retained the oxidation stability of corn oil for 16 weeks at 30 °C, compared to 12 weeks for the oil in a control package (without antioxidant) [28]. In a further study, effect of BHT on oat flakes packaged in HDPE packs were studied by Han et al. [32]. In the cereal, migration of BHT was 25 and 70%. Extra polymer layer with low permeability could be used to prevent loss. However, BHT has hazardous effect on human health. Therefore, utilization of natural antioxidants (such as vitamins C and E) in the packaging films can diminish the oxidative reactions includes rancidity, change in color, and flavor in high-fat products. Vitamin E is also considered safe and effective in cereal and snack food products, which have low to medium water activity. It has also stability during processing of film due to high solubility in packaging material (polyolefins) [18].

10.4.2 ANTIMICROBIAL PACKAGING

Antimicrobial packaging is designed to retard or suppress the growth of microorganism during the storage that may be present in the food or packaging material [2]. To control the growth of undesirable microorganisms, antimicrobial compounds can be added in or coated onto the food packaging materials. Several natural and artificial antimicrobials are used in packaging materials, which include natural, organic acids, chemical compounds, and nanoparticles. Recent years have focused on reduction on the use of synthetic petroleum-based additives compared to more focus on natural ingredients [1].

Essential oils with strong antimicrobial properties play vital role in plant defense. Among the essential oils, cinnamon oil is most effective in the use of active packaging. It was found that packaging material containing 18% and 10% of cinnamaldehyde is highly effective against *E. coli* and *Saccharomyces cerevisiae* [25]. Carvacrol is used in the development of bioactive packaging material. Carvacrol has synergistic antimicrobial effects against *Botryi scinera* when used in combination with thymol in the HDPE/modified montmorillonite nanocomposite films [14].

Chitosan (also known as chito oligosaccharides) has antifungal and antimicrobial properties against different kinds of microorganisms. It is used in combination with synthetic polymers, which include LDPE, HDPE, and natural polymers, such as carboxymethylcellulose [87]. It is also used on plastic films. It is effective against mold, yeast, and many kinds of bacteria; hence, it can increase shelf-life of the resultant product [40].

10.5 NANOCOMPOSITE-BASED FOOD PACKAGINGS

Nanocomposites comprise of single or mixture of polymers, which contain at least one organic or inorganic filler of dimensions <100 nm that further can be prepared by using nanoparticles (Figure 10.2). Integration of nanoparticles (such as SiO_2, clay, TiO_2, $KMnO_4$, nanocellulose, SiC, nanofibrillated cellulose, carbon nanotubes) into synthetic polymers and biopolymers can upsurge their mechanical and barrier properties [76]. Nanocomposite polymers are stronger, are of better thermal properties, and are more flame-resistant than the commercial polymers.

Traditional food packages have poor barrier performance, poor mechanical strength, low thermal stability and poor biodegradable. Nanocomposites packaging provides all these properties in addition to cost-effective [87]. The advancement of nanotechnology has potential to lessen the negative environmental impacts caused by synthetic polymer packaging by replacing with biodegradable or edible materials, such as plant-based nanocomposite materials [33]. Two types of nanocomposite particles are used: (1) inorganic and metal nanoparticles (nanoclay, halloysite nanotubes, and montmorillonite nanoparticles); (2) plant extracts (milk thistle extract, and green tea extract) mixtures of biopolymers (cellulose, chitosan, starch, and others).

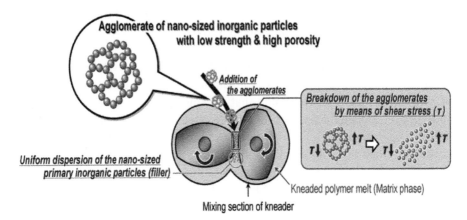

FIGURE 10.2   Development of polymer nanocomposites using nanoparticles.

Source: Reprinted from Ref. [76]. https://creativecommons.org/licenses/by/3.0/

Nanocomposites polymers are either single or mixture of compounds, which contain minimum one organic and inorganic polymer. The polymers having diameter <100 nm are being used for the development of packaging material. Various studies have reported that the addition of nanoparticles and biopolymers can enhance the barrier and mechanical properties [38].

10.6   SUMMARY

Adoption of suitable smart packaging material by the bakery industry can be useful to enhance the shelf-life, nutritional value, improve quality, safety, and provide information on the product. The innovation of bioactive packaging and nanopackaging has brought many changes in food preservation, storage, distribution, and consumption. These biopackaging materials can enhance the shelf-life of foods and are cost-effective, eco-friendly, degradable, and renewable packaging materials. However, there is a need to have fundamental studies on toxicity, migration assays, and risk assessment of nanocomposites materials.

KEYWORDS

- active packaging
- bioactive packaging
- intelligent packaging
- nanocomposite packaging

REFERENCES

1. Alves-Silva, J. M.; dos Santos, S. M. D.; Pintado, M. E. Chemical Composition and *In Vitro* Antimicrobial, Antifungal and Antioxidant Properties of Essential Oils Obtained from Some Herbs Widely Used in Portugal. *Food Control*, **2013**, *32*, 371–378.
2. Appendini, P.; Hotchkiss, J. H. Review of Antimicrobial Food Packaging. *Innovative Food Science & Emerging Technologies*, **2002**, *3*, 113–126.
3. Ashwar, B. A.; Shah, A.; Gani, A.; Shah, U. Rice Starch Active Packaging Films Loaded with Antioxidants: Development and Characterization. *Starch-Stärke*, **2015**, *67*, 294–302.
4. Axel, C.; Brosnan, B.; Zannini, E.; Furey, A.; Coffey, A.; Arendt, E.K. Antifungal Sourdough Lactic Acid Bacteria (LAB) as Biopreservation Tool in Quinoa and Rice Bread. *International Journal of Food Microbiology*, **2016**, *239*, 86–94.
5. Bai, Y. P.; Guo, X. N.; Zhu, K. X.; Zhou, H. M. Shelf-life Extension of Semi-Dried Buckwheat Noodles by the Combination of Aqueous Ozone Treatment and Modified Atmosphere Packaging. *Food Chemistry*, **2017**, *237*, 553–560.
6. Balaguer, M. P.; Lopez-Carballo, G.; Catala, R.; Gavara, R. Antifungal Properties of Gliadin Films Incorporating Cinnamaldehyde and Application in Active Food Packaging of Bread and Cheese Spread Foodstuffs. *International Journal of Food Microbiology*, **2013**, *166*, 369–377.
7. Baldwin, E. A.; Hagenmaier, R.; Bai, J. (Eds.). *Edible Coatings and Films to Improve Food Quality*. Boca Raton, FL: CRC Press; **2011**; p. 460.
8. Baleizao, C.; Nagl, S.; Schäferling, M.; Berberan-Santos, M. N.; Wolfbeis, O. S. Dual Fluorescence Sensor for Trace Oxygen and Temperature with Unmatched Range and Sensitivity. *Analytical Chemistry*, **2008**, *80*, 6449–6457.
9. Berenzon, S.; Saguy, I. S. Oxygen Absorbers for Extension of Crackers Shelf-Life. *LWT-Food Science and Technology*, **1998**, *31*, 1–5.
10. Biji, K. B.; Ravishankar, C. N.; Mohan, C. O.; Gopal, T. S. Smart Packaging Systems for Food Applications: A Review. *Journal of Food Science and Technology*, **2015**, *52*, 6125–6135.
11. Borchert, N. B.; Kerry, J. P.; Papkovsky, D. B. The CO_2 Sensor Based on Pt-Porphyrin Dye and FRET Scheme for Food Packaging Applications. *Sensors and Actuators B: Chemical*, **2013**, *176*, 157–165.

12. Brody, A.L; Strupinsky, E.R; Kline, L.R. Odor removers. In: *Active Packaging for Food Applications*; Brody, A.L.; Strupinsky, E.R.; Kline, L.R. (Eds.); Lancaster, PA: Technomic Publishing Company Inc.; **2001**; pp. 107–117.

13. Campos-Requena, V. H.; Rivas, B. L.; Pérez, M. A.; Figueroa, C. R. The Synergistic Antimicrobial Effect of Carvacrol and Thymol in Clay/Polymer Nanocomposite Films Over Strawberry Gray mold. *LWT—Food Science and Technology*, **2015**, *64,* 390–396.

14. Chen, C. W.; Xie, J.; Yang, F. X.; Zhang, H. L. Development of Moisture Absorbing and Antioxidant Active Packaging Film Based on Poly(Vinyl Alcohol) Incorporated With Green Tea Extract and Its Effect on the Quality of Dried Eel. *Journal of Food Processing and Preservation*, **2018**, *42,* e13374; https://doi.org/10.1111/jfpp.13374;

15. Cruz, R. S.; Soares, N. D. F; Andrade, N. J. D. Evaluation of Oxygen Absorber on Antimicrobial Preservation of Lasagna-Type Fresh Pasta under Vacuum Packed. *Ciência e Agrotecnologia*, **2006**, *30,* 1135–1138.

16. Cruz, R. S; Camilloto, G. P. Oxygen Scavengers: An Approach on Food Preservation. *Structure and Function of Food Engineering.* **2012**; *2012*, p. 21; https://www.intechopen. com/books/structure-and-function-of-food-engineering/oxygen-scavengers-an-approach-on-food-preservation; Accessed on December 31, 2019.

17. Day, B.P.F. Active Packaging of Food. In: *Smart Packaging Technologies for Fast Moving Consumer Goods*; Kerry, J., Butler, P.J. (Eds.); West Sussex, England: Wiley & Sons Ltd.; **2008**; pp. 1–18.

18. Degirmencioglu, N.; Göcmen, D.; Inkaya, A. N. Influence of Modified Atmosphere Packaging and Potassium Sorbate on Microbiological Characteristics of Sliced Bread. *Journal of Food Science and Technology*, **2011**, *48,* 236–241.

19. Floros, J.D.; Dock, L.L.; Han, J.H. Active Packaging Technologies and Applications. *Food Chemistry Drug Packaging*, **1997**, *20,* 10–17.

20. Franke, I.; Wijma, E.; Bouma, K. Shelf-Life Extension of Pre-Baked Buns by an Active Packaging Ethanol Emitter. *Food Additive and Contamination*, **2002**, *19,* 314–322.

21. Fuentes, E.; Palomo, I. Antiplatelet Effects of Natural Bioactive Compounds by Multiple Targets: Food and Drug Interactions. *Journal of Functional Foods*, **2014**, *6,* 73–81.

22. Gaikwad, K.K; Singh, S; Lee, Y.S Oxygen Scavenging Films in Food Packaging. *Environmental Chemistry Letters*, **2018**, *16,* 523–538.

23. Gaikwad, K. K.; Singh, S.; Ajji, A. Moisture Absorbers for Food Packaging Applications. *Environmental Chemistry Letters*, **2019**, *17,* 609–628.

24. Gherardi, R.; Becerril, R.; Nerin, C.; Bosetti, O. Development of a Multilayer Antimicrobial Packaging Material for Tomato Puree Using an Innovative Technology. *LWT—Food Science Technology*, **2016**, *72,* 361–366.

25. Giannakourou, M. C.; Koutsoumanis, K.; Nychas, G. J. E.; Taoukis, P. S. Field Evaluation of the Application of Time Temperature Integrators for Monitoring Fish Quality in the Chill Chain. *International Journal of Food Microbiology*, **2005**, *102,* 323–336.

26. Graciano-Verdugo, A. Z.; Soto-Valdez, H. Migration of α-Tocopherol from LDPE Films to Corn Oil and Its Effect on the Oxidative Stability. *Food Research International*, **2010**, *43,* 1073–1078.

27. Granda-Restrepo, D. M.; Soto-Valdez, H.; Peralta, E. Migration of α-Tocopherol from an Active Multilayer Film into Whole Milk Powder. *Food Research International*, **2009**, *42,* 1396–1402.

28. Guo, X.; Liu, L.; Hu, Y.; Wu, Y. Water Vapor Sorption Properties of TEMPO Oxidized and Sulfuric Acid Treated Cellulose Nanocrystal Films. *Carbohydrate Polymers*, **2018**, *197*, 524–530.

29. Gutierrez, L.; Sanchez, C. New Antimicrobial Active Package for Bakery Products. *Trends Food Science and Technology*, **2009**, *20*, 92–99.

30. Guynot, M. E.; Sanchis, V.; Ramos, A. J.; Marin, S. Mold free Shelf Life Extension of Bakery Products by Active Packaging. *Journal of Food Science*, **2003**, *68*, 2547–2552.

31. Han, J.K.; Miltz J.; Harte, B.R. Loss of 2-Tertiarybuty-l-4-Methoxy Phenol (BHA) from HDPE Film. *Polymer Engineering & Science*, **1987**, *27*, 934–938.

32. Han, J.W.; Ruiz-Garcia, L.; Qian, J.P.; Yang, X.T. Food Packaging: Comprehensive Review and Future Trends. *Comprehensive Review of Food Science and Food Safety*, **2018**, *17*, 860–877.

33. Hata T.; Sato T.; Ichikawa T.; Morimitsu Y. Antifungal Activity of Protamine Salmine Hydrochloride and Ɛ-Poly-L-Lysine in Actual Food Systems, Rice- or Wheat-Based Confectioneries. *Journal of Food Processing and Preservation*, **2016**, *40*, 1180–1187.

34. Heising, J. K. Simulations on the Prediction of Cod (*Gadusmorhua*) Freshness from an Intelligent Packaging Sensor Concept. *Food Packaging and Shelf-life*, **2015**, *3*, 47–55.

35. Hempel, A.W.; O'Sullivan, M.G.; Papkovsky, D.B.; Kerry, J.P. Use of Smart Packaging Technologies for Monitoring and Extending the Shelf-Life Quality of Modified Atmosphere Packaged (MAP) Bread: Application of Intelligent Oxygen Sensors and Active Ethanol Emitters. *European Food Research Technology*, **2013**, *237*, 117–124.

36. Hogan, S. A.; Kerry, J. Smart Packaging of Meat and Poultry Products. Chapter 3; In: *Smart Packaging Technologies for Fast Moving Consumer Goods*; J. Kerry and P. Butler (Eds.); West Sussex, UK: John Wiley & Sons Ltd.; **2008**; pp. 33–60.

37. Huang, Y.; Mei, L.; Chen, X.; Wang, Q. Recent Developments in Food Packaging Based on Nanomaterials. *Nanomaterials*, **2018**, *8*, 830–835.

38. Hutton, T. Food Packaging: An introduction. In: *Key Topics in Food Science and Technology*; Gloucestershire, UK: Campden and Chorleywood Food Research Association Group; **2003**; volume *7*; pp. 108–113.

39. Joerger, R.D.; Sabesan, S. Antimicrobial Activity of Chitosan Attached to Ethylene Copolymer Films. *Packaging Technology*, **2009**, *22*, 125–138.

40. Juneja, V. K.; Dwivedi, H. P. Novel Natural Food Antimicrobials. *Annual Review of Food Science and Technology*, **2012**, *3*, 381–403.

41. Kerry, J.P.; O'Grady, M.N. Past, Current and Potential Utilization of Active and Intelligent Packaging Systems for Meat and Muscle-Based Products: A Review. *Meat Science*, **2006**, *74*, 113–130

42. Kotsianis, I. S.; Giannou, V. Production and Packaging of Bakery Products using MAP Technology. *Trends in Food Science & Technology*, **2002**, *13*, 319–324.

43. Lang, C.; Hubert. Color Ripeness Indicator for Apples. *Food and Bioprocess Technology*, **2012**, *5*, 3244–3249.

44. Latou, E.; Mexis, S.F. Shelf-life Extension of Sliced Wheat Bread using Either an Ethanol Emitter or an Ethanol Emitter Combined with an Oxygen Absorber as Alternatives to Chemical Preservatives. *Journal Cereal Science*, **2010**, *2010*, 52–57.

45. Lee, D. S. Carbon Dioxide Absorbers for Food Packaging Applications. *Trends in Food Science & Technology*, **2016**, *57*, 146–155.

46. Leites, C., Fernando, L. Bioactive Compounds Incorporation into the Production of Functional Biodegradable Films—A Review. *Polymers from Renewable Resources*, **2017**, *8*, 151–176.

47. Llobet, E. Gas Sensors Using Carbon Nanomaterials: A Review. *Sensors and Actuators B: Chemical*, **2013**, *179*, 32–45.

48. Lopes, F. A.; Ferreira-Soares, N.F. Conservation of Bakery Products Through Cinnamaldehyde Antimicrobial Films. *Packaging Technology and Science*, **2014**, *27*, 293–302.

49. Lopez-Rubio A.; Almenar, E.; Hernandez-Munoz, P. Overview of Active Polymer-Based Packaging Technologies for Food Applications. *Food Review International*, **2004**, *20*, 357–387.

50. Luchese, C. L.; Brum, L. F. W.; Piovesana, A. Bioactive Compounds Incorporation into the Production of Functional Biodegradable Films: A Review. *Polymers from Renewable Resources*, **2017**, *8*, 151–176.

51. Luz, C.; Calpe, J. Antimicrobial Packaging Based on Ɛ Polylysine Bioactive Film for the Control of Mycotoxigenic Fungi *In Vitro* and in Bread. *Journal of Food Processing and Preservation*, **2018**, *42*, e13370; online; doi: 10.1111/jfpp.13370.

52. Mateo, E. M.; Gómez, J. V.; Domínguez, I. Impact of Bioactive Packaging Systems Based on EVOH Films and Essential Oils in the Control of Afla Toxigenic Fungi and Aflatoxin Production in Maize. *International Journal of Food Microbiology*, **2017**, *254*, 36–46.

53. Mbuge, D.O.; Negrini, R.; Nyakundi, L.O. Application of Superabsorbent Polymers (SAP) as Desiccants to Dry Maize and Reduce Aflatoxin Contamination. *Journal of Food Science and Technology*, **2016**, *53*, 3157–3165.

54. Mehauden, K.; S. Bakalis, P.; Cox, P. Use of Time Temperature Integrators for Determining Process Uniformity in Agitated Vessels. *Innovative Food Science and Emerging Technologies*, **2008**, *9*, 385–395.

55. Melikoglu, M.; Webb, C. Food Industry Wastes. Chapter 4; In: *Use of Waste Bread to Produce Fermentation Products*; San Diego, CA: Academic Press; 2013; pp. 63–76.

56. Melini, V.; Melini, F. Strategies to Extend Bread and GF Bread Shelf-Life: From Sourdough to Antimicrobial Active Packaging and Nanotechnology. *Fermentation*, **2018**, *4*, 9–12.

57. Mihaly-Cozmuta, A.; Peter, A. Active Packaging System Based on Ag/TiO$_2$ Nanocomposite Used for Extending the Shelf-Life of Bread: Chemical and Microbiological Investigations. *Packaging Technology and Science*, **2015**, *28*, 271–284.

58. Mitsubishi Gas Chemical. *Ageless Eye Oxygen Indicator*; **2014**; online; http://www.mgccojp/eng/products/abc/ageless/eye.html; Accessed September 24, 2019.

59. Mohammadmoradi, S.; Javidan, A.; Kordi, J. Boom of Probiotics: This Time Non-Alcoholic Fatty Liver Disease: A Mini Review. *Journal of Functional Foods*, **2014**, *11*, 30–35.

60. Muratore, F.; Barbosa, S. E. Development of Bioactive Paper Packaging for Grain-Based Food Products. In: *Food Packaging*; **2019**; volume 20; online; https://doi.org/10.1016/j.fpsl.2019.100317; Accessed December 31, 2019.

61. Neethirajan, S.; Jayas, D. S. Carbon Dioxide (CO$_2$) Sensors for the Agri-Food Industry: A Review. *Food and Bioprocess Technology*, **2009**, *2*, 115–121.

62. Oliveira, P.M.; Zannini, E.; Arendt, E.K. Cereal Fungal Infection, Mycotoxins and Lactic Acid Bacteria Mediated Bioprotection: From Crop Farming to Cereal Products. *Food Microbiology*, **2014**, *37*, 78–95.

63. Otoni, C. G.; Pontes, S. F. O.; Medeiros, E. A. A. Edible Films from Methylcellulose and Nano Emulsions of Clove Bud (*Syzygium aromaticum*) and Oregano (*Origanum vulgare*) Essential Oils as Shelf-Life Extenders for Slice Bread. *Journal of Agricultural and Food Chemistry*, **2014**, *62*, 5214–5219.

64. Pant, A.; Sängerlaub, S.; Müller, K. Gallic Acid as an Oxygen Scavenger in Bio-Based Multilayer Packaging Films. *Materials,* **2017** *10,* 489–493.
65. Pereira de Abreu, D. A.; Cruz, J. M. Active and Intelligent Packaging for the Food Industry. Food Reviews International, **2012**, *28,* 146–187; online; https://doi.org/10.10 80/87559129.2011.595022;
66. Pereira, V. A.; de Arruda, I. N. Q. Active chitosan/ PVA Films with Anthocyanins from *Brassica oleraceae* (Red cabbage) as Time-Temperature Indicators for Application in Intelligent Food Packaging. *Food Hydrocolloids,* **2015**, *43,* 180–188.
67. Ramos, M.; Beltrán, A.; Peltzer, M.; Valente, A. J. Release and Antioxidant Activity of Carvacrol and Thymol from Polypropylene Active Packaging Films. *LWT-Food Science and Technology,* **2014**, *58,* 470–477.
68. Realini, C. E.; Marcos, B. Active and Intelligent Packaging Systems for a Modern Society. *Meat Science,* **2014**, *98,* 404–419.
69. Rodriguez-Aguilera, R.; Oliveira, J. C. Review of Design Engineering Methods and Applications of Active and Modified Atmosphere Packaging Systems. *Food Engineering Reviews,* **2009**, *1,* 66–83.
70. Rooney, M. L. Introduction to Active Food Packaging Technologies. In: *Innovations in Food Packaging*; Han, J. H. (Ed.); London, UK: Elsevier Ltd.; **2005**; pp. 63–69.
71. Shankar, S.; Rhim, J. W. Polymer Nanocomposites for Food Packaging Applications. *Functional and Physical Properties of Polymer Nanocomposites,* **2016**, *2016,* 29–31.
72. Sivertsvik, M. Lessons from Other Commodities: Fish and Meat. In: *Intelligent and Active Packaging for Fruits and Vegetables*; Wilson, C.L. (Ed.); Boca Raton, FL: CRC Press; **2007**; pp. 151–161.
73. Smiddy, M.; Fitzgerald, M.; Kerry, J.P. Use of Oxygen Sensors to Non-Destructively Measure the Oxygen Content in Modified Atmosphere and Vacuum-Packed Beef: Impact of Oxygen Content on Lipid Oxidation. *Meat Science,* **2002**, *61,* 285–290.
74. Soares, N. F. F.; Rutishauser, D. M. Inhibition of Microbial Growth in Bread Through Active Packaging. *Packaging Technology and Science: An International Journal,* **2002**, *15,* 129–132.
75. Sohail, M.; Sun, D. W. Recent Developments in Intelligent Packaging for Enhancing Food Quality and Safety. *Critical Reviews in Food Science and Nutrition,* **2018**, *58,* 2650–2662.
76. Tanahashi, M. Development of Fabrication Methods of Filler/Polymer Nanocomposites: With Focus on Simple Melt-Compounding-Based Approach Without Surface Modification of Nano-Fillers. *Materials,* **2010**, *3,* 1593–1619.
77. Tripathi, M. K.; Giri, S. K. Probiotic Functional Foods: Survival of Probiotics During Processing and Storage. *Journal of Functional Foods,* **2014**, *9,* 225–241.
78. Ünalan, İ. U.; Arcan, I.; Korel, F. Application of Active Zein-Based Films with Controlled Release Properties to Control *Listeria Monocytogenes* Growth and Lipid Oxidation in Fresh Kashar Cheese. *Innovative Food Science & Emerging Technologies,* **2013**, *20,* 208–214.
79. Upasen, S.; Wattanachai, P. Packaging to Prolong Shelf-life of Preservative-Free White Bread. *Heliyon,* **2018**, *4,* e00802; online; doi: 10.1016/j.heliyon.2018.e00802;
80. Vermeiren, L.; Devlieghere, F.; Van Beest, M. Developments in the Active Packaging of Foods. *Trends Food Science and Technology,* **1999**, *10,* 77–86.
81. Vu, C. H. T.; Won, K. Novel Water-Resistant UV-Activated Oxygen Indicator for Intelligent Food Packaging. *Food Chemistry,* **2013**, *140,* 52–56.

82. Vulpoi, A.; Baia, L.; Falup, A. Changes in the Microbiological and Chemical Characteristics of White Bread During Storage in Paper Packages Modified with Ag/ TiO2 -SiO$_2$, Ag/N-TiO$_2$ or Au/TiO$_2$. *Food Chemistry*, **2016**, *197*, 790–798.

83. Wang, S.; Marcone, M. F.; Barbut, S. Fortification of Dietary Biopolymers-Based Packaging Material with Bioactive Plant Extracts. *Food Research International,* **2012**, *49,* 80–91.

84. Yildirim, S.; Röcker, B. Active Packaging Applications for Food. *Comprehensive Review of Food Science and Food Safety,* **2018**, *17*, 165–199.

85. Youssef, A.M.; El-Sayed, S.M. Bionanocomposite Materials for Food Package Applications: Concepts and Future Outlook. *Carbohydrate Polymerization,* **2018**, *193*, 19–27.

86. Youssef, A.M.; El-Sayed, S.M.; El-Sayed, H.S. Enhancement of Egyptian Soft White Cheese Shelf-Life Using a Novel Chitosan/Carboxymethyl Cellulose/Zinc Oxide Bionanocomposite Film. *Carbohydrate Polymerization,* **2016**, *151,* 9–19.

CHAPTER 11

CEREAL-BASED NUTRIENTS AND PHYTOCHEMICALS: TRADITIONAL AND NOVEL PROCESSING TECHNIQUES FOR INCREASING BIOAVAILABILITY

HARINDERJEET KAUR BHULLAR, RESHU RAJPUT, AMARJEET KAUR, and KAMALJIT KAUR

ABSTRACT

Epidemiological studies have suggested that whole grain possesses several health benefits that are linked to the combined impacts of micronutrients, phytochemicals, and dietary fiber. Bioactive compounds in cereal grains are covalently linked to cellular matrix that hinders the bioavailability. Several strategies and processing techniques (such as germination, fermentation, and extrusion) have been implemented to develop functional cereal-based foods with increased content of bioactive compounds and dietary fiber. Also, baking, roasting, and microwave cooking can influence the amounts of phytochemicals in cereals. This chapter highlights the techniques to improve the bioavailability of nutrients and phytocompounds in cereal and cereal-based foods.

11.1 INTRODUCTION

Cereals (such as wheat, rice, maize, oats, millets, rye, sorghum, and barley) have been an integral part of human diet because of rich source of carbohydrates, proteins, vitamins, and minerals. Cereal grains constitute three edible parts: bran, germ and endosperm. During the refining process, bran and germ portions are removed and remaining endosperm fractions are used in the preparation of various products because of exclusive functionality of wheat protein. However, regular consumption of such products has increased the risk of pharynx, larynx, thyroid, and digestive tract disorders. Thus nowadays, consumption of whole cereal grain products has gained preference owing to their richness in functional components, such as dietary fiber and phytochemicals. These functional components are referred to as bioactive compounds. The bran layer of grains contains higher levels of bioactive components, such as phenolic compounds, phytosterols, carotenoiods, γ-aminobutyric acid, β-glucans, and arabinoxylans than those in the endosperm parts. The phenolic compounds (such as lignans, alkyl-resorcinols, and phenolic acids) are major compounds responsible for protective effects in cereals [32].

Although these compounds are present in small amounts, yet their regular consumption may impart several health benefits that are attained through complex physiological mechanisms including enhancement of immune system and antioxidant activity, production of short-chain fatty acids during fermentation in colon, increased rate of bile excretion, mediation of hormones, alleviated transition of substance through the digestive tract, and absorption of substances in the gut. Epidemiological studies have proposed that the complex mixture of bioactive components in whole-grain foods may have more beneficial impacts because of possible additive and synergistic effects of the bioactive compounds than the individual isolated components [98]. In last decade, the whole-grain cereals have gained scientific, governmental, and commercial interest as studies have increasingly shown their protective role against many diseases, such as reduction in the risk of type-2 diabetes, cardiovascular diseases, various cancers, hypertension, gastrointestinal disorders; helps in weight management; influences satiety; and normalization of bowel movement [59].

Several factors as a source of food, interactions among the phytochemicals and with food and gastrointestinal environment interfere the bioavailability of nutrients and bioactive components [15], and not necessarily relies on the number of active metabolites in target tissues and concentration of bioactive compounds in food [4]. Bioavailability of bioactive components is determined by their bioaccessibility that liberates nutrients from complex food matrix, and transport across the gut wall, thus entering the bloostream and making nutrients bioavailable.

Various processing techniques (such as germination, fermentation, extrusion) result in an increase in the biological activity of grains thus increasing the bioavailability of nutritional health-promoting compounds. Processing possesses both positive and negative impacts on bioactive compounds, for example, during thermal treatments due to high-temperature thermolabile compounds get decomposed or polymerization of phenolic compounds occurs due to the interaction of phenols with Millard reaction by-products resulting in loss in antioxidant activity of phenolic compounds [5]. In contrast, milling, extrusion, fermentation, and the action of hydrolytic enzymes during germination facilitate the release of bound phenolic compounds by breaking the cell wall and releasing cellular constituents [8].

This chapter reviewed bioactive compounds in cereals, processing techniques of cereal grains, and the impact of processing on the bioavailability of nutrients and bioactive components in cereals.

11.2 BIOACTIVE COMPOUNDS IN CEREALS

Bioactive compounds are classified into polyphenolic compounds, organo-sulfur compounds, terpenoids, and phytosterols, which vary in chemical structure, concentration, and nature of distribution in foods (Figure 11.1). These phytochemicals possess antimutagenic, antioxidant, and various other biological activities [98]. Concentration of these compounds in cereals depends on the genus, species, and the site of localization, as majority of bioactive components are present in the bran layer of cereals. Bioactive compounds present in whole-grain cereals are listed in Table 11.1.

Polyphenolic compounds are characterized as flavonoids and nonflavonoids. Major polyphenols present in cereals are phenolic acids, alkylresorcinols, and avenanthramides. Phenolic acids are aromatic compounds widely present in cereals, oil seeds, legumes, fruits, and vegetables; and these occur in free, conjugated, and ester-bound forms. Bound phenolic acids are predominant than the free phenolic acids and they form bridges between the polymer chains in the kernel. Two groups of phenolic acids in the bound form are:

- Derivatives from dihydroxybenzoic acids comprise of gallic, proto-catechuic, vanillic, and syringic acids;
- Derivatives from hydroxycinnamic acids comprise of ferulic, caffeic, p-coumaric, and sinapic acids.

Hydroxybenzoic acid derivatives are chief constituents of tannins and lignins, whereas hydroxycinnamic acid derivatives form ester-linkages with cell-wall components, for example, cellulose, arabinoxylans, and proteins [37]. Free phenolic acids are concentrated in the bran layer of grains, whereas bound phenolics are present beyond the cell wall [37], hence whole-grain cereals are richer in phenolic compounds. Bound phenolics are not digested

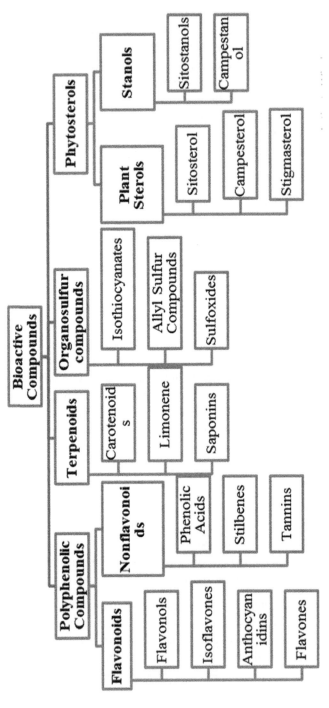

FIGURE 11.1   Classification of bioactive compounds [42].

in the small intestine but can enter portal circulation [89] being absorbed slowly by the enzymes (β-glucosidases, esterases) present in the colon and show preventive action against colon cancer. Consequently, the enhancement of bound phenolics during processing exerts profound beneficial health effects throughout the digestive tract after absorption [92].

TABLE 11.1 Bioactive Components in Whole Cereal Grains.

Cereal Grain	Bioactive Components	Ref.
Barley	Ferulic acid, *p*-hydroxybenzoic acid, vanillic acid, caffeic acid, salicylic acid, α-tocotrienols, β-glucans, arabinoxylans, 3-deoxyanthocyanidins, γ-aminobutyric acid.	[10, 59, 98]
Maize	*p*-Coumaric acid, syringic acid, lignan, carotenoid, γ-aminobutyric acid.	[32, 59]
Millet	Cinnamic acid, *p*-coumaric acid, *p*-hydroxybenzoic acid, vanillic acid.	[32, 59, 98, 105]
Oat	Ferulic acid, *p*-coumaric acid, caffeic acid, vanillic acid, *p*-hydroxybenzoic acid, avenanthramides, β-glucans, lignans, phytosterols.	[19, 59, 98]
Rice	*p*-Coumaric acid, sinapic acid, caffeic acid, *p*-hydroxybenzoic acid, gallic acid, protocatechuic acid, γ-oryzanol, γ-aminobutyric acid, phytosterols.	[32, 59, 98]
Rye	Ferulic acid, sinapic acid, *p*-coumaric acid, caffeic acid, alkylresoricinols, α-tocopherols, phytosterols, arabinoxylans.	[6, 32, 59, 98]
Sorghum	*p*-Coumaric acid, caffeic acid, salicylic acid, syringic acid, protocatechuic acid.	[32, 59, 98, 105]
Wheat	Ferulic acid, caffeic acid, salicylic acid, syringic acid, vanillic acid, alkylresorcinols, β-tocotrienol, arabinoxylans, lutein, zeaxanthin, phytosterols.	[32, 54, 59]

However, the bioavailability and the impact of processing on phenolic acid content depends on the type of phenolic acids that have ability to chelate ions of transition metals thus reduce the generation of free radicals [1], prevent the oxidation of methyl linoleate [48], decrease the low-density lipoproteins (LDL) in human blood [2]. However, stability and bioavailability of individual phenolic compounds are greatly influenced by the processing methods and the chemical nature of phenolic acids. The predominant phenolic acid in the whole grain is ferulic acid that is present at all stages of grain development [62]. It acts as a potent antioxidant, anti-inflammatory, plays a role as chemo-protectant by inhibiting the formation of N-nitroso compounds and by preventing peroxynitrite-mediated nitration of tyrosine residues in collagen and suppresses lipid peroxidation in the microsomal membrane [90, 97]. Other phenolic acids found in grains are vanillic acid, caffeic acid, p-coumaric, sinapic, and syringic acid (Table 11.1).

Avenanthramides are the nitrogen-containing polyphenols in oats (Table 11.1) and have anti-inflammatory, antipruritic, antioxidation, anti-proliferation, and antiatherogenic activities [60]. Avenanthramides are classified into two groups: amides of different cinnamic acids and cinnamic acids substituted with anthranilic acid. Out of 40 known structures of avenanthramides, most common forms are avenanthramide A (2f), B (2p) and C (2c) of group 2. Avenanthramide C is thermostable, shows higher antioxidant activities, and is abundant in oats than A and B forms, whereas avenanthramide A possesses 4-folds higher bioavailability than the others. However, their levels in oats depend on the variety, environmental stress, and geographical parameters. Research works have shown that avenanthramides are present mainly in outer layer than the whole oats, whereas few researchers found higher concentration in oat flakes of whole-grain oats (27 mg/kg) than the oat bran (13 mg/kg) [60]. The concentration of A-, B- and C-avenanthramides reported in oat flakes of whole oats was 9.0, 8.6, and 9.0 mg/kg, respectively; of which, oat bran contains 4.4, 4.1, 4.3 mg/kg of fresh weight, respectively [60]. Several health benefits of avenanthramides are enhancing cognitive abilities, reduce inflammation, relax blood vessels, interferes cancer proliferation, and regulates blood sugar [95].

Alkylresorcinols are the phenolic lipids that are composed of resorcinol type phenolic ring to which long odd-numbered aliphatic chain is attached at the 5th position. These phenolic compounds are abundant in whole grains (such as: rye (0.36–3.2 mg/g), triticale (0.58–1.63 mg/g), wheat (0.32–1.01 mg/g) and barley (0.04–0.5 mg/g) (see Table 11.1). According to Ross et al. [81], alkylresorcinols metabolites (3,5-dihydroxyphenylpropionic acid and 3,5-dihydroxybenzoic acid (DHBA)) act as nutritional biomarkers to identify the intake of whole wheat and rye. Ross et al. [82] also reported higher concentrations of alkylresorcinols in plasma on ingestion of whole grains than the refined grains; therefore, it could be used as a marker to detect the bran in flour and other cereal products. These compounds interfere cancer progression, reduce mutagenic activity, inhibit metabolic enzyme in-vitro, have antimicrobial and antioxidant activities. Some researchers reported loss of alkylresorcinols on processing (such as extrusion, fermentation and baking) [99]. However, Ross et al. [82] found no impact of baking on the concentration of alkylresorcinols. Besides polyphenolic compounds, other bioactive components in cereals are γ-aminobutyric acid, phytosterols, carotenoids, arbinoxylans, and β-glucans.

The γ-aminobutyric acid (GABA) is a naturally occurring nonprotein amino acid in whole grains (brown rice, maize, barley), lentils, soy, nuts, beans, fish, fruits, and vegetables (Table 11.1). GABA is synthesized by α-decarboxylation of glutamic acid catalyzed by GAD (glutamate decarboxylase) [61]. It acts as inhibitory neurotransmitter, induces diuretic, and hypotensive effects [73] and inhibits the proliferation of cancer cells [69]. Concentration of GABA in grains and legumes could be increased by the biological activation of grains and legumes. Several health benefits exhibited by GABA are prevention of headache, regulation of blood sugar level, relief of constipation, reduction in anxiety, and risk of developing colon cancer and Alzheimer disease [39].

Phytosterols can lower LDL and the risk of coronary heart diseases and also exhibit antioxidant activity [47]. Phytosterols are mainly present in whole cereal grains (rice, wheat, oat), vegetable oil, legumes, nuts, and vegetables (Table 11.1). In general, out of all plant sterols, β-sitosterol is higher in concentration (62%), followed by campesterol (21%) and stigmasterol (4%), whereas stanols as β-sitostanol (4%) and campestanol (2%) are present in very small amounts [47]. β-sitosterol shows antipyretic, anti-inflammatory properties and also control blood sugar and exhibits immune-modulating effects.

Arabinoxylans are found in all cereal grains, but are abundant in rye followed by wheat, barley, oat, and rice. Arabinoxylans are localized in the husk, bran layer, and cell wall of endosperm in different concentrations. However, the concentration and structure of these compounds depend on the cereal variety and type, and geographical variation and location. These polysaccharides are diverse in nature and arbinoxylans of endosperm are soluble in water, whereas those present in bran layer form covalent linkages with ferulic acid, resulting in water-insoluble arabinoxylans [29], which exhibit high antioxidant and antimicrobial activities. However, consumption of cereal arabinoxylans showed significant impact on health by lowering serum cholesterol, serum insulin, by production of short-chain fatty acids, reducing postprandial and fasting glucose, increasing absorption of minerals (magnesium and calcium), enhancing immunity, thus preventing from chronic diseases [87].

The β-glucan is naturally present polysaccharide comprising of β-D-glucose units. It is found in cereals, cell wall of bacteria, yeast, and fungi. Among cereals, it is predominantly present in the bran layer of oats and barley. β-Glucans present in cereals comprises of β-1,3 and β-1,4 linkages of β-D-glucose units and are water soluble, whereas those existing in cell wall of bacteria, yeast and fungi constitute β,1–3 linkages of β-D-glucose units and are water insoluble. These polysaccharides are not digested by the enzymes of intestine, but get fermented in the colon by the gut bacteria thus acting as substrate for the gut microflora. On anaerobic fermentation, short-chain fatty acids are produced that exhibit several health benefits, such as reduce risk of heart diseases, type-2 diabetes, and helps in weight management [35].

All bioactive compounds undergo various processes in the body after consumption, for their effective absorption and metabolism (Figure 11.2). However, the oral bioavailability of these compounds is considerably affected due to their poor solubility and less stability in gastrointestinal environment, inadequate gastric residence time, and limited diffusion across lipid-bilayer cell membrane of the intestine [42].

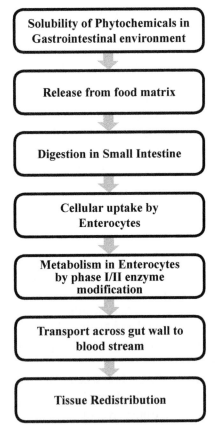

FIGURE 11.2   Summary of the process of absorption of bioactive compounds in our body [42].

Novel technologies, such as nanotechnology can be used to enhance the bioaccessibility and protect these compounds from degradation after their isolation from the cereals, but this may add up to the cost of the products prepared from nanoparticles of bioactive compounds [42]. Therefore, it is recommended to subject the whole cereal grains to various processing techniques, such as germination, malting, thermal treatments, fermentation, high hydrostatic pressure processing, ultrasonication, or the combination of these techniques to improve the bioaccessibility and bioavailability of these compounds.

11.3 PROCESSING OF CEREALS

11.3.1 GERMINATION

Germination is a traditional natural biological process, in which the cereals or legumes are harvested at the initial growth stages of the plant. During germination, complex biochemical changes take place, such as transformation of nutritionally undesirable compounds into essential constituents, synthesis of hydrolytic enzymes, and generation of health-promoting compounds (i.e., phenolic acids, γ-oryzanol, γ-aminobutyric acid, arbinoxylans with antioxidant activities [87]. Germination improves the nutritional and sensorial quality and reduces the antinutritional compounds thus maximizing the bioavailability of nutrients. Germination at room temperature for 48 h showed significant increase in protein content, calcium, iron, and vitamin C content in finger millet [23]. During germination due to the activation of hydrolytic and amylolytic enzymes, saccharification takes place, resulting in increase in simple sugars to improve carbohydrate digestibility [66]. Zhang et al. [106] observed a significant increase in total protein and its bioavailability during germination of buckwheat flour, whereas Bhathal and Kaur [14] found decrease in protein content of germinated quinoa flour, which may be due to the degradation of proteins by proteases. Due to increase in digestibility and bioavailability of proteins and carbohydrates, germinated cereals can be used for the preparation of infant foods. Reduction of antinutritional factors after processing led to the enhancement of bioavailability of minerals.

Ogbonna et al. [68] observed notable drift in the iron, calcium, magnesium, phosphorus, sodium, and potassium contents after germination of sorghum at 50 °C for 96 h. However, Laxmi et al. [58] reported 8% increase in iron content but substantial decrease in calcium and phosphorus on germination of foxtail millet for 48 h. The shift in the level of nutrition and bioactive components depends on the time and temperature of steeping and germination, and on the species of grains. At several instances, higher temperature and longer germination time exhibited higher antioxidant activity of phenolic compounds, γ-aminobutyric acid and flavonoids, which might be due to the degradation of cell wall polysaccharides and also enzymes convert the storage proteins into transportable amides [30] or due to the activation of phenylpropanoid metabolic pathway during germination that synthesis flavonoids [92]. Table 11.2 illustrates the effect of germination time and temperature on bioactive compounds of cereals.

Caceres et al. [17] observed prominent increase in antioxidant activity and phenolic acids in brown rice as germination progressed. Comparable results for increase in antioxidants and total phenols were reported in cereals by them indicating positive relation between these parameters. Pal et al. [75] also reported a notable increase in protein content (5%–20%) and total phenolics (9%–75%) in different cultivars of germinated brown rice. Furthermore, it was anticipated that the increase in free phenolics is due to the hydrolysis of cell-wall polysaccharides that liberate bound phenolics or polymerization and oxidation of phenolics resulted in the increment of free or bound phenolics, thus attributing to the increase in total phenols.

Salinity in soaking water also promoted antioxidant compounds in rice. Umnajkitikorn et al. [96] reported 41% increase in the antioxidant activity of rice germinated on agar supplemented with NaCl than the control. However, few researchers have reported decrease in antioxidant activity and total phenols content of cereals (i.e., wheat, corn, finger millet, and sorghum) with increase in the germination time (Table 11.2) due to the increase in the activity of polyphenol oxidase and other catabolic enzymes. Furthermore, increase in other polyphenols was also reported [49]. Within this context, a study by Bryngelsson et al. [16] reported increase in transferase enzyme (hydroxycinnamoyl-CoA: hydroxyanthranilate N-hydroxycinnamoyl transferase) during germination that is involved in the biosynthesis of avenanthramides to increase the concentration of this phenolic compound in germinated oats.

Besides polyphenols, effect of germination on other bioactive compounds has also been investigated. An increase in γ-aminobutyric acid was reported in all germinated cereals due to the activation of enzyme, GAD that catalyzes the γ-decarboxylation of L-glutamic acid to γ-aminobutyric acid [17, 30, 53, 63]. Oh [70] observed 13-times higher γ-aminobutyric acid than in the ungerminated brown rice. Sie-Cheong et al. [86] reported reduction of γ-oryzanol content in five cultivars of brown rice and increase in 3 cultivars of brown rice after germination at room temperature for 24 h. Therefore γ-oryzanol content is influenced by species, genus, and environmental factors.

TABLE 11.2 Effects of Germination on Bioactive Compounds of Cereals

Cereal	Germination Time and Temperature	Bioactive Compound	Ref.
Barley	25 °C for 1 day	Increased antioxidant activity (AOX) (%2,2 diphenyl 1 picrylhydrazyl [DPPH] scavenging activity) by 91.37%; decreased total polyphenolic content (TPC) (mg gallic acid equivalent [GAE]/100 g) by 18.18%.	[84]
	16.5 °C for 5days	Increased γ-aminobutyric acid (mg/g) by 188.8%; decreased arabinoxylans (g/100 g) by 21.74%.	[30]
	25 °C for 2 days	Increased AOX (%DPPH scavenging activity) by 174.83%; TPC (mg/g) by 217.92%.	[38]
Buckwheat	16.5 °C for 5days	Increased arabinoxylans (g/100 g) by 125%; γ-aminobutyric acid (mg/g) by 131.3%.	[30]
	25 °C for 9 days	Increased rutin (mg/g) by 3005.8%; Total free phenylalanine (%) by 3136.8%.	[79]
Buckwheat, common	25 °C for 7 days	Increased AOX (%DPPH scavenging activity) by 631.85%;	[79]
	25 °C for 9 days	Increased TPC by 1174.3%; Total flavonoids (mg/g) by 580.7%.	[79]
Buckwheat, tertiary	25 °C for 7 days	Increased AOX (%DPPH scavenging activity) by 493.45%.	[79]
	25 °C for 9 days	Increased TPC by 2148.8%; Total flavonoids (mg/g) by 1077.4%.	[79]
Finger millet	28 °C for 2 days	Increased TPC by 122.17%; Total flavonoids (mg%) by 66.76%; Decreased AOX by 45.39%.	[49]
Oat	Room temperature for 10 h	Increased avenanthramides (mg/kg) by 250%;	[16]
	15 °C for 3 days	Increased sterols (%) by 20%;	[52]
	16 °C for 6 days	Increased TPC (mg GAE/g) by 355%;	[94]
	16.5 °C for 5 days	Decreased arabinoxylans (g/100 g) by 90.20%.	[30]
Rice, Rough	28–30 °C for 2 days	Increased TPC (mg GAE/100 g) by 40.25%; γ-oryzanol (mg/100 g) by 57.58%; γ-minobutyric acid (mg/100 g) by 383.2%;	[63]
	15 °C for 3 days	Increased γ-oryzanol (mg/g) by 11.76%; γ-aminobutyric acid (mg/100 g) by 116.7%.	[53]
	28 °C for 5 days	Increased AOX (μmol of Trolox equivalent/g) by 41.67%;	[96]
	37 °C for 5 days	Increased arabinoxylans (%) by 150%; γ-oryzanol (mg/100 g) by 56.10%;	[55]
	37 °C for 4 days	Increased γ-aminobutyric acid (mg/100 g) by 241.1%.	[55]

TABLE 11.2 (Continued)

Cereal	Germination Time and Temperature	Bioactive Compound	Ref.
Rice, brown	30 °C for 3 days	Increased γ-oryzanol (mg/100 g) by 4.6%; γ-aminobutyric acid (mg/100 g) by 2367%;	[72]
	35°C for 1 day	Increased γ-aminobutyric acid (mg/100 g) by 38.6%;	[56]
	35 °C for 2 days	Increased γ-aminobutyric acid (mg/100 g) by 1095.4%;	[21]
	25 °C for 1 day	Increased γ-oryzanol (mg/g) by 22.30%;	[86]
	35 °C for 2 days	Increased γ-aminobutyric acid (mg/100 g) by 3090.5%;	[22]
	28–30 °C for 1 day	Increased TPC (mg GAE/100 g) by 19.91%; γ-oryzanol (mg/100 g) by 27.27%; γ-aminobutyric acid (mg/100 g) by 187.4%;	[63]
	30 °C for 4 days	Increased γ-aminobutyric acid (mg/g) by 1710%;	[80]
	16.5 °C for 5 days	Increased arabinoxylans (g/100 g) by 400%;	[30]
	15 °C for 3 days	Increased γ-oryzanol (mg/g) by 18.97%; γ-aminobutyric acid (mg/100 g) by 143.5%;	[53]
	28 °C for 4 days	Increased AOX (mg trolox equivalent /100 g) by 7.98%; TPC (mg GAE/100 g) by 15.54%.	[17]
	28 °C for 4 days	Increased AOX (mg trolox equivalent /100 g) by 200.66%; TPC (mg GAE/100 g) by 337.40%.	[17]
	34 °C for 4 days	Increased AOX (mg trolox equivalent /100 g) by 344.61%; γ-aminobutyric acid (mg/100 g) by 815.8%.	[17]
	20 °C for 2 days	Increased AOX (μmol of trolox equivalent/g) by 87.86%; TPC (mg/100 g) by 63.16%; Total flavonoids (mg catechin/100 g) by 21.22%.	[92]
	35°C for 3 days	Increased γ-aminobutyric acid (mg/100 g) by 442.9%.	[27]
	25 °C for 3 days	Increased γ-aminobutyric acid (mg/100 g) by 1180%.	[80]
	30 °C for 4 days	Increased AOX (μmol of trolox equivalent/g) by 31.2%; TPC (mg/100 g) by 30.1%.	[26]
			[26]
Rye	18 °C for 6 days	Increased TPC (mg/100 g) by 32.35%.	[51]
	16.5 °C for 5 days	Increased γ-aminobutyric acid (mg/g) by 783%; decreased arabinoxylans (g/100 g) by 40.00%.	[30]

TABLE 11.2 *(Continued)*

Cereal	Germination Time and Temperature	Bioactive Compound	Ref.
Sorghum	16.5 °C for 5 days	Increased arabinoxylans (g/100 g) by 140%; γ-aminobutyric acid (mg/g) by 150%.	[30]
	Room temperature for 1 day	Decreased AOX (%DPPH scavenging activity) by 16.85%; TPC (mg GAE/100 g) by 14.9%.	[77]
	Room temperature for 2 days	Decreased AOX (%DPPH scavenging activity) by 30.79%; TPC (mg GAE/100 g) by 35.93%.	[77]
Wheat	16.5 °C for 8 days	Increased AOX (α-tocopherol μg/g) by 149.89%; TPC (μg ferulic acid equivalent /g) by 58.03%.	[102]
	37 °C for 2 days	Increased AOX (%DPPH scavenging activity) by 97.44%; TPC (μg ferulic acid equivalent/g) by 43.33%.	[43]
	21 °C for 1 day	Increased AOX (% DPPH scavenging activity) by 64.10%; decreased TPC (μg ferulic acid equivalent/g) by 13.78%.	[43]
	16.5 °C for 5 days	Increased Arabinoxylans (g/100 g) by 22%.	[30]
	Room temperature for 1 day	Decreased AOX (% DPPH scavenging activity) by 6.85%; TPC (mg GAE /100 g) by 15.18%.	[77]
	Room temperature for 2 days	Decreased AOX (%DPPH scavenging activity) by 21.6%; TPC (mg GAE/100 g) by 41.09%.	[77]
	25 °C for 3 days	Increased arabinoxylans (%) by 11.39%; γ-aminobutyric acid (mg/100 g) by 2023.6%. Decreased β-glucan (%) by 30.36%;	[88]
	20 °C for 4 days	Increased AOX (mg trolox equivalent/g) by 117.54%; TPC (mg/g) by 73.31%; total flavonoids (mg/g) by 26.36%	
	25 °C for 2 days	Increased TPC (mg/g) by 23.31%; Total flavonoids (mg/g) by 17.73%; Increased AOX (mmol of trolox equivalent/kg) by 27.9%; TPC (g GAE/kg) by 23.8%.	[91]
	Room temperature for 5 days	Increased AOX (mmol of trolox equivalent/kg) by 82.5%; TPC (g GAE/kg) by 92.9%.	[71]

The activation of hydrolytic enzymes (D-xylanase and cell wall hydrolase) during germination resulted in increase in soluble arabinoxylans in wheat, rice, buckwheat and sorghum (Table 11.2). However, Donkor et al. [30] reported 90% degradation of arabinoxylans in oats. Also, a reduction in β-glucan content has also been observed due to increase in the β-glucanase activity during germination [88]. Therefore optimum time and temperature is required to enhance the nutritional value of cereals during germination.

11.3.2 EXTRUSION COOKING

Extrusion is a high temperature short-time cooking process, in which simultaneous action of temperature, pressure, and shear is applied to produce ready-to-eat or ready-to-cook food products. Extrusion cooking causes biochemical changes in food, such as gelatinization of starch, protein denaturation, modification of lipids, and browning reactions, which led to the development of certain flavors, increase soluble dietary Fiber, enhance nutrient bioavailability, improves protein and starch digestibility and inactivate heat labile toxic compounds.

Extrusion cooking inactivates enzymes (lipase and lipoxidase) thus increasing the food stability and reducing antinutritional factors. Huth et al. [45] optimized extrusion conditions to maintain the structure of β-glucan and to increase resistant starch by 6% in barley extrudates.

In addition, extrusion breaks the cell-wall matrices thereby liberates covalently bound phenols, while it also degrades heat labile phenolic compounds; therefore phenolic concentration of extruded cereals depends upon the chemical changes and thermal effects during extrusion [33]. Antioxidant capacity and retention of total phenols in extrudates also depends on feed moisture and screw speed. The low moisture content and high screw speed extensively retained total phenols on extrusion of brown rice [20] and the increase in moisture content at low temperature showed greater retention in extruded rice flour [93]. Extrusion cooking of wheat, barley, rye, and oat at 120–160–200°C, constant screw speed of 500 rpm and mass flow rate of 225 g/min was able to increase the 200%–300% of free and ester-bound phenolic acid content [107]. Extrusion led to a significant increase in ferulic acid, vanillic acid, syringic acid and coumaric acid, except for sinapic and caffeic acids, which might be due to the conversion of phenolic acids from one kind to another during processing or due to thermolability of phenolic acids.

The increment in coumaric acid was predominant in oats, whereas increase in ferulic acid was more pronounced in wheat, barley, and rye than other phenolic compounds. Furthermore, there was intense increase in the free phenolic acid content in all four grains when compared to conjugated form of phenolic compounds. Ramos-Diaz et al. [78] revealed the increase in total phenolics on substitution of corn-based extrudates with 20%–50% of amaranth and quinoa. They also reported the decrease in α-, β- and γ-tocopherols and fatty acids due to the formation of amylose–lipid complex.

Altan et al. [5] reported a noticeable reduction in antioxidant activity (60%–68%) and total phenolic content (46%–60%) in extruded barley flour in comparison to the unprocessed barley flour. Similarly, extrusion process reduced the antioxidant activity and phenolic content of brown rice and germinated brown rice extrudates. This decrease might be attributed to the decarboxylation of phenols at high temperature and feed moisture [20].

In another study, Sensoy et al. [83] observed no effects of extrusion on antioxidant activity of buckwheat flour. Extruded cereal flour may be used to make products higher in phytonutrients. Mora-Rochin et al. [65] prepared tortillas with extruded flour of pigmented (yellow, white, red, and blue) Mexican maize that retained 76%–93% phenolic compounds, 58%–97% ferulic acid, and exhibited higher antioxidant activity than the tortillas made from traditional processed flour. They also revealed a significant decrease in anthocyanins, which is due to the degradation of heat labile flavonoids during extrusion. Therefore optimal extrusion conditions are required to improve the bioactive components concentration in cereals.

11.3.3 FERMENTATION

Fermentation can enhance the flavor and acceptability, increases nutritional value by activation of endogenous enzymes, and improves functional properties of cereals. During fermentation, starch hydrolyzing enzymes (α-amylase and maltase) get activated that hydrolyze complex carbohydrates into dextrins and simple sugar, thus improves digestibility of carbohydrates. Zhai et al. [104] observed the increase in reducing sugars on fermentation of cereals (i.e., wheat, millet, oat, corn, and sorghum) by *Agaricus subrufescens*, whereas in rice these sugars decrease due to the utilization of sugars by mycelia of fungi. Fermentation also increases the biological value and in-vitro protein digestibility by activation of proteolytic enzymes to cleave protein-tannin complex, thereby liberate peptides and amino acids and by degradation of storage proteins. El-Hag et al. [34] observed 82%–84% increase in protein digestibility of fermented flour in contrast to the unfermented flour. Researchers found negative relation between protein content and natural fermentation of pearl millet flour due to the utilization of amino acids by microbes during fermentation. Furthermore, it was anticipated that microbial fermentation loosens the complex food matrix that releases embedded minerals, lowers pH that converts ferrous ions to absorbable ferric ions, and reduces antinutritional that interferes minerals bioavailability, thereby increases mineral content in cereals. Interestingly, fermentation of rice by baker's yeast led to the increase in vitamins (riboflavin, nicotininc acid, pyridoxine, and tocopherols) [46].

Microbial breakdown of chemical constituents releases cellular bound compounds, such as flavonoids, phenols thus influencing the antioxidant activity of cereals and improving the bioavailability of nutrients. Fermentation of cereals by different strains of microbes exhibited different effects on the bioaccessibility of bioactive compounds especially phenolic compounds and antioxidant activity [98]. Fermentation of rye, rice, and wheat by baker's yeast showed profound effect on the antioxidant activity and phenolic contents with 14%–28% increment in antioxidant activity and 11%–100% in total phenols [31, 46]. Dordevic et al. [31] found that bacterial fermentation of cereals is more effective than the fermentation by yeast in releasing bioactive compounds (Table 11.3).

In another study, Bhanja and Kuhad [11] found that antioxidant property of cereals (wheat, brown rice, maize, and oat) fermented by *Rhizopus oryzae, Aspergillus oryzae, Aspergillus awamori*, and *Rhizopus oligosporus* depends on the species and variety of grains; as maximum antioxidant activity was found in wheat fermented by *R. oryzae*. Interestingly, fermentation of wheat by *A. oryzae* led to a significant increase in protocatechuic acid, caffeic acid, ferulic acid and cinnamic acid, but decreases the amount of gallic acid due to the conversion of phenolic acids from one kind to another during processing [11].

Zhai et al. [104] reported significant increase in total phenols when fermentation of cereals (millet, oat, wheat and rice) progressed up to 30 days, whereas in corns total phenols was increased up to 20 days then decreases as fermentation continued, therefore the influence of fermentation on phytonutrients is also time-dependent.

Oladeji et al. [74] observed loss in antioxidant activity and total phenols during natural fermentation of maize flour, due to the utilization of phytochemicals by microorganisms. Thus, nutritional value of cereals depends on the microbial strain along with time of fermentation. Table 11.3 indicates effects of fermentation by different microbes on bioactive compounds in cereals.

TABLE 11.3 Effects of Fermentation on Bioactive Compounds in Cereals

Cereal	Strain	Bioactive Components	Ref.
Barley	Lactobacillus rhamnosus	Increased AOX (%DPPH scavenging activity) by 138.46%; TPC (mg GAE/g) by 22.56%.	[31]
	Saccharomyces cerevisiae	Increased AOX (%DPPH Scavenging activity) by 92.30%; TPC (mg GAE/g) by 12.80%.	[31]
Buckwheat	Lactobacillus rhamnosus	Increased AOX (%DPPH scavenging activity) by 26.72%; TPC (mg GAE/g) by 17.15%.	[31]
	Saccharomyces cerevisiae	Increased AOX (%DPPH scavenging activity) by 8.62%; TPC (mg GAE/g) by 5%.	[31]
Maize	Rhizopus oryzae	Increased AOX (μmol trolox equivalent /g) by 210.32%.	[11]
	Aspergillus awamori	Increased AOX (μmol trolox equivalent/g) by 197.82%.	[11]
	Natural fermentation	Increased total flavonoids (mg quercetin equivalent /100 g) by 94.32%; decreased TPC (mg GAE/g) 2.92%.	[74]
Oats	Aspergillus oryzae	Increased AOX (μmol of trolox equivalent/g) by 46.52%; TPC (mg GAE/g) by 140%; γ-aminobutyric acid (μg/g) by 662.17%.	[18]
	Rhizopus oryzae	Increased AOX (μmol of trolox equivalent/g) by 76.08%; TPC (mg GAE/g) by 260%; γ-aminobutyric acid (μg/g) by 119.6%.	[18]
Pearl millet	Rhizopus oryzae	Increased AOX (μmol trolox equivalent /g) by 222.14%.	[12]
	Natural fermentation	Decreased TPC (mg /100 g) 59.86%.	[34]
Rice	Aspergillus oryzae	Increased AOX (%DPPH scavenging activity) by 205.78%; TPC (mg GAE/g) by 300.86%.	[13]
	Saccharomyces cerevisiae	Increased AOX (mmol TE/g) by 14.56%; TPC (mg GAE/g) by 11.71%; Decreased γ-oryzanol (μg/g) by 22.43%.	[46]
Rye	Lactobacillus rhamnosus	Increased AOX (%DPPH scavenging activity) by 42.45%; TPC (mg GAE/g) by 39.39%.	[31]
	Saccharomyces cerevisiae	Increased AOX (%DPPH scavenging activity) by 28.30%; TPC (mg GAE/g) by 22.72%.	[31]
Wheat	Saccharomyces cerevisiae	Increased AOX (%DPPH scavenging activity) by 25%; TPC (μmol GAE/g) by 100%.	[64]
	Aspergillus oryzae	Increased AOX (vitamin C equivalent antioxidant capacity [VCEAC] scavenging activity) by 8341.6%; TPC (μmol GAE/g) by 2088.87%.	[12]
	Aspergillus awamori	Increased AOX (VCEAC scavenging activity) by 4621.78%; TPC (μmol GAE/g) by 1610.41%.	[12]
	Lactobacillus rhamnosus	Increased AOX (%DPPH scavenging activity) by 50.74%; TPC (mg GAE/g) by 27.77%.	[31]
	Saccharomyces cerevisiae	Increased AOX (%DPPH scavenging activity) by 22.38%; TPC (mg GAE/g) by 13.58%.	[31]
	Rhizopus Oryzae	Increased AOX (μmol trolox equivalent /g) by 562.01%.	[11]

11.3.4   OTHER PROCESSING TECHNIQUES

Besides, germination, fermentation, and extrusion, other processing techniques (such as milling, dehulling, thermal treatment of cereals) also have significant effect on the amounts of bioactive compounds. Milling is the separation of bran and germ from endosperm by mechanical methods. Bioactive components are mostly concentrated in the bran layer of the cereals, thus removal of bran during milling decreases the phytonutrient concentration in the milled flour. Hung and Morita [44] reported higher phenolic content in the hull of buckwheat than other buckwheat flour fractions. In addition, higher concentration of bioactive components and antioxidant activity was reported in germ of milled wheat followed by bran, shorts, whole grain, and flour [9]. Ultra-fine grinding of bran enhances free and conjugated phenolic acid contents [41]. The milled fractions of cereals (such as bran or germ) can be incorporated into certain products to enhance their nutritional value.

Thermal treatments (such as baking, cooking, and roasting) add flavor increase nutritional value and reduce heat-labile toxic compounds. During the thermal treatments, chemical oxidation of phenols, Millard browning or caramelization takes place that improves the total phenolic content of processed cereals. Thermal processing at high temperature also decomposes ferulic acid to vanillin and vanillic acid; p-coumaric acid to p-hydroxybenzaldehyde [76]. Heating up to 100 °C could also degrade conjugated phenolics, thus increasing free phenolic acid contents (ferulic, vanillic, syringic, and p-coumaric acids) in wheat flour [25]. Furthermore, baking increases free phenolic acids in whole wheat bread, muffin, and cookie; while bound phenolic acid content is decreased in bread, but is slightly altered in cookie and muffin. Also, higher phenolic compounds were found in the crust of bread than its crumbs. Abdel-Aal [1] concluded that the impact of baking on phytochemicals depends on the baking conditions, form of phenolic acid, type of cereals, and also on the type of baked product.

In another study, Konopka et al. [57] reported an increase in antioxidant activity and free and bound phenolic acids in sourdough fermented whole wheat and rye bread in comparison to yeast fermented whole wheat and rye bread. Therefore, fermentation along with baking releases bound phenolics due to simultaneous hydrolysis, polymerization, oxidation and degradation of compounds. Oboh et al. [67] reported a significant decrease in total phenolic content and antioxidant capacity on roasting of yellow and white sorghum, whereas sand roasting of different cultivars of barley increased the antioxidant capacity from 27.4% to 39.8% but phenolic concentration was decreased [85]. Similarly, Gallegos-Infante [36] observed increase in phenolic compounds and antioxidant activity in microwave roasting of barley grains at 600 W power for 8.5 min but Sharma and Gujral [85] reported significant reduction in antioxidant capacity and phenolic compounds among all microwave roasted cultivars of barley (at 900 W power for 2 min) in comparison to traditional roasted barley. This decrease can be attributed to the loss of heat-sensitive phenolics due to longer heat exposure in the microwave oven.

Nevertheless, all processes alone are not enough to enhance bioavailability of bioactive compounds and reduce antinutritional factors; certain prerequisites to these techniques may prove more beneficial. Fermentation in combination with grinding promoted mineral bioavailability by exposing more grain surface to microbes due to mechanical breakage of cellular-structure [40]. In another study, Katina et al. [51] stated that germination prior to fermentation of rye by *Saccharomyces cerevisiae* enhanced total phenolic content by 11-times and free phenolic acids by 3-times due to the activity of hydrolytic enzymes that caused structural breakdown of cell wall before the fermentation. Similarly, fermentation along with enzymatic treatment (α-amylase, β-glucanase, xylanase, ferulic acid esterase, and cellulase) of wheat bran increased the free phenolic acid content by 1.6 times in comparison to fermentation alone [7]. Furthermore, high hydrostatic pressure treatment (100–500 MPa for 10 min) of germinated brown rice increased bioaccessibility of minerals (calcium and copper), amino acids (mainly indispensable amino acids), γ-aminobutyric acid and also increased the antioxidant activity of brown rice [101]. Similarly, high-pressure processing (30 MPa for 24 or 48 h of germination) of germinated rough rice (37 °C for 2 days) increased the contents of γ-aminobutyric acid, arabinoxylans, γ-oryzanol, vitamins; due to the increased exposed surfaces for enzymatic action [55]. Recently, Xia et al. [100] reported pressure-dependent increase in γ-aminobutyric acid, exhibiting 25% improvement in 50 MPa-stressed germinated brown rice than the other stressed grains in comparison to the control. These results recommend the use of high-pressure treatment to enhance the chemical composition of germinated cereals.

In addition, exposure of brown rice to 3 kV low-pressure plasma reduces the germination time up to 24 h, increases α-amylases activity by 162%, γ-aminobutyric acid by 42.6%, total phenols by 25% and antioxidant activity by 34% than the unexposed germinated brown rice [24]. Interestingly, ultrasound treatment (at 0, 30, 45 kHz for 0, 5, 10, 15, 20, 25, and 30 min) prior to germination (for 16 and 24 h) of brown rice significantly improved the concentration of γ-aminobutyric acid in contrast to the untreated germinated brown rice [103]. The authors reported frequency and time dependent increase in production of γ-aminobutyric acid, with maximum accumulation when ultrasound treatments were applied at 30 kHz for 15 min and afterward there was a decrease.

Increase in the endogenous enzymes activity after the ultrasound treatment promoted γ-aminobutyric acid production, whereas the decrease on further progression of treatment might be due to the increase in temperature. In addition, maximum accumulation of γ-aminobutyric acid was observed in ultrasound treated brown rice germinated for 16 h than that the germinated rice for 24 h. Thereby, ultrasound treatment can stimulate the endogenous enzyme activation, as well as the degradation of biopolymeric compounds, hence accelerates the germination time.

11.4 SUMMARY

In order to make the nutrients bioaccessible, cereals need to be preprocessed through germination, fermentation, extrusion, or roasting before the preparation of a final product. Processing of cereals breaks down complex carbohydrates into dextrins, improves the biological value and digestibility of carbohydrates and proteins, reduces the antinutritional factors that interferes mineral bioavailability, breaks cellular matrices to release bound phenolic compounds. The effect of processing on phytonutrients depends on the processing techniques and conditions, the species and genus of cereal grains. In addition, a time-dependent effect of fermentation and germination on bioactive compounds has been reported in cereals. The high hydrostatic pressure, ultrasonication, low-pressure plasma processing in combination with traditional technologies might attribute to the better retention of nutrients and health-promoting compounds.

KEYWORDS

- **bioactive compounds**
- **bioavailability**
- **cereals**
- **processing techniques**
- **whole grain**

REFERENCES

1. Abdel-Aal, E. S. M.; Choo, T. M.; Dhillon, S.; Rabalski, I. Free and Bound Phenolic Acids and Total Phenolic in Black, Blue, and Yellow Barley and Their Contribution to Free Radical Scavenging Capacity. *Cereal Chemistry,* **2012,** *89,* 198–204.
2. Abdel-Aal, E. S. M.; Gamel, T. H. Effects of Selected Barley Cultivars and their Pearling Fractions on the Inhibition of Human LDL Oxidation *In Vitro* Using a Modified Conjugated Dienes Method. *Cereal Chemistry,* **2008,** *85,* 730–737.
3. Abdel-Aal, E. S. M.; Rabalski, I; Effect of Baking on Free and Bound Phenolic acids in Wholegrain Bakery Products. *Journal of Cereal Science,* **2013,** *57,* 312–318.
4. Acosta, E. Bioavailability of Nanoparticles in Nutrient and Nutraceutical Delivery. *Current Opinion in Colloid and Interface Science,* **2009,** *14,* 3–15.
5. Altan, A.; McCarthy, K. L.; Maskan, M. Effect of Extrusion Process on Antioxidant Activity, Total Phenolics and B-Glucan Content of Extrudates Developed from Barley-Fruit and Vegetable By-Products. *International Journal of Food Science and Technology,* **2009,** *44,* 1263–1271.
6. Andreasen, M. F.; Christensen, L. P.; Meyer, A. S.; Hansen, A. Content of Phenolic Acids and Ferulic Acid De-Hydrodimers in 17 Rye (*Secale Cereale*) Varieties. *Journal of Agricultural and Food Chemistry,* **2000,** *48,* 2837–2842.
7. Anson, N. M.; Hemery, Y. M.; Bast, A.; Haenen, G. R. Optimizing the Bioactive Potential of Wheat Bran by Processing. *Food and Function,* **2012,** *3,* 362–375.
8. Awika, J. M.; Dykes, L.; Gu, L.W.; Rooney, L.W.; Prior, R. L. Processing of Sorghum (Sorghum bicolor) and Sorghum Products Alters Procyanidin Oligomer and Polymer Distribution and Content. Journal of Agricultural and Food Chemistry, 2003, 51, 5516–5521.
9. Barron, C.; Hemery, Y.; Rouau, X.; Lullien-Pellerin, V.; Abecassis, J. Dry Processes to Develop Wheat Fractions and Products with Enhanced Nutritional Quality. *Journal of Cereal Science.* **2007,** *46,* 327–347.
10. Bartłomiej, S.; Justyna, R.; Ewa, N. Bioactive Compounds in Cereal Grains Occurrence, Structure, Technological Significance and Nutritional Benefits: A Review. *Food Science and Technology International,* **2011,** *18,* 559–568.
11. Bhanja, D. T.; Kuhad, R. C. Upgrading the Antioxidant Potential of Cereals by Their Fungal Fermentation under Solid-State Cultivation Conditions. *Letters in Applied Microbiology,* **2014,** *59* (5), 493–499.
12. Bhanja, T.; Kumari, A.; Banerjee, R. Enrichment of Phenolics and Free Radical Scavenging Property of Wheat Koji Prepared with Two Filamentous Fungi. *Bioresource Technology,* **2009,** *100,* 2861–2866.
13. Bhanja, T.; Rout, S.; Banerjee, R.; Bhattacharyya, B. C. Studies on the Performance of a New Bioreactor for Improving Antioxidant Potential of Rice. *LWT—Food Science and Technology,* **2008,** *41,* 1459–1465.
14. Bhathal, S.; Kaur, N. Effect of Germination on Nutrient Composition of Gluten Free Quinoa (*Chenopodium quinoa*). *International Journal of Scientific Research,* **2015,** *4,* 423–425.
15. Bohn, T. Dietary Factors Affecting Polyphenol Bioavailability. *Nutrition Reviews,* **2016,** *72,* 429–452.
16. Bryngelsson, S.; Ishihara, A.; Dimberg, L.H. Levels of Avenanthramides and Activity of Hydroxy-Cinnamoyl-Coa: Hydroxy-Anthranilate N-Hydroxy-Cinnamoyl Transferase (HHT) in Steeped or Germinated Oat Samples. *Cereal Chemistry,* **2003,** *80,* 356-360.

17. Caceres, P. J.; Martınez-Villaluenga, C.; Amigo, L.; Frias, J. Maximizing the Phyto-chemical Content and Antioxidant Activity of Ecuadorian Brown Rice Sprouts Through Optimal Germination Conditions. *Food Chemistry*, **2014**, *152*, 407–414.

18. Cai, S.; Gao, F.; Zhang, X.; Wang, O.; Wu, W.; Zhu, S.; Ji, B. Evaluation of Γ-Aminobutyric Acid, Phytate and Antioxidant Activity of Tempeh-Like Fermented Oats (*Avena sativa L.*) Prepared with Different Filamentous Fungi. *Journal of Food Science and Technology,* **2014**, *51*, 2544–2551.

19. Cai, S.; Wang, O.; Wu, W. Comparative Study of the Effects of Solid-State Fermentation with Three Filamentous Fungi on the Total Phenolics Content (TPC), Flavonoids, and Antioxidant Activities of Subfractions from Oats (*Avena sativa*). *Journal of Agricultural and Food Chemistry*, **2012**, *60*, 507–513.

20. Chalermchaiwat, P.; Jangchud, K.; Jangchud, A. Antioxidant Activity, Free Gamma-Aminobutyric Acid Content, Selected Physical Properties and Consumer Acceptance of Germinated Brown Rice Extrudates as Affected by Extrusion Process. *LWT Food Science and Technology*, **2015**, *64*, 490–496.

21. Charoenthaikij, P.; Jangchud, K.; Jangchud, A. Germination Conditions Affect Physicochemical Properties of Germinated Brown Rice Flour. *Journal of Food Science*, **2009**, *74*, 658–665.

22. Charoenthaikij, P.; Jangchud, K.; Jangchud, A. Germination Conditions Affect the Quality of Composite Wheat-Germinated Brown Rice Flour and Bread Formulations. *Journal of Food Science*, **2010**, *75*, 321–318.

23. *Chauhan, E. S. Effects of Processing (Germination and Popping) on the Nutritional and Anti-Nutritional Properties of Finger Millet (Eleusine coracana). Current Research in Nutrition and Food Science*, **2018**, *6*, 566–572.

24. Chen, H. H.; Chang, H. C.; Chen, Y. K.; Hung, C. L.; Lin, S. Y.; Chen, Y. S. An Improved Process for High Nutrition of Germinated Brown Rice Production: Low-Pressure Plasma. *Food Chemistry*, **2016**, *191*, 120–127.

25. Cheng, Z.; Su, L.; Moore, J.; Zhou, K.; Luther, M.; Yin, J. J.; Yu, L. Effects of Postharvest Treatment and Heat Stress on Availability of Wheat Antioxidants. *Journal of Agricultural and Food Chemistry*, **2006**, *54*, 5623–5629.

26. Cho, D.H.; Lim, S.T. Germinated Brown Rice and Its Bio-Functional Compounds. *Food Chemistry*, **2016**, *196*, 259–271.

27. Chungcharoen, T.; Prachayawarakorn, S.; Tungtrakul, P.; Soponronnarit, S. Effects of Germination Process and Drying Temperature on Gamma-Aminobutyric Acid (GABA) and Starch Digestibility of Germinated Brown Rice. *Drying Technology*, **2014**, *32*, 742–753.

28. Cornejo, F.; Caceres, P. J.; Martinez-Villaluenga, C.; Rosell, C. M.; Frias, J. Effects of Germination on the Nutritive Value and Bioactive Compounds of Brown Rice Breads. *Food Chemistry*, **2015**, *173*, 298–304.

29. Cui, S. W.; Wu, Y.; Ding, H. *The Range of Dietary Fiber Ingredients and a Comparison of their Technical Functionality*. In: *Fiber-Rich and Wholegrain Foods*. London, UK: Woodhead Publisher; Series in Food Science, Technology and Nutrition; **2013**; pp. 96–119.

30. Donkor, O. N.; Stojanovska, L.; Ginn, P.; Ashton, J.; Vasiljevic, T. Germinated Grains—Sources of Bioactive Compounds. *Food Chemistry*, **2012**, *135*, 950–959.

31. Đorđević, T. M.; Šiler-Marinković, S. S.; Dimitrijević-Branković, S. I. Effect of Fermentation on Antioxidant Properties of Some Cereals and Pseudo Cereals. *Food Chemistry*, **2010**, *119*, 957–963.

32. Dykes, L.; Rooney, L. W. Phenolic Compounds in Cereal Grains and their Health Benefits. *Cereal Foods World*, **2007**, *52*, 105–111.

33. Eastman, J.; Orthoefer, F.; Solorio, S. Using Extrusion to Create Breakfast Cereal Products. *Cereal Foods World*, **2001**, *46*, 468–471.

34. El-Hag, M. E.; El-Tinay, A. H.; Yousif, N. E. Effect of Fermentation and Dehulling on Starch, Total Polyphenols, Phytic Acid Content and *In Vitro* Protein Digestibility of Pearl Millet. *Food Chemistry*, **2002**, *77*, 193–196.

35. El-Khoury, D.; Cuda, C.; Luhovyy, BL.; Anderson, G. H. Beta-Glucan: Health Benefits in Obesity and Metabolic Syndrome. *Journal of Nutrition and Metabolism*, **2012**, *2012*, Article ID: 851362.

36. Gallegos-Infante, J. A.; Rocha, Guzman, N. E. Quality of Spaghetti Pasta Containing Mexican Common Bean Flour (*Phaseolus vulgaris* L.). *Food Chemistry*, **2010**, *119*, 1544–1549.

37. Gani, A.; Wani S. M.; Masoodi, F. A.; Hameed, G. Whole-grain Cereal Bioactive Compounds and Their Health Benefits: A Review. *Food Processing and Technology*, **2012**, *3*, 1–10.

38. Ha, K. S.; Jo, S. H.; Mannam, V.; Kwon, Y. I.; Apostolidis, E. Stimulation of Phenolics, Antioxidant and *A*-Glucosidase Inhibitory Activities During Barley (*Hordeum vulgare*) Seed Germination. *Plants Foods for Human Nutrition*, **2016**, *71*, 211–217.

39. Hagiwara, H.; Seki, T.; Ariga, T. The Effect of Pre-Germinated Brown Rice Intake on Blood Glucose and PAI-1 Levels in Streptozotocin Induced Diabetic Rats. *Bioscience, Biotechnology, and Biochemistry*, **2004**, *68*, 444–447.

40. Hemalatha, S.; Platel, K.; Srinivasan, K. Influence of Germination and Fermentation on Bioaccessibility of Zinc and Iron from Food Grains. *European Journal of Clinical Nutrition*, **2007**, *61*, 342–348.

41. Hemery, Y. M.; Anson, N. M.; Havenaar, R. Dry-Fractionation of Wheat Bran Increases the Bioaccessibility of Phenolic Acids in Breads Made from Processed Bran Fractions. *Food Research International*, **2010**, *43*, 1429–1438.

42. Hu, B.; Liu, X.; Zhang, C.; Zeng, X. Food Macro Molecule-Based Nano-Delivery Systems for Enhancing the Bioavailability of Polyphenols. *Journal of Food and Drug Analysis*, **2016**, *2016*, 1–13.

43. Hung, P. V.; Hatcher, D. W.; Barker, W. Phenolic Acid Composition of Sprouted Wheats by Ultra-Performance Liquid Chromatography (UPLC) and Their Antioxidant Activities. *Food Chemistry*, **2011**, *126*, 1896–1901.

44. Hung, P.; Morita, N. Distribution of Phenolic Compounds in the Graded Flours Milled from Whole Buckwheat Grains and their Antioxidant Capacities. *Food Chemistry*, **2008**, *109*, 325–331.

45. Huth, M.; Dongowski, G.; Gebhardt, E.; Flamme, W. Functional Properties of Dietary Fiber Enriched Extrudates from Barley. *Journal of Cereal Science*, **2000**, *32*, 115–128

46. Ilowefah, M.; Bakar, J.; Ghazali, H. M.; Mediani, A.; Muhammad, K. Physicochemical and Functional Properties of Yeast Fermented Brown Rice Flour. *Journal of Food Science and Technology*, **2014**, *52*, 5534–5545.

47. Jiang, Y.; Wang, T. Phytosterol in Cereal By-Product. *Journal of the American Oil Chemists Society* **2005**, *82*, 439–444.

48. Kahkonen, M. P.; Hopia, A. I.; Vuorela, H. J.; Rauha, J. P. Antioxidant Activity of Plant Extracts Containing Phenolic Compounds. *Journal of Agricultural and Food Chemistry*, **1999**, *47*, 3954–962.

49. Karki, D. B.; Kharel, G. P. Effect of Finger Millet Varieties on Chemical Characteristics of Their Malts. *African Journal of Food Science*, **2012**, *6*, 308–316.

50. Katina, K.; Liukkonen, K. H.; Kaukovirta-Norja, A. Fermentation-Induced Changes in the Nutritional Value of Native or Germinated Rye. *Journal of Cereal Science*, **2007**, *46*, 348–355.

51. Kaukovirta-Norja, A.; Wilhemson, A.; Poutanen, K. Germination: A Means to Improve the Functionality of Oat. *Agricultural and Food Science*, **2004**, *13*, 100–112.

52. Kauklasapathy, K.; Masondole, L. Survival of Free and Microencapsulated *Lactobacillus acidophilus* and *Bifidobacterium lactis* and their effect on Texture of Feta Cheese. *Australian Journal of Dairy Technology*, **2005**, 60, 252–258.

53. Kim, H. Y.; Hwang, I. G.; Kim, T. M.; Woo, K. S. Chemical and Functional Components in Different Parts of Rough Rice (*Oryza Sativa*) before and after Germination. *Food Chemistry,* **2012**, *134*, 288–293.

54. Kim, K. H.; Tsao, R.; Yang, R.; Cui, S. W. Phenolic Acid Profiles and Antioxidant Activities of Wheat Bran Extracts and the Effect of Hydrolysis Conditions. *Food Chemistry*, **2006**, *95*, 466–473.

55. Kim, M. Y.; Lee, S. H.; Jang, G. Y. Effects of High Hydrostatic Pressure Treatment on the Enhancement of Functional Components of Germinated Rough Rice (*Oryza sativa* L.). *Food Chemistry,* **2015**, *166*, 86–92.

56. Komatsuzaki, N.; Tsukahara, K.; Toyoshima, H. Effect of Soaking and Gaseous Treatment on GABA Content in Germinated Brown Rice. *Journal of Food Engineering,* **2007**, *78*, 556–560.

57. Konopka, I.; Tańska, M.; Faron, A.; Czaplicki, S. Release of Free Ferulic Acid and Changes in Antioxidant Properties during the Wheat and Rye Bread Making Process. *Food Science and Biotechnology*, **2014**, *23*, 831–840.

58. Laxmi, G.; Chaturvedi, N.; Richa, S. The Impact of Malting on Nutritional Composition of Foxtail Millet, Wheat and Chickpea. *Journal of Nutrition and Food Sciences*, **2015**, *5*, 407–410.

59. Liu, R. H. Whole Grain Phytochemicals and Health. *Journal of Cereal Science*, **2007**, *46*, 207–219.

60. Mattila, P.; Pihlava, J.; Hellström, J. Contents of Phenolic Acids, Alkyl- and Alkenyl resorcinols, and Avenanthramides in Commercial Grain Products. *Journal of Agricultural and Food Chemistry,* **2005**, *53*, 8290–8295.

61. Mayer, R.; Cherry, J.; Rhodes, D. Effects of Heat Shock on Amino Acid Metabolism of Cowpea Cells. *Plant Physiology*, **1990**, *94*, 796–810.

62. McKeehen, J.D.; Busch, R.H.; Fulcher, R.G. Evaluation of Wheat (*Triticum Aestivum* L.) Phenolic Acids During Grain Development and Their Contribution to Fusarium Resistance. *Journal of Agricultural and Food Chemistry,* **1999**, *47*, 1476–1482.

63. Moongngarm, A.; Saetung, N. Comparison of Chemical Compositions and Bioactive Compounds of Germinated Rough Rice and Brown Rice. *Food Chemistry.* **2010**, *122*, 782–788.

64. Moore, J.; Cheng, Z.; Hao, J.; Guo, G.; Liu, J. G.; Lin, C.; Yu, L. Effects of Solid-State Yeast Treatment on the Antioxidant Properties and Protein and Fiber Compositions of Common Hard Wheat Bran. *Journal of Agricultural and Food Chemistry,* **2007**, *55*, 10173–10182.

65. Mora-Rochin, S.; Gutiérrez-Uribe, J. A.; Serna-Saldivar, S. O. Phenolic Content and Antioxidant Activity of Tortillas Produced from Pigmented Maize Processed by

Conventional Nixtamalization or Extrusion Cooking. *Journal of Cereal Science*, **2015**, *2*, 502–508.

66. Nkhata, S. G.; Ayua, E.; Kamau, E. H.; Shingiro, J. B. Fermentation and Germination Improve Nutritional Value of Cereals and Legumes Through Activation of Endogenous Enzymes. *Food Science and Nutrition*, **2018**, *2018*, 1–13.

67. Oboh, G.; Ademiluyi, A. O.; Akindahunsi, A. A. The Effect of Roasting on the Nutritional and Antioxidant Properties of Yellow and White Maize Varieties. *International Journal of Food Science and Technology*, **2010**, *45*, 1236–1242.

68. Ogbonna, A. C.; Abuajah, C. I.; Ide, E. O.; Udofia, U. S. Effect of Malting Conditions on the Nutritional and Anti-Nutritional Factors of Sorghum. *Food Technology*, **2012**, *36*, 64–72.

69. Oh, C. H.; Oh, S. H. Effect of Germinated Brown Rice Extracts with Enhanced Levels of GABA on Cancer Cell Proliferation and Apoptosis. *Journal of Medicinal Food*, **2004**, *7*, 19–23.

70. Oh, S. H.; Stimulation of Gamma-Aminobutyric Acid Synthesis Activity in Brown Rice by a Chitosan/Glutamic Acid Germination Solution and Calcium/Calmodulin. *Journal of Biochemistry and Molecular Biology*, **2003**, *36*, 319–325.

71. Ohm, J.; Lee, C. W.; Cho, K. Germinated Wheat: Phytochemical Composition and Mixing Characteristics. *Cereal Chemistry*, **2016**, *93*, 612–617.

72. Ohtsubo, K. I.; Suzuki, K.; Yasui, Y.; Kasumi, T. Bifunctional Components in the Processed Pre-Germinated Brown Rice by a Twin Screw Extruder. *Journal of Food Composition and Analysis*, **2005**, *18*, 303–316.

73. Okada, T.; Sugishita, T.; Murakami, T.; Murai, H. Effect of the Defatted Rice Germ Enriched with GABA for Sleepless, Depression, Autonomic Disorder by Oral Administration. *Nippon Shokuhin Kagaku Kougaku Kaishi*, **2000**, *47*, 596–603 (in Japanese).

74. Oladeji, B. S.; Akanbi, C. T.; Gbadamosi, S. O. Effects of Fermentation on Antioxidant Properties of Flours of a Normal Endosperm and Quality Protein Maize Varieties. *Journal of Food Measurement and Characterization*, **2017**, *11*, 1148–1158.

75. Pal, P.; Singh, N.; Kaur, P.; Kaur, A.; Virdi, A. S.; Parmar, N. Comparison of Composition, Protein, Pasting, and Phenolic Compounds of Brown Rice and Germinated Brown Rice from Different Cultivars. *Cereal Chemistry*, **2016**, *93*, 584–592.

76. Ragaee, S.; Seetharaman, K.; Abdel-Aal, E. S. M. The Impact of Milling and Thermal Processing on Phenolic Compounds in Cereal Grains. *Critical Reviews in Food Science and Nutrition*, **2014**, *54*, 837–849.

77. Ramadan, B. R.; Sorour, M.; Abdel, H.; Kelany, M. A. Changes in Total Phenolics and DPPH Scavenging Activity During Domestic Processing in Some Cereal Grains. *Annual Review of Food Science and Technology*, **2012**, *13*, 190–196.

78. Ramos, Diaz, J, M.; Sundarrajan, L.; Kariluoto, S. Effect of Extrusion Cooking on Physical Properties and Chemical Composition of Corn-Based Snacks Containing Amaranth and Quinoa: Application of Partial Least Squares Regression. *Journal of Food Process Engineering*, **2017**, *40*, 1–15.

79. Ren, S. C.; Sun, J. T. Changes in Phenolic Content, Phenylalanine Ammonia-Lyase (PAL) Activity, and Antioxidant Capacity of Two Buckwheat Sprouts in Relation to Germination. *Journal of Functional Foods*, **2014**, *7*, 298– 304.

80. Roohinejad, S.; Omidizadeh, A.; Mirhosseini, H. Effect of Pre-germination Time on Amino Acid Profile and Gamma Amino Butyric Acid (GABA) Contents in Different Varieties of Malaysian Brown Rice. *International Journal Food Properties*, **2011**, *14*,1386–1399.

81. Ross, A. B.; Åman, P.; Kamal-Eldin, A. Identification of Cereal Alkyl resorcinol Metabolites in Human Urine: Potential Biomarkers of Wholegrain Wheat and Rye Intake. *Journal of Chromatography B*, **2004**, *809* (1), 125–130.

82. Ross, A. B.; Shepherd, M. J.; Schupphaus, M. Alkyl resorcinols in Cereals and Cereal Products. *Journal of Agricultural and Food Chemistry*, **2003**, *51*, 4111–4118.

83. Şensoy, İ.; Rosen, R. T.; Ho, C, T.; Karwe, M. V. Effect of Processing on Buckwheat Phenolics and Antioxidant Activity. *Food Chemistry*, **2006**, *99*, 388–393.

84. Sharma, P.; Gujral, H. S. Antioxidant and Polyphenol Oxidase Activity of Germinated Barley and Its Milling Fractions. *Food Chemistry.* **2010**, *120*, 673–678.

85. Sharma, P.; Gujral, H. S. Effect of Sand Roasting and Microwave Cooking on Antioxidant Activity of Barley. *International Food Research Journal*, **2011**, *44*, 235–240.

86. Sie-Cheong, K.; Pang-Hung, Y.; Amartalingam, R.; Sie-Chuong, W. Effect of Germination on Γ-Oryzanol Content of Selected Sarawak Rice Cultivars. *American Journal of Applied Science*, **2009**, 6, 1658–1661.

87. Singh, A.; Sharma, S. Bioactive Components and Functional Properties of Biologically Activated Cereal Grains: A Bibliographic Review. *Critical Reviews in Food Science and Nutrition*, **2017**, *57*, 3051–3071.

88. Singkhornart, S.; Edou-ondo, S.; Ryu, G. H. Influence of Germination and Extrusion with CO_2 Injection on Physicochemical Properties of Wheat Extrudates. *Food Chemistry,* **2014**, *143*, 122–131.

89. Slavin, J. Whole Grains and Human Health. *Nutrition Research Reviews*, **2004**, *17*, 99–110.

90. Srinivasan, M.; Sudheer, A.R.; Menon, V.P. Ferulic Acid: Therapeutic Potential Through its Antioxidant Property. *Journal of Clinical Biochemistry and Nutrition*, **2007**, *40*, 92–100.

91. Swieca, M.; Dziki, D. Improvement in Sprouted Wheat Flour Functionality: Effect of Time, Temperature and Elicitation. *International Journal of Food Science and Technology*, **2015**, *50*, 2135–2142.

92. Ti, H.; Zhang, R.; Zhang, M.; Li, Q.; Wei, Z. Dynamic Changes in the Free and Bound Phenolic Compounds and Antioxidant Activity of Brown Rice at Different Germination Stages. *Food Chemistry.* **2014**, *161*, 337–344.

93. Ti, H.; Zhang, R.; Zhang, M.; Wei, Z.; Chi, J.; Deng, Y.; Zhang, Y. Effect of Extrusion on Phytochemical Profiles in Milled Fractions of Black Rice. *Food Chemistry*, **2015**, *178*, 186–194.

94. Tian, S.; Nakamura, K.; Kayahara, H. Analysis of Phenolic Compounds in White Rice, Brown Rice and Germinated Brown Rice. *Journal of Agricultural and Food Chemistry*, **2004**, *52*, 4808–4813.

95. Tripathi, V.; Singh, A.; Ashraf, M. T. Avenanthramides of Oats: Medicinal Importance and Future Perspectives. *Pharmacognosy Reviews*, **2018**, *12*, 66–71.

96. Umnajkitikorn, K.; Faiyue, B.; Saengnil, K. Enhancing Antioxidant Properties of Germinated Thai Rice (*Oryza sativa* L.) with Salinity. *Journal of Rice Research*, **2013**, *1*, 1–8.

97. Verma, B.; Hucl, P.; Chibbar, R.N. Phenolic Acid Composition and Antioxidant Capacity and Alkali Hydrolyzed Wheat Bran Fractions. *Food Chemistry*, **2009**, *116*, 947–954.

98. Wang, T.; He, F.; Chen, G. Improving Bioaccessibility and Bioavailability of Phenolic Compounds in Cereal Grains Through Processing Technologies: A Concise Review. *Journal of Functional Foods*, **2014**, 7, 101–111.

99. Winata, A.; Lorenz, K. Effects of Fermentation and Baking of Whole Wheat and Whole Rye Sourdough Breads on Cereal Alkylresorcinols. *Cereal Chemistry*, **1997**, *74*, 284–287.

100. Xia, Q.; Wang, L.; Li, Y. Exploring High Hydrostatic Pressure-Mediated Germination to Enhance Functionality and Quality Attributes of Wholegrain Brown Rice. *Food Chemistry*, **2018**, *249*, 104–110.

101. Xia, Q.; Wang, L.; Xu, C.; Mei, J.; Li, Y. Effects of Germination and High Hydrostatic Pressure Processing on Mineral Elements, Amino Acids and Antioxidants in Vitro Bio-Accessibility, as well as Starch Digestibility in Brown Rice (*Oryza sativa* L.). *Food Chemistry*, **2017**, *214*, 533–542.

102. Yang, F.; Basu, T. K.; Ooraikul, B. Studies on Germination: Conditions and Antioxidant Contents of Wheat Grain. *Journal of Food Science and Nutrition*, **2001**, *52*, 319–330.

103. Yi, Z.; Ting-ting, Z.; Juan-li, S.; Xin, W. Study on Effect of Ultrasonic Treatment on GABA Accumulation and Antioxidant Capacity in Germinated Brown Rice. *Journal of Food Science and Technology*, **2016**, *2*, 130–137.

104. Zhai, F. H.; Wang, Q.; Han, J.R. Nutritional Components and Antioxidant Properties of Seven Kinds of Cereals Fermented by the Basidiomycete *Agaricus blazei*. *Journal of Cereal Science*, **2015**, *65,*202–208.

105. Zhang, G.; Hamaker, B. *Nutraceutical and Health Properties of Sorghum and Millet*. In: *Cereals and Pulses*; London: Wiley-Blackwell; **2012**; pp. 165–186.

106. Zhang, G.; Xu, Z.; Gao, Y.; Huang, X.; Yang, T. Effects of Germination on the Nutritional Properties, Phenolic Profiles, and Antioxidant Activities of Buckwheat. *Journal of Food Science*, **2015**, *80*, 1111–1119.

107. Zielinski, H.; Kozowska, H.; Lewczuk, B. Bioactive Compounds in the Cereal Grains Before and after Hydrothermal Processing. *Innovative Food Science and Emerging Technologies*, **2001**, *2*, 159–169.

PART IV
Role of Cereals in Disease Management

CHAPTER 12

FOODS FOR GLUTEN INTOLERANCE

ASHWANI KUMAR, SARABJIT SINGH, VIDISHA TOMER, and
RASANE PRASAD

ABSTRACT

The gluten intolerant individuals may encounter three types of medical conditions associated with gluten, that is, celiac disease, allergy to wheat and nonceliac gluten sensitivity. The major gluten sources are wheat, rye, barley, and oats. Management through gluten-free diet using alternative sources (such as rice, millets, pseudo-cereals, legumes, and others) is the only solution for the gluten intolerants. The traditional Gluten-free foods like *idli, dosa, paddu, uthapam, tofu, and others* are traditional alternatives to the gluten-containing foods. Various formulations of gluten-free bread, biscuits, cake, cookies, pasta, and spaghetti have been standardized by the food scientists. Consumer requirements have suggested the food industry to continuously improve the formulations and processing methodologies used for gluten-free foods.

12.1 INTRODUCTION

Gluten is a protein complex, which is formed during the dough preparation from flours of wheat (*Triticum aestivum* L.), rye (*Secale cereal* L.), and barley (*Hordeum vulgare* L.) and is responsible for the rheological properties required for the leavened bread production. The hybrids and the cross-breeds of these cereals [such as kamut and triticale (×*Triticosecale*)] also contain gluten. In addition to these, although inherently gluten free, oats (*Avena sativa* L.) may also contain gluten due to the contamination with wheat during cultivation or processing.

The proteins responsible for gluten formation are monomeric gliadins and polymeric glutenins. The gliadins (30–80 kDa: α-, β-, γ-, and ω-gliadins) are protein polymorphic mixtures, which are soluble in 70% alcohol. Among these, α-gliadins are considered the most immunogenic and mainly responsible for gluten intolerance. The γ-gliadins and glutenins are comparatively less responsible for gluten intolerance. Glutenins have a molecular weight in the range of 30–120 kDa and are broadly classified as low molecular weight glutenins (30–74 kDa) and high molecular weight glutenins (75–120 kDa) [95]. The gliadin is believed to control the viscosity of dough while glutenins provide elasticity and strength to the dough. This complex is crucial in the preparation of leavened bakery products. However, it can result in negative health effects in genetically susceptible individuals.

The susceptible individuals may encounter three types of conditions on consumption of gluten, that is, celiac disease (CD), allergy to wheat and nonceliac gluten sensitivity (NCGS) [7]. CD is a genetic, autoimmune disorder, where the ingestion of gluten triggers an immune response in genetically susceptible individuals. This signals the immune cells to target the lining of the small intestine causing atrophy of the intestinal villi leading to the annihilation of the cell lining. This reaction interferes with the absorption of nutrients and can lead to a wide range of symptoms and deficiency disorders, such as anemia and bone diseases.

Wheat allergy is a rapid immune response to several proteins in wheat that takes place within minutes or a few hours. In the individuals sensitive to wheat, T-cells trigger the release of immunoglobulin E (IgE) antibodies to attack the wheat protein and the local tissues alert the problem to other parts of the body by sending natural chemical messengers. As a response, body produces a range of symptoms, such as nausea, abdominal pain, itching, swelling of the lips and tongue, troubled breathing or anaphylaxis [5].

NCGS (also known as gluten sensitivity) involves an immune reaction to gluten, but it does not involve the production of antibodies to damage the intestine [36]. Initial symptoms from NCGS reactions can be observed as early as 48 h after ingestion of gluten and continue for long periods. The mechanism is still not well understood and the symptoms include nausea, abdominal cramps, and diarrhea.

The 20th session report of the *Codex Alimentarius* (1996) limits gluten consumption to not more than 10 mg/day for celiac patients [23]. Till now, the lifelong avoidance of gluten diet is the only observed solution for the gluten intolerants, which limits their food choices. Limited food choice is another factor that results in nutrient deficiency among these patients. The patients suffering from celiac disease especially children, who follow gluten-free diet, are associated with malnutrition with the reduced energy intake and the deficiency of minerals like calcium and iron. The limited knowledge on gluten-free alternatives and their products also limit the food choices.

Hence, it becomes very important to develop gluten-free food products with enhanced nutritive values [21].

This chapter discusses the gluten-free alternatives, their nutritional value, traditional gluten-free food products, status of research on the development of gluten-free products, and the gluten-free products commercially available.

12.2 HISTORY AND CURRENT STATUS OF GLUTEN INTOLERANCE

It is reported that the CD occurs at a frequency of 1:100 people, whereas the frequency of occurrence of wheat allergy is 0.1%. The data lacks accuracy because most of the cases are self-reported and have not undergone a clinical diagnosis [91]. The frequency of NCGS is unknown.

12.3 PATHOPHYSIOLOGY OF GLUTEN INTOLERANCE

12.3.1 SYMPTOMS

CD, wheat allergy, and gluten sensitivity show gastrointestinal symptoms, which range from chronic diarrhea to intestinal pain [79]. Extraintestinal symptoms can also occur as listed in Table 12.1. It should be noted that the symptoms of glutens sensitivity closely resemble those of CD; however, it gives negative results when tested for CD or wheat allergy.

12.3.2 MECHANISM OF ACTION

Gastrointestinal proteases are responsible for digestion of gluten proteins. Glutamines and prolamins are hydrolyzed partially in the gastrointestinal tract. This results in upregulation of zonulin peptides present in the intestine. Zonulin is a tight junction regulator, which builds gradient for efficient absorption of nutrients. It is also involved in efficient permeability in the gut. The peptides obtained from gluten are transported to lamina propria (mucosa) and are reformed into transglutaminase (TG) present in the tissues. The modification enhances their similarity to MHC-II molecules. Patients containing HLA-DQ2/DQ8 protein subunits account for toxic and immunogenic effects. In such individuals, the peptide sequence presented by HLA-DQ2/DQ8 subunits on coming in contact with gluten-specific T-cells triggers a

TABLE 12.1 Comparison of Frequency, Pathogenesis, Clinical Presentation, Biological Markers, and Treatments for Different Types of Gluten Intolerances

Parameter	Celiac Disease	Ref.	Wheat Allergy	Ref.	Gluten Sensitivity	Ref.
Frequency	1%–3%	[91]	0.1%		Unknown	[91]
Pathogenesis	Autoimmune		Allergic		Not autoimmune, not allergic	[79]
Clinical presentations/ reactions	Chronic diarrhea, weight loss, osteoporosis, anaemia, neurological disturbances		Wheat-dependent, exercise-induced anaphylaxis (WDEIA), occupational asthma, rhinitis, contact urticarial, atopic dermatitis		Symptoms usually resemble those associated with CD, along with extraintestinal symptoms like weight loss, fatigue, joint pain, muscle pain or cramps, numbness etc.	
Biological markers	TG2A, endomysium antibody (EMA), deamidated gliadin peptides antibody (DGPA)		Wheat specific IgE		No biological markers, exclusion of CD and wheat allergy is crucial.	[91]
Treatment	Exclude wheat, barley and rye (gluten-free diet)		Exclude wheat, barley and rye (gluten-free diet)		Avoid any kind of contact with wheat	[71]
Duration of treatment	Lifelong		Lifelong		Annual evaluation is recommended	[71]

Notes: *CD*, celiac disease.

prompt immune reaction in the form of both innate and adaptive responses. As a result, interferon and Interleukin-15 are produced. This whole process leads to enteropathy, inflammation in the intestine, degeneration of villi, crypt hyperplasia, and amplified penetration by intraepithelial lymphocytes [7].

12.3.3 IDENTIFIESD SOURCES OF GLUTEN IN HUMAN DIET

Wheat, barley, and rye are the major gluten-containing cereals (GCC). Among these cereals, the highest gluten content (g/100 g) is in wheat (8.92 ± 0.11) followed by barley (4.23 ± 0.08) and rye (3.08 ± 0.04). In addition to these cereals, oats may also contain gluten due to cross contamination in the fields. The gluten content in oats has been reported to be (1.29 ± 0.03 g/100 g) [80]. As per the data for worldwide production of grains (2019), wheat, barley, oats, and rye rank 2nd, 4th, 6th, and 7th with a production of 734.74, 140.6, 22.22, and 10.57 million metric tons, respectively [84]. Wheat and rye are major cereals used in bakery industry: barley is majorly used for the beer production, whereas oats have recently gained popularity as a health grain. These cereals are also potent sources of proteins and minerals (Table 12.2).

TABLE 12.2 Nutritional Composition of Gluten Containing Cereals

Parameter	Wheat	Barley	Rye	Oat
Ash (g/100 g)	1.63 ± 0.26	2.16 ± 0.75	1.11 ± 0.73	2.14 ± 0.65
Calcium (mg/100 g)	43.41 ± 3.69	55.5 ± 47.92	80	47.16 ± 17.81
Carbohydrates (g/100 g)	69.88 ± 1.66	66.73 ± 4.56	65.54 ± 9.10	63.07 ± 3.83
Crude fiber (g/100 g)	1.77 ± 0.15	9.19 ± 5.06	11.16 ± 6.80	10.93 ± 2.04
Fat (g/100 g)	2.81 ± 0.18	2.32 ± 0.83	1.90 ± 0.72	25.1 ± 20.50
Glutelins (g/100 g)	2.98 ± 0.04	1.10 ± 0.02	1.10 ± 0.02	1.01 ± 0.05
Gluten (g/100 g)	8.92 ± 0.11	4.23 ± 0.08	3.08 ± 0.04	1.29 ± 0.03
Iron (mg/100 g)	5.24 ± 0.80	4.93 ± 2.69	7 ± 2.83	4.31 ± 0.48
Prolamins (g/100 g)	5.94 ± 0.07	3.13 ± 0.06	2.53 ± 0.03	1.29 ± 0.03
Proteins (g/100 g)	13.78 ± 1.40	14.5 ± 3.74	10.65 ± 0.83	13.28 ± 2.79
Zinc (mg/100 g)	2.9	2.69 ± 0.52	4 ± 1.42	3.64 ± 0.63

Source: [6, 58, 64, 80].

Wheat is consumed by approximately one-third of the total world population [60]. However, being a rich source of gluten, the products prepared from wheat cannot be consumed by gluten-intolerant individuals. Hence, the omission of this cereal from the diet poses a big challenge to the bakery industry. This limits the availability of food types to gluten intolerants and may also result in malnutrition among them.

12.4 GLUTEN-FREE ALTERNATIVES

Many gluten-free alternatives are available in the form of staple cereals, millets, pseudocereals, and legumes. The details of GCC, their alternatives, and processed products have been indicated in Figure 12.1.

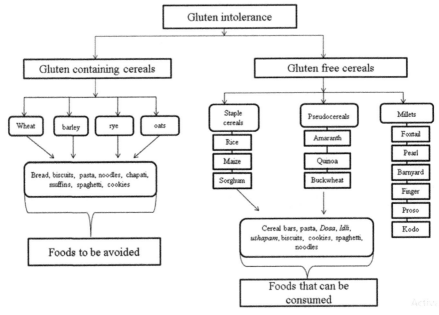

FIGURE 12.1 Schematic diagram of the gluten-containing cereals, their alternatives, and processed products.

12.4.1 STAPLE CEREALS

The major gluten-free staple cereals (GFSC) are corn (*Zea mays*), rice (*Oryza sativa*), and sorghum (*Sorghum bicolor*). Among these cereals, corn ranked first with a worldwide production of 1099.61 million metric tons followed by rice (495.87 million metric tons) and sorghum (58.4 million metric tons) with the 1st, 3rd, and 5th rank among all major cereals [84].

The carbohydrate content of these cereals is comparable to the GCC (i.e., wheat, barley, oats, and rye) (Table 12.3). However, the GFSC contain lower protein compared to GCC. The fat content of GFSC is also similar to the GCC, but the GFSC contain lower crude fiber and ash compared to GCC. The GCC have also been reported to contain a higher amount of calcium and a comparable amount of iron and zinc to GFSC.

12.4.2 PSEUDOCEREALS

Pseudocereals are nongrass family fruits or seeds, consumed in a similar fashion as cereals [19]. Commonly consumed pseudocereals are buckwheat (*Fagopyrum esculentum*), quinoa (*Chenopodium quinoa*), and amaranth (*Amaranthus spp*). The worldwide production of buckwheat and quinoa has been reported to be 3,827,748 tons and 146,735 tons, respectively [30]. However, the data for the world total amaranth production is not available. Asia is the largest producer of buckwheat with a 49% share in global production. The major buckwheat producing countries are China, Russia, Ukraine, and France. America is the largest producer of quinoa with Peru, Bolivia, and Ecuador as its largest producer. Amaranth originated in America and is presently grown in Mexico, Central America, and Nepal [27]. These cereals are rich source of macro- and micronutrients (Table 12.3).

The amount of calcium content in amaranth (180.07 ± 0.06) is higher in comparison with major cereals like rice (0.12 ± 0.07), maize (12.06 ± 1.08), and sorghum (35.23 ± 7.42). Amount of the iron content in quinoa (7.88 ± 4.75) is also significantly higher than the major cereals. Moreover, these crops do not contain gluten as a source of protein [3]. Apart from being gluten-free crops, these are also rich in bioactive compounds, which confer health benefits to the human body. The high antioxidant capacity of these crops due to the presence of phenolic compounds, which is effective in treating chronic diseases [87]. Therefore these pseudocereals can be utilized as a substitute to GCC to develop a variety of nutrient-dense food products for the individuals suffering from gluten intolerance [4].

12.4.3 MILLETS

Millets are cereal grasses belonging to the family *Poaceae*. The worldwide millet production was estimated to be 27.8 million tons in 2018 [67] with India being the largest producer with a global market share of 41.01%, followed

TABLE 12.3   Alternatives of Gluten-Containing Cereals Along With Their Nutritional Composition

Parameter	Carbohydrates	Proteins	Fat	Crude Fiber	Ash	Calcium	Iron	Zinc
	(g/100 g)						(mg/100 g)	
Major Cereals								
Maize	73.35 ± 6.77	10.01 ± 0.84	3.27 ± 1.70	3.02 ± 1.40	1.39 ± 0.18	12.06 ± 1.08	3.67 ± 1.88	2.43 ± 1.57
Rice	82.86 ± 7.53	4.99 ± 1.38	1.90 ± 1.03	1.63 ± 0.42	0.99 ± 0.42	0.12 ± 0.07	1.25 ± 0.78	0.5
Sorghum	72.97 ± 2.25	10.82 ± 2.45	3.23 ± 1.60	1.97 ± 0.35	1.70 ± 0.66	35.23 ± 7.42	5.29 ± 1.28	3.01 ± 0.89
Pseudocereals								
Amaranth	62.7 ± 10.75	16.05 ± 0.97	7.25 ± 1.79	3.57 ± 0.76	3.05 ± 0.30	180.07 ± 8.76	9.2	2.17 ± 1.14
Buckwheat	69.29 ± 4.94	11.38 ± 2.51	2.10 ± 0.78	2.39 ± 1.76	2.10 ± 0.08	39.8 ± 18.1	3.84 ± 1.75	2.13 ± 0.44
Quinoa	66.51 ± 5.06	14.51 ± 0.59	5.17 ± 0.09	1.97 ± 0.40	2.68 ± 0.57	46.25 ± 26.7	7.88 ± 4.75	2.35 ± 1.1
Millets								
Barnyard millet	56.88 ± 6.86	10.76 ± 1.11	3.53 ± 1.19	12.8 ± 2.4	4.30 ± 0.26	18.33 ± 6.0	17.47 ± 2.0	57.45 ± 1.9
Finger millet	71.52 ± 3.59	7.44 ± 0.87	1.43 ± 0.12	3.60	2.63 ± 0.06	348 ± 3.5	4.27 ± 0.6	36.6 ± 3.7
Foxtail millet	67.30 ± 5.70	11.34 ± 0.91	3.33 ± 0.76	8.23 ± 1.66	3.37 ± 0.12	31 ± 11	3.5 ± 1.2	60.6
Kodo millet	63.82 ± 7.94	9.94 ± 1.6	3.03 ± 1.03	8.20 ± 2.3	2.83 ± 0.40	32.33 ± 4.6	3.17 ± 1.3	32.7 ± 2.2
Pearl millet	69.10 ± 1.52	11.4 ± 0.8	4.87 ± 0.12	2.0 ± 0.55	2.13 ± 0.21	35 ± 8.9	10.3 ± 7.0	—
Proso millet	67.09 ± 4.79	11.74 ± 0.86	3.09 ± 1.18	8.47 ± 3.4	2.73 ± 0.72	10 ± 3.5	2.2 ± 1.2	—
Legumes								
Bengal gram	60.3 ± 0.7	17.33 ± 0.49	5.15 ± 0.21	3.95 ± 0.87	3.15 ± 0.23	205 ± 4.24	4.8 ± 0.28	—
Black gram	56.1 ± 3.65	26.35 ± 2.25	3.76 ± 2.12	0.9	5.23 ± 1.76	130 ± 21.50	3.6 ± 0.52	—
Chickpea	63.93 ± 4.76	23.78 ± 3.72	7.02 ± 1.23	3.97 ± 0.83	3.61 ± 0.59	99 ± 60.3	6.75 ± 5.26	3.68 ± 2.03

TABLE 12.3 *(Continued)*

Parameter	Carbohydrates	Proteins	Fat	Crude Fiber	Ash	Calcium	Iron	Zinc
			(g/100 g)				(mg/100 g)	
Kidney beans	58.86 ± 8.55	22.22 ± 5.30	3.41 ± 1.5	3.83 ± 0.57	3.79 ± 0.43	100.3 ± 65.60	8.11 ± 4.96	7.33 ± 9.57
Lentil	62.70 ± 8.36	27.68 ± 2.29	1.55 ± 0.97	5.42 ± 1.23	2.96 ± 0.31	66.18 ± 33.59	6.92 ± 2.54	3.30 ± 1.40
Peas	60.41 ± 2.30	23.81 ± 3.45	0.99 ± 0.94	2.89 ± 1.60	2.89 ± 0.21	82.53 ± 0.37	3.79 ± 0.23	3.97 ± 0.18
Soybean	19.8 ± 0.12	42.9 ± 0.20	19.8 ± 0.30	5.1 ± 0.30	5.6 ± 0.08	245 ± 50.17	9.6 ± 0.10	—
Alternative Sources								
Green banana flour	75.75 ± 10.26	2 ± 0.71	0.53 ± 0.04	—	2.13 ± 0.54	—	—	—
Tapioca starch	99	0.1	0.1	0.2	—	—	—	—
Unripe plantain flour	39.14 ± 0.212	3.15 ± 0.042	0.21 ± 0.028	3.74 ± 0.27	2.30 ± 0.09	—	—	—

Source: [3, 13, 58, 66, 94].

by Nigeria (16.7%), Niger (9.89%), and China (5.4%) [92]. Majorly grown millets throughout the world are as follows:

1. Barnyard millet (*Echinochloa frumentacea*);
2. Finger millet (*Elusine coracana*);
3. Foxtail millet (*Setaria italica*);
4. Kodo millet (*Paspalum scrobiculatum*);
5. Pearl millet (*Pennisetum glaucum*);
6. Proso millet (*Panicum miliaceum*).

The largest produced millet is pearl millet (50%), followed by the proso- and foxtail millet (30%), finger millet (10%), barnyard millet, and kodo millet (10%).

Nutritionally, millets are considered comparable to major staple cereals. The total amount of carbohydrate in millets ranges from 56.88 to 72.97 g/100 g [58]. The highest amount of carbohydrate in the range of 66.5 to 72 g/100 g has been stated in finger millet while the lowest carbohydrate content in the range of 49–65 g/100 g is reported in the barnyard millet [83]. A large variation has been reported in the protein content of millets; however, the average amount of protein in the millets is reported in the range of 8 to 12 g/100 g. The proteins of all the millets are gluten free and are considered superior to major grains particularly in the finger millet, where 44.7% of the total protein constitutes of the essential amino acids [63].

Finger millet protein is high-quality protein and is rich in essential amino acids (such as methionine, valine, and lysine). The protein content of proso millet is analogous to wheat and has a higher amount of essential amino acids (viz. leucine, isoleucine) in comparison to wheat. The lipid content of millets ranges from 1.43 to 6 g/100 g and is comparable to wheat. The lowest amount of lipid has been reported in finger millet (1.43 g/100 g). Among millets, the highest amount of lipid is found to be in pearl millet (6 g/100 g) [70, 82]. Millets are rich in fiber with crude fiber content in the range of 2 to 12.8 g/100 g.

The minimum crude fiber content is found in pearl millet (2 g/100 g), whereas the maximum content in millets is possessed by barnyard millet (12.8 g/100 g) [78]. In addition to the macronutrients, millets are also a rich source of micronutrients, like minerals and vitamins. The millets have higher ash content (2.13–4.30 g/100 g) compared to GCC (1.11–2.16 g/100 g). Finger millets possess highest amount of calcium (348 mg/100 g), barnyard millet and pearl millet have the highest content of iron (17.47 mg/100 g and 10.3 mg/100 g, respectively) and foxtail millet and barnyard millet have the highest content of zinc (60.6 mg/100 g and 57.45 mg/100 g, respectively). The content of these minerals is several-folds higher than those in GCC. Millets are also exceptionally high in vitamins B_1, B_3, and B_9. The highest content of thiamine (0.60 mg/100 g) is reported in foxtail millet, B_1 (4.20 mg/100 g) in barnyard millet, folate content (39.49 µg/100 g) in kodo millet [20, 58].

12.4.4 LEGUMES

Many legumes have high nutritive value and contain on an average about the double proteins in comparison to cereals [52]. The legume proteins are gluten free and hence make these a healthy choice for gluten intolerant individuals. The nutritional profiles of some of the commonly consumed legumes have been presented in Table 12.3. The carbohydrate content of legumes varies from 19.8 g/100 g to 63.93 g/100 g (Table 12.3). The highest content of carbohydrate has been reported in chickpea, whereas the lowest carbohydrate content is reported in soybean. Soybean is the richest source of proteins among legumes with a protein content of 42.9 mg/100 g. Its protein provides all the eight essential amino acids; however, the content of methionine and tryptophan is low.

The lipids content varies a lot among legumes and the highest lipid content is found in soybean (19.8 mg/100 g), whereas the lowest amount of lipid has been found in peas (0.99 mg/100 g). Legumes are also a good source of crude fiber with average crude fiber content in the range of 2.89 to 5.42 mg/100 g, except black gram that contains 0.9 mg of crude fiber per 100 g. The ash content of legumes ranges from 2.96 to 5.6 mg/100 g and are rich source of essential minerals like calcium, iron, and zinc. Being an inexpensive source of nutrients, legumes can perform an important role in improving the nutritional status of malnourished persons [68]. These are also good source of health-protective compounds, such as inositol phosphates and phenolics.

12.4.5 OTHER SOURCES

In addition to above-mentioned sources, the use of other food crops (like tapioca, green banana flour, and unripe plantain flour) has been also reported by various researchers for developing of gluten-free food products.

Tapioca is among the important tropical root crops [11, 16] found in the regions between the Tropic of Cancer and Tropic of Capricorn. It is rich in carbohydrates and is one of the major sources of dietary energy. Tapioca starch is extracted from the root and has been used in various food preparations. According to the United Nation Food and Agriculture Organization, tapioca ranks fourth after rice, maize, and wheat in developing countries. Tapioca starch is an exclusive source of gluten-free energy and is reported to contain only traces of lipid, protein, and phosphorus [11]. Green banana flour is currently being researched for the development of gluten-free products. The mature green banana pulp is a fair source of starch, protein, ash, and soluble fiber [17].

12.5 TRADITIONAL GLUTEN-FREE FOOD PREPARATIONS

12.5.1 DHOKLA

Dhokla is traditionally made using rice and Bengal gram dal and is a popular fermented food in Western India, particularly Gujarat. Rice and Bengal gram *dal* are grounded coarsely and a batter is prepared by adding sufficient amount of potable water. The batter is allowed to ferment overnight at low temperature. On completion of fermentation, curd along with other ingredients and spices are mixed with the batter. The batter is steamed for 15–10 min in a pie-dish, and cut into cubes, and is seasoned [12]. *Dhokla* has a tangy flavor with a slight sweet taste. It can be eaten as a breakfast, main course, side dish, or a snack. The traditional ingredients of *dhokla*, that is, rice and Bengal gram dal can also be replaced with other cereals, legumes, millets, chickpea [61].

In some places, semolina is also used to replace rice in *dhokla*. Semolina is prepared from wheat and the *dhokla* prepared using semolina can cause the symptoms of gluten intolerance. Food researchers have also developed the recipes, where the rice has been replaced with millets and Bengal gram dal is replaced with lentils. Roopa et al. [77] standardized a recipe for *dhokla* preparation, where besan and rice semolina were substituted with kodo rice (10%–50%) and lentil (10%).

Pathak et al. [74] developed an instant powder of *dhokla* mix using foxtail and barnyard millet (55%), legumes (35%), and fenugreek seeds (10%). Gupta et al. [35] optimized the recipe for the preparation of *dhokla* with the incorporation of finger millet flour (20%), pearl millet flour (20%), and drumstick leaves (10%).

12.5.2 DOSA

Dosa is made using rice and pulse. The two ingredients in the proportions ranging from 6:1 to 10:1 are soaked, grounded, and the batter is fermented.

The ingredients for *dosa* preparation are not limited to just rice and black gram instead pure rice or a mixture of rice, wheat, sorghum, maize, or millets can also be used. Black-gram dal may also be substituted by other pulses like green gram or Bengal gram.

The ingredients are finely grounded for the *dosa* batter preparation. Gluten-sensitive individuals are advised to avoid the wheat-based *dosa* and special precautions must be taken in case of *rawa dosa*, where wheat semolina is used as one of the ingredients. Krishnamoorthy et al. [56] formulated *dosa* dry mix using rice flour, millet mix flour and decuticuled black gram in equal proportions (33:33:33). Narayanan et al. [69] prepared the millet dosa using unpolished foxtail millet (90 g) and Black-gram dal (20 g).

12.5.3 IDLI

Idli is a popular steamed pudding in South India, prepared from a thick fermented batter of rice (*Oryza sativa*) and split dehusked black gram (*Phaseolus mungo*) dal. Traditionally, the ingredients are soaked in water for overnight followed by grinding them separately on a stone pestle and mortar. In *idli* preparation, rice is ground coarsely while the black gram dal is ground into a fine paste. The two pastes are then mixed together to form a batter, salted to taste, and allowed to ferment for 12–18 h. The fermented mixture is steamed in flat plates or perforated cups to obtain soft, spongy, tasty, and easily digestible *idlis*.

The rice and black gram *idli* can be enjoyed by the gluten intolerant individuals; however, semolina based *idli* should be avoided by such individuals. Many new variants of the *idli* have been also developed by substituting rice with millets like finger millet, pearl millet, barnyard millet, and foxtail millet [14, 35, 90]. The prepared products have been reported to have improved nutritional value in comparison to the traditional *idli*.

12.5.4 PADDU

Paddu is a soft, shallow fried product of a mixture of fermented rice and pulses, and is a common breakfast in South India. The pulse used in its preparation may be similar as of *idli* and *dosa*; alternatively, lentils can be used. It is round in shape and golden brown in appearance.

12.5.5 ROTI

Millets are mainly consumed in the form of *"roti"*: Indian flatbread. These *rotis* were mainly prepared using pearl millet (*bajra*) and finger millet (*koda*). In addition to millets, the *rotis* of gluten-free cereals like corn and sorghum are also popular in many states of India. Pearl millet *roti* (*bajre ki roti*) and sorghum *roti* (*jowar ki roti*) are very popular in Rajasthan. Finger millet *roti* (*kode ki roti/ mandal ki roti/ nchani ki roti*) is very popular in Himachal Pradesh and Uttarakhand and is still a staple in some hilly regions. Corn flour *rotti*, famously known as *"makke di rotti,"* is consumed in Punjab and bordering states. All these cereals are gluten free and provide an inexpensive healthier option to the gluten-intolerant individuals.

12.5.6 TOFU

Tofu is a product prepared by coagulating the soya milk and is also known as soya curd. Many variants of tofu, that is, firm tofu/cotton tofu, silken tofu/ soft tofu, sufu/fermented tofu are being produced. Coagulants (such as citric acid ($C_6H_8O_7$), calcium chloride ($CaCl_2$), calcium sulfate ($CaSO_4$), Nigari ($MgCl_2$), and glucono-D-lactone (GDL)) are majorly used in Japan for the preparation of tofu [26].

A soy protein gel matrix is formed during the production of the tofu, which entraps the water, lipids, and constituents present in the matrix to form curd. This curd is further pressurized to form solids. This product is a rich source of soya proteins and lipids. This also provides a healthy option to gluten intolerant people as the soybean proteins are gluten free. However, soybean itself can also act as an allergen; hence, it is recommended to conduct the tests for such allergy.

12.5.7 UTHAPAM

Uthapam, uttapam, uttappa, or *oothapam* is another healthy breakfast food made from fermented batter of common rice and Bengal gram dal. The ingredients are same as that of *idli* and *dosa*; however, *uthapam* batter is little thicker and is topped with different types of veggies like onion, tomato, capsicum, carrot, green chilli, sweet corn, and others, which make it evenmore healthier and tastier. Like *idli* and *dosa*, the basic ingredients for preparation of *uthapam* can also be replaced with semolina; however, the gluten-intolerant individuals are advised to avoid such *uthapam*.

12.6 NOVEL DEVELOPMENTS IN GLUTEN-FREE PRODUCTS

The omission of gluten-containing food from the diet is the only treatment for gluten intolerance; hence, it is very important to develop alternatives for the products that are traditionally prepared using wheat, rye, barley, and oats. Therefore, the food scientists have developed various gluten-free alternatives like gluten-free flour, bread, cookies, pasta, and spaghetti. A list of major gluten-free food preparations is shown in Table 12.4.

TABLE 12.4 Recent Developments in Gluten-Free Products

Product Name	Ingredients Used	Ref.
Amaranth bread	Popped amaranth flour (60%); raw amaranth flour (40%)	[21]
Amaranth cookies	Whole popped amaranth flour (18.75%), rice flour (6.25%), ground pop-amaranth (12.50%), maize starch (4.20%)	[21]
Biscuits with rice and chick pea flour	Rice flour (78.13 g), chickpea flour (21.17 g), xanthan gum (1.5%), ammonium bicarbonate (0.93 g), sodium bicarbonate (0.46 g), salt (0.75 g), sugar (18.12 g)	[9]
Cake	Rice flour (11%), corn flour (11%), soya oil (5%), pasteurized milk (18%), xanthum gum (0.4%)	[75]
Carrot cake	Gluten-free flour (170 g), xanthan gum (7 g), finely shredded carrot (220 g), coconut chips (50 g), raisins (80 g), and walnuts (45 g)	[33]

TABLE 12.4 *(Continued)*

Product Name	Ingredients Used	Ref.
Cereal bar	Quinoa (40%), brown rice (35%), flaxseed (10%), almonds (5%), raisins (5%), figs (5%), and honey (50%)	[54]
Gluten-free bread	Chestnut flour (30%), rice flour (70%)	[22]
Okara cookies	Okara flour (30%), manioc flour/tapioca flour, sugar, inulin, eggs, vanilla essence, baking powder, and blend of xantic (xanthan) guar and guar gum	[72]
Rice pasta with legumes flour	Rice flour, chickpea flour (10 g/100 g), yellow pea (20 g/100 g), and lentil flour (30 g/100 g)	[10]
Rice plus maize cookies	Maize flour (50%), rice flour (50%)	[76]

12.6.1 BREAD

Bread is traditionally prepared using wheat, as the wheat gluten has the unique property of forming the viscoelastic and cohesive dough. This elasticity is responsible for retention of Carbon dioxide at the time of fermentation and provides porous structure and appearance to bread. Gluten-free bread has been prepared from tapioca, amaranth, quinoa, buckwheat, rice, potato starch, and a mixture of chestnut flour and rice flour.

Turkut et al. [81] prepared gluten-free bread using quinoa, buckwheat, rice flour, and potato starch. They prepared the bread consisting of rice flour (25%), potato starch (25%), buckwheat flour (50%) along with the ingredients like salt, sugar, yeast, oil, xanthan gum, and water (2%, 3%, 3%, 6%, 0.5%, and 87%, respectively on the flour mixture basis) was taken as control. In the test study, the buckwheat was replaced with quinoa at the rate of 12.5%, 25%, 37.5%, and 50%. The rheological study of the samples revealed that quinoa flour made the bread batter more elastic and improved the structure of the bread similar to gluten. Incorporation of quinoa up to 25% reduced instrumental hardness significantly. It also increased hardness and chewiness of the crumb. The sensorial analysis of the prepared breads revealed that the control sample had the lowest scores while the gluten-free bread prepared using quinoa flour (25%) scored highest for flavor and overall acceptability.

Tapioca is one of the naturally occurring gluten-free ingredients, which requires phytochemicals as proteins and hydrocolloids for the development of viscoelasticity that provides the structure and helps in retention of gas. Milde et al. [65] prepared bread from a mixture of tapioca starch (80%) and corn flour (20%) with the base ingredients like yeast, salt, sugar, and water. In this study, the level of vegetable fat (10–30 g), whole hen egg (1–2 units), and soybean flour (0–50 g) were optimized per 500 g of Tapioca starch and corn flour blend. It was found that the bread prepared using 30 g of vegetable fat, one unit of hen egg, and 50 g of soybean flour was most acceptable. The prepared bread was readily accepted by habitual wheat bread consumers (84% overall acceptability) and celiac patients (100% acceptability).

De le Barca et al. [21] formulated the gluten-free bread using popped and raw amaranth flours. The best formulation for bread preparation was reported to be popped (60%–70%) and raw (30%–40%) amaranth flour. The baked product with this composition had homogeneous crumb and greater specific volume (3.5 mL/g). The bread prepared using 60:40 of popped: raw amaranth flour had 73.30% carbohydrates, 17.65% proteins, 5.20% fat, and 3.85% ash on a dry basis. This formulation also had better sensorial characteristics.

Demirkesen et al. [22] optimized the recipe for preparation of the gluten-free bread making use of chestnut flour and rice flour. Chestnut and rice flours in different proportions ranging from 0:100, 10:90, to 100/0 were mixed. Influence of hydrocolloid blend (xanthan–locust bean gum, xanthan–guar gum blend) and emulsifier diacetyl tartaric acid ester of mono- and diglycerides on the rheological and quality characteristics of dough formulations breads respectively was analyzed. The 30:70 blend (chestnut: rice) with xanthan–guar blend and emulsifier was found to possess the best traits including hardness, specific volume, and sensory scores. A chestnut flour level of above 30% was reported to deteriorate the quality of bread as it produced a lower volume, hard texture, and was found to be darker in color.

12.6.2   BISCUITS

Benkadri et al. [9] studied the effect of xanthan gum (0, 0.5%, 1%, and 1.5%) on biscuits prepared using a blend of rice (78.13 g) and chickpea flour (21.87 g) and compared them with wheat cookies. The other ingredients, such as sugar (18.12 g), hydrogenated vegetable fat (13.36 g), ammonium bicarbonate (0.93 g), sodium bicarbonate (0.46 g), salt (0.75 g), and water (33 g) were added according to the standard recipe. The rheological evaluation showed that the dough prepared using the composite of rice and chickpea flour was sticky in comparison to the control. The addition of gum significantly improved the dough handling during kneading. The textural analysis showed that the hardness and elasticity of dough got increased and a decrease in the springiness, cohesiveness, and adhesiveness was found with increase in gum concentration. The physicochemical evaluation showed that there was an improvement in the thickness and the specific volume of biscuits. The sensory scores for the prepared biscuits were at par with the control biscuits with no significant changes.

12.6.3   CAKE

Preichardt et al. [75] analyzed the effects of xanthan gum on the acceptability of gluten-free cake. In this study, the cake was prepared using a gluten-free cake batter by using rice flour (11%), corn flour (11%), refined sugar (31%), soya oil (5%), pasteurized milk (18%), baking powder (2%), and eggs. Effect of xanthan gum @ 0.2%, 0.3%, and 0.4% was studied on the chemical, physical, and sensorial parameters of cake prepared using this recipe. In this study, two control samples were used: (1) cake without xanthan gum (2) cake with wheat flour and without incorporation of xanthan gum. The incorporation of xanthan gum resulted in altered qualitative properties, such as increased specific volume. The improved texture was observed with a decrease in firmness and late staling. Cakes with incorporated xanthan gum had superior and uniform internal structure. The sensory analysis revealed that 56% of respondents preferred the cake prepared using 0.4% of xanthan gum. A total of 44% preferred the cake prepared using 0.3% of xanthan gum. The average preference test score grades for these formulas were 7.7 and 7.1 as given by the consumers.

Gambus et al. [33] conducted a study on the impact of amaranth flour on the corn flour and potato starch-based sponge cakes. The standard gluten-free sponge cake was prepared using eggs (150 g), sucrose (120 g), gluten-free baking powder (5 g), and gluten-free flour (60:60 corn flour: potato starch). Effect of amaranth flour was studied by substituting parts of corn flour and potato starch with amaranth flour in the ratios (20:40:60) and (0:60:60). When 50% of corn starch was replaced with amaranth flour in cake batter, the end product was superior in quality and quantity. This led to a 40% increase in protein and three-times increase in dietary fiber when compared to corn flour only sponge cake. However, the sensory scores were maximum when a blend of corn flour and potato starch (60:60) or a blend of potato starch and amaranth (60:60) was used. The use of a blend of three cereals, that is, corn flour, potato starch, and amaranth in the ratios 20:40:60 resulted in the least sensory scores for the sponge cake.

In the same study by Gambus et al. [33], the effect of linseed meal in a carrot and coconut-based cake was also studied. The control carrot cake was prepared using 170 g of dough along with other ingredients like xanthan gum (7 g), gluten-free baking powder (9 g), cinnamon (7 g), ginger (1 g), salt (6 g), eggs (100 g), sucrose (250 g), milk (150 g), oil (60 g), vanilin (5 g), finely shredded carrot (220 g), coconut chips (50 g), raisins (80 g), and finely ground walnuts (45 g); and 110 g of the corn flour was replaced with linseed meal in the test carrot cake and rest of the ingredients were same as that of control. The incorporation of linseed cake in place of corn flour resulted in the increased protein, fat, and fiber content without affecting the sensorial properties of the product as the linseed meal-based carrot cake exhibited the sensorial properties similar to control. The control coconut cake was prepared using rice paste 250 g along with other ingredients like margarine (250 g), sucrose powder (185 g), gluten-free baking powder (10 g), shredded coconut (100 g), and vaniline (5 g). In the coconut cake, 125 g of the rice flour was replaced with the linseed meal and it resulted in increased protein, fat, and dietary fiber content. An increase was also reported in the content of amino acids, fatty acids, and minerals with the addition of amaranth and linseed meal.

12.6.4 COOKIES

Ostermann-Porcel et al. [72] prepared cookies with the incorporation of okara flour (@ 15%, 30%, 50%) in manioc flour/tapioca flour. The other ingredients like sugar, margarine, eggs, vanilla essence, baking powder, and blend of xantic guar and guar gum were added as per the standard recipe. The physicochemical analysis revealed that incorporation of okara flour in the cookies increased the protein, fat, and fiber content but decreased the carbohydrates and ash content. The rheological study revealed that the addition of okara flour reduced the gas retention capacity as the fibers present in the okara flour may interfere with the matrix structure that decreased the gas retention capacity in the dough. Textural evaluation showed a decrease in spread ratio and height, whereas the hardness increased with increasing amount of okara flour. On sensory evaluation, it was revealed that the cookies prepared using 30% of okara flour were highly acceptable while the incorporation of 50% okara flour had negative effects on the color, aroma, taste, and overall acceptability of the cookies.

Altindag et al. [2] optimized the recipe for gluten-free cookies using blend of buckwheat flour with rice and corn flour in varying proportions. The effect of the enzyme TG was also studied on the quality of cookies. In this study, the cookies prepared using buckwheat (100%) were used as control and three other mixes of the flour, that is, buckwheat: rice (1:1), buckwheat: corn (1:1), and buckwheat: corn: rice flour (1:0.5:0.5) were used for the preparation of cookies. The cookies were also prepared by adding 0.002% of TG in this mix. The physicochemical analysis of the prepared cookies revealed that the control cookies (cookies prepared from 100% buckwheat) had the highest protein and ash, whereas the buckwheat-corn cookies had high lipid and crude fiber amount. The lowest protein content was also in buckwheat-corn cookies, which resulted in the lowest moisture content in the cookies prepared using this mix.

In the above study, the addition of transglutaminase significantly increased the moisture content in cookies. The enhancement in moisture content on the addition of TG was believed to be because of the deamidation of glutamine resulting in the production of glutamic acid residues. These residues are negatively charged and are known to enhance the capacity of proteins to bind with water. The study also revealed that the control cookies had the maximum thickness, hardness, and fracturability among all the cookies. The maximum width was obtained in the cookies prepared using a mix of buckwheat, corn, and rice and the maximum spread ratio was obtained in cookies prepared from the mix of buckwheat and corn. There was a nonsignificant impact of TG on width, thickness, and spread ratio of cookies; however, the addition of TG significantly reduced the hardness and increased the fracturability of all the prepared cookies.

Chauhan et al. [13] investigated formulation of the cookies with raw and germinated amaranth grain flour and compared them with the wheat-based cookies. The cookies were prepared using flour (100 g), sodium bicarbonate (1 g), salt (1 g), sugar (40 g), shortening (50 g), skim milk powder (20 g), and water (20 mL). The physicochemical analysis of the prepared cookies revealed that the highest spread ratio was shown in the decreasing trend in raw amaranth flour cookies, germinated amaranth flour, and wheat flour cookies, respectively. The highest amount of antioxidant activity (21.43 g/100 g) and dietary fiber (13.97 g/100 g) was shown in the germinated amaranth cookies in comparison to the raw amaranth and wheat flour cookies. The cookies prepared using raw amaranth flour showed harder cookies than the germinated amaranth flour and wheat flour. The change in the hardness could be the result of the degradation in the structure of protein and starch brought due to germination. The cookies prepared by using germinated flour from amaranth possessed the maximum overall scores for the sensory points than raw amaranth and wheat flour cookies.

De-Simas et al. [24] conducted a study to formulate cookies using king palm (*Archontophoenix Alexander*) flour with rice flour and corn starch. A mix of rice flour (70%) and corn starch (30%) was taken as control and it was further replaced with the king palm flour at the concentration of 0% to 30% (w/w). The addition of king palm flour in the cookies increased the lipid and moisture content and decreased the crude protein content. It also decreased the carbohydrate contents and an increased the minerals like calcium, magnesium, potassium, iron, zinc, and manganese. The hardness of the cookies was also amplified with enhanced concentration of king palm flour. A consumer-based sensory analysis by the celiac patients revealed that the cookies fortified with 10% and 20% of the king palm flour were the most acceptable.

Sharma et al. [81] prepared the gluten-free cookies using blends of nongerminated and germinated minor millets using foxtail, barnyard, and kodo millets. Mixture of foxtail, barnyard, and kodo millet flour consisted of ratio with 80:15:5, 70:20:10, 60:25:15, 50:30:20, 40:35:25, and 35:35:30, respectively. The physicochemical analysis of the flours indicated that the carbohydrates, fat, ash was decreased and protein and crude fiber were significantly increased on germination. The millet-based cookies had more thickness in comparison to the control cookies made from wheat. Cookies prepared from the blend of raw millet flours had a greater spread ratio as compared to germinated millet flour and control cookies. The hardness value of the cookies made with the raw millet flour was greater than the cookies prepared from the germinated millet flours. Germination results in the structural degradation of starch and proteins which might be responsible for the soft texture of cookies prepared from germinated flour. The prepared cookies were also evaluated for sensory characteristics on 9-point hedonic scale. Among all the treatments, the cookies made by the incorporation of raw and germinated foxtail, barnyard, and kodo millets in the ratio of 70:20:10 had the highest acceptability. However, germinated samples had more acceptability among the germinated and nongerminated samples.

12.6.5   CEREAL BARS

Kaur et al. [54] prepared a cereal bar using quinoa as a main ingredient. Four cereal blends were prepared using quinoa, brown rice, and flaxseed in the ratios 35:25:25, 35:30:20, 40: 30: 15, 40:35:10 with the addition of 5% almonds, raisins, and dried figs. The sensory evaluation revealed that the bar prepared with 25% of flaxseed was bitter in taste. Further, the optimization of honey (40%–60%) was done in the prepared formulations of the flour. Honey was used as a binding agent and sweetener. The textural evaluation showed that significant changes were observed in the texture with the addition of honey. The bars were brittle when 40% of honey was used, whereas the bars prepared using 60% honey were soft, sticky, and sweet in taste. The bar prepared using 40% quinoa, 35% brown rice, 10% flaxseed, 5% of almonds–raisins–dried figs, and 50% honey was highly acceptable and was a good source of protein (10.50%), fat (2.89%), and minerals (1.34%).

12.6.6   PASTA

Bouasla et al. [10] developed gluten-free precooked rice pasta enriched with legume-flour. The pasta was prepared from rice flour (control) and rice flour with blend of yellow pea flour, chickpea flour and red lentil flour in concentrations of 10 g/100 g, 20 g/100 g, and 30 g/100 g. The protein, ash, and fiber contents were improved on the incorporation of legume flour. However, the addition of the legumes weakened the network of starch and a significant cooking loss was observed. Textural analysis revealed that the hardness value was higher for control pasta as compared to legumes fortified pasta. The sensory evaluation revealed that all the pasta preparations were acceptable.

Ferreira et al. [31] prepared pasta using different flours like sorghum, rice, corn flours with potato starch. The ingredients used for pasta preparation were gluten-free flour mix using sorghum (40%–60%), 15%–30% of rice and/or 10%–20% of corn flour, and 10%–40% of potato starch, eggs (50 g), soybean oil (4 mL), and water. In total 15 blends of sorghum, rice, and corn flour were prepared in this study. The flour mixture was blended in extrusion equipment and dough was prepared. The dough was cut into spaghetti strips and dehydrated in an oven at 50 °C for 60 min. It was followed by rising the temperature to 60 °C and dehydrating the pasta for 30 more minutes. The samples prepared by using higher concentrations of sorghum were rejected as they produced bitter pasta. Among all the prepared pasta samples, samples with a sorghum/ rice/ potato ratio of 5:2.5:2.5, 4:2:4, 4:3:3 were selected on the basis of attributes like odor, residual bitterness, stickiness, grittiness, and overall quality. The Chemical analysis and cooking evaluation of the selected pasta treatments revealed that the mix containing sorghum, rice, and potato flours (40:20:40) had the highest quality; as it had the best yield, density, and lowest loss of solids.

12.6.7   SPAGHETTI

Flores Silva et al. [32] developed gluten-free spaghetti using chickpea flour, unripe plantain, and maize flours. Durum wheat pasta made from semolina was used as control, whereas eight formulations were prepared using different percentages of chickpea flour (70%, 65%, 60%, and 50%), unripe plantain flour (30%, 25%, 20%, and 15%), and white maize flour (20%, 15%, 10%, and 0%). To these formulations, 0.5% of carboxymethyl cellulose was added as a binder to replace gluten. The physicochemical evaluation of the prepared samples showed that the protein content was increased from 43.6% to 60.3% on the addition of chickpea and plantain flours. The fat content was also increased in spaghetti as maize is a good source of fat. Gluten-free spaghetti also had the more cooking loss as compared to the control sample. However, the loss of solids in the gluten-free pasta was in the range of 10.04%–10.91%, which is considered to be an acceptable range as a cooking loss of 12% is considered as the maximum acceptable limit for good quality pasta. The result of this study indicates the potential for the formulation of gluten-free spaghetti using the mixtures of chickpea, unripe plantain, and maize flours.

12.7   GLUTEN-FREE ALCOHOLIC BEVERAGES

Alcoholic beverages have been consumed traditionally throughout the world. In addition to the traditional beverages, some well-established alcoholic beverages like rum, beer, whiskey, wine, vodka, and brandy are also consumed:

1. Rum is a product prepared from molasses [25],
2. Beer is prepared from barley [93],
3. Whiskey is prepared from premium grains of wheat [1],
4. Wine is prepared from grapes and other fruits and brandy are the distilled form of wine [88].

Gluten intolerant individuals need to avoid the beverages prepared using the barley and wheat. However, the beers based on rice and millets are gluten free and can be consumed by such individuals. *Sake* is a rice beer produced and consumed traditionally in Japan [85], *Lugdi* is indigenous rice-based alcoholic drink of Himachal Pradesh [57], *Sur* is an alcoholic drink prepared using finger millet produced in Himachal Pradesh [59], *kodo ka jannanr* is a traditional finger millet-based alcoholic beverage used in Northeast India (Tamang). In addition to the gluten-free beer, recipes have been also standardized by the food researchers. Khandelwal et al. [55] investigated the suitability of the finger millet and pearl millet for developing low-alcoholic beverages with good acceptability. The process for the production of beer from a blend of finger millet, barnyard millet, and paddy was optimized by Bano et al. [8]. A millet-based gluten-free whiskey known as Koval single barrel millet whiskey is also produced by Koval Breweries Ltd., Chicago, IL, United States. The type of millet used for this preparation is not revealed by the company [62].

12.8   COMMERCIALLY AVAILABLE GLUTEN-FREE ALTERNATIVES

A list was prepared (Table 12.5) for the gluten-free products available on the online amazon store [37–51]. The online store is full of the food items ranging from the flour of brown rice, buckwheat, quinoa, amaranth to the processed products like gluten-free multigrain energy bar, spaghetti, cake rusks, pancakes, biscuits, and chips. Along with gluten-free claim, they also provide additional benefits like free from cholesterol, suitable for diabetes, rich source of proteins, vital minerals, and vitamins.

12.9   SUMMARY

The Avoidance of the GCC (wheat, rye, barley, and oats) and their products is a recommended remedy for the gluten-intolerant persons. The quantity of gluten consumed by the celiac patients should not exceed 10 mg/day. The

TABLE 12.5 Gluten-Free Products Available in the Market

Gluten-Free Product	Main Ingredients	Nutritional Composition	Special Features
Amaranth flour	Amaranth	Energy (420 calories), total fat (10 g), protein (14 g), total carbohydrates (68 g), dietary fiber (8 g), sodium (40 mg), calcium (78 mg), iron (3.6 mg)	Rich in lysine and good source of minerals like potassium, calcium, manganese, iron, and phosphorus
Brown rice flour	Brown rice	Total energy (383 kcal), total carbohydrates (82.5 g), dietary fiber (17 g), proteins (11.5 g), sodium (24 mg), total fat (1 g)	Suitable for diabetes, cardiac health, celiac disease
Coconut cookies	Rice flour, tapioca, gram flour, coconut	Energy (528 kcal), total fat (50 g), carbohydrates (18 g), dietary fiber (2 g), protein (1 g), sodium (852 mg)	Eggless
Gulab jamun	Brown rice flour, sorghum flour, tapioca (*sabudana*) flour, rice flour	Energy (751.3 kcal), total fat (20.95 g), dietary fiber (1.38 g), sodium (112.98 mg), cholesterol 6.9 g), protein (7.09 g), calcium (86.49 mg), iron (1.148 mg)	Soya free
Hulled buckwheat	Natural buckwheat	Energy (384 kcal), carbohydrates (30 g), protein (4 g), fiber (10 g), fat (0.60 g), magnesium (40 g), vitamin B3 (5 g)	Rich in fiber and magnesium
Kale crisps	Kale, onion, cheese powder, dehydrated garlic	Energy (430 kcal), carbohydrates (45 g), total fat (19.9 g), protein (17.7 g), dietary fiber (7.4 g), iron (4.25 mg), (calcium 460 mg)	Cholesterol free, rich sources of proteins, calcium, folic acid, iron, vitamin A, K, and C
Keto atta biscuits	Almond, flax seeds, pumpkin seeds, sunflower seeds, watermelon seeds	Energy (437 calories), total carbohydrates (19.79 g), protein (13 g), sodium (42.5 mg), zinc (4.5 mg), magnesium (263 mg), potassium (478 mg), phosphorus (426 mg), calcium (76.3 mg)	Essential vitamin and minerals like vitamin A, C, niacin, calcium, magnesium
Multi grain energy bar	Oats, millets (saamai, foxtail), almonds, and honey	Energy (166 kcal), proteins (5.3 g), dietary fiber (3 g), total carbs (21.8%) and fat 7.14 g, and sugars (7.3 g)	High protein content and fiber
Pancake and waffle-almond flour mix	Almond flour, arrowroot, organic coconut flour, organic coconut sugar	Energy (160 cal), total carbohydrates (18 g), protein (3 g), dietary fiber (2 g), sodium (180 mg)	Free of grains, soy, corn, dairy and gums/emulsifiers

TABLE 12.5 *(Continued)*

Gluten-Free Product	Main Ingredients	Nutritional Composition	Special Features
Pasta spinach	Rice starch, spinach powder (3%)	Energy (354 kcal), protein (5.83 g), carbohydrates (82.39 g), fat (0.163 g), dietary fiber (2.77 g)	High fiber
Quinoa flour	Stone-grinded quinoa	Energy (110 kcal), protein (4 g), fats (15 g), fiber (2 g), carbohydrates (18 g)	Good source of iron, potassium, manganese, and phosphorus
Ragi chips	Ragi, rock salt, and pepper	Energy (362.75 kcal), total fat (0.932 g), total carbohydrates (80.68 g), dietary fiber (5.05 g), protein (7.91 g), sodium (1378.12 mg), calcium (363.38 mg), iron (1.29 mg)	Contains calcium, iron and helps in the prevention of anaemia
Rice spaghetti	Brown rice	Energy (200 calories), cholesterol (0 mg), sodium (4.2 mg), potassium (9.2 mg), protein (3.1 g)	Cholesterol free, gluten-free
Tapioca starch	Tapioca starch	Energy (346 kcal), fat (0.22 g), protein (11), carbohydrates (93 g), sodium (1.1 mg)	Contains low level of saturated fats, cholesterol
Wheat-free cake rusk	Bengal gram, black gram, maize, egg, rice	Energy (495.84 kcal), protein (9.90 g), total fats (21.68 g), carbohydrates (65.28 g)	Contains egg, lactose free, and soy free

Source: [amazon.com].

major gluten-free alternatives include rice, maize, sorghum. In addition to these, the gluten-free food preparations can also be developed from millets, pseudocereals, and legumes. Some exotic alternatives like green banana flour and unripe plantain flour can also commercially be used for the development of gluten-free products. The future food scientists must come with new low cost, ready-to-cook/ready-to-eat food recipes, which are within the reach of the poor segment. The development of household recipes using locally available raw materials should also be encouraged.

KEYWORDS

- celiac disease
- gluten-free foods
- gluten intolerance
- millets
- pseudocereals
- staple cereals
- wheat allergy

REFERENCES

1. Agu, R. C.; Bringhurst, T. A.; Brosnan, J. M. Production of Grain Whisky and Ethanol from Wheat, Maize and Other Cereals. *Journal of the Institute of Brewing*, **2006**, *112*, 314–323.
2. Altındag, G.; Certel, M.; Erem, F. İlknur Konak, Ü. Quality Characteristics of Gluten-Free Cookies from Buckwheat, Corn, and Rice Flour with/without Trans glutaminase. *Food Science and Technology International*. **2015**, *21*, 213–220.
3. Alvarez-Jubete, L.; Arendt, E. K.; Gallagher, E. Nutritive Value of Pseudocereals and Their Increasing Use as Functional Gluten-Free Ingredients. *Trends in Food Science and Technology*, **2010**, *21*, 106–113.
4. Alvarez-Jubete, L.; Arendt, E. K.; Gallagher, E. Nutritive Value and Chemical Composition of Pseudocereals as Gluten-Free Ingredients. *International Journal of Food Sciences and Nutrition*, **2009**, *60*, 240–257.
5. Anonymous. Celiac Disease, Non-Celiac Gluten Sensitivity or Wheat Allergy: What is the Difference? Gluten Intolerance Group; GIG Educational Bulletin, **2014** https://gluten.org/wp-ontent/uploads/2015/01/EDU_AllrgyIntlrnc_5.28.141.pdf; Accessed on May 26, 2019.

6. Arendt, E. K.; Moore, M. M. Gluten-Free Cereal-Based Products. *Bakery Products: Science and Technology*, **2006**, *2006*, 471–496.

7. Balakireva, A. V; Zamyatnin, A. A. Properties of Gluten Intolerance: Gluten Structure, Evolution, Pathogenicity and Detoxification Capabilities. *Nutrients*, **2016**, *8*, 644–650.

8. Bano, I; Gupta, K; Singh, A; Shahi, N.C. Finger Millet: A Potential Source for Production of Gluten-free Beer. *International Journal of Applied Engineering Research*, **2015**, *5* (7), 74–77.

9. Benkadri, S.; Salvador, A.; Zidoune, M. N.; Sanz, T. Gluten-Free Biscuits Based on Composite Rice–Chickpea Flour and Xanthan Gum. *Food Science and Technology International*, **2018**, *24*, 607–616.

10. Bouasla, A.; Wójtowicz, A.; Zidoune, M. N. Gluten-Free Precooked Rice Pasta Enriched with Legumes Flours: Physical Properties, Texture, Sensory Attributes and Microstructure. *LWT-Food Science and Technology*, **2017**, *75*, 569–577.

11. Breuninger, W. F.; Piyachomkwan, K.; Sriroth, K. Tapioca/Cassava Starch: Production and Use. In: *Starch*; New York: Academic Press; **2009**; pp. 541–568.

12. Chandra, A.; Singh, A. K.; Mahto, B. Processing and Value Addition of Finger Millet to Achieve Nutritional and Financial Security. *International Journal of Current Microbiology and Applied Sciences*, **2018**, *7* (Special issue), 2901–2910.

13. Chauhan, A.; Saxena, D.C.; Singh, S. Total Dietary Fiber and Antioxidant Activity of Gluten-free Cookies Made from Raw and Germinated Amaranth (*Amaranthus* Spp.) Flour. *LWT—Food Science and Technology*, **2015**, *63*, 939–945.

14. Chelliah, R.; Ramakrishnan, S. R.; Premkumar, D.; Antony, U. Accelerated Fermentation of Idli Batter Using *Eleusine coracana* and *Pennisetum glaucum*. *Journal of Food Science and Technology*, **2017**, *54* (9), 2626–2637.

15. Chinma, C. E.; Igbabul, B. D.; Omotayo, O. O. Quality Characteristics of Cookies Prepared From Unripe Plantain and Defatted Sesame Flour Blends. *American Journal of Food Technology*, **2012**, *7*, 398–408.

16. Corbishley, D. A.; Miller, W. Tapioca, Arrowroot, and Sago Starches: Production. In: *Starch: Chemistry and Technology*; online; https://www.sciencedirect.com/book/9780127462707/starch-chemistry-and-technology#book-description; **1984**; 2nd Edition; pp. 469–478.

17. Daramola, B.; Osanyinlusi, S. A. Production, Characterization and Application of Banana (Musa Spp) Flour in Whole Maize. *African Journal of Biotechnology*, **2006**, *5*, 992–995.

18. Da Mota, R. V.; Lajolo, F. M.; Cordenunsi, B. R.; Ciacco, C. Composition and Functional Properties of Banana Flour from Different Varieties. *Starch Stärke*, **2000**, *52*, 63–68.

19. Das, S. Pseudocereals: An Efficient Food Supplement. In: *Amaranthus: A Promising Crop of Future*; Singapore: Springer; **2016**; pp. 5–11.

20. Dayakar Rao, B.; Bhaskarachary, K. *Nutritional and Health Benefits of Millets*. Rajendranagar, Hyderabad: ICAR Indian Institute of Millets Research (IIMR); **2017**; p. 112.

21. De la Barca, A. M. C.; Rojas-Martínez, M. E. Gluten-Free Breads and Cookies of Raw and Popped Amaranth Flours with Attractive Technological and Nutritional Qualities. *Plant Foods for Human Nutrition*, **2010**, *65*, 241–246.

22. Demirkesen, I.; Mert, B.; Sumnu, G.; Sahin, S. Utilization of Chestnut Flour in Gluten-Free Bread Formulations. *Journal of Food Engineering*, **2010**, *101*, 329–336.

23. Denery-Papini, S.; Nicolas, Y.; Popineau, Y. Efficiency and Limitations of Immunochemical Assays for The Testing of Gluten-Free Foods. *Journal of Cereal Science*, **1999**, *30*, 121–131.

24. De Simas, K. N.; Vieira, L. D. N.; Podestá, R. Effect of King Palm (*Archontophoenix Alexandrae*) Flour Incorporation on Physicochemical and Textural Characteristics of Gluten-Free Cookies. *International Journal of Food Science & Technology*, **2009**, *44*, 531–538.

25. De Souza; Maria, D. C. A.; Vásquez, P. Characterization of Cachaça and Rum Aroma. *Journal of Agricultural and Food Chemistry*, **2006**, *54*, 485–488.

26. Dey, A.; Prasad, R.; Kaur, S.; Singh, J.; Lawang, M. D. Tofu: Technological and Nutritional Potential. *Indian Food Industry*, **2017**, 36, 8–24.

27. Directorate Agricultural Information Services. *Amaranthus-Production Guideline.* Department of Agriculture, Forestry and Fisheries; **2010**; Republic of South Korea; https://www.nda.agric.za/docs/brochures/amaranthus.pdf; Accessed on June 12, 2016.

28. Eleazu, C. O.; Okafor, P. N.; Amajor, J. Chemical Composition, Antioxidant Activity, Functional Properties and Inhibitory Action of Unripe Plantain (M. *Paradisiacae*) Flour. *African Journal of Biotechnology*, **2011**, *10*, 16937–16947.

29. Eleazu, C. O.; Okafor, P. N.; Ahamefuna, I. Total Antioxidant Capacity, Nutritional Composition and Inhibitory Activity of Unripe Plantain (*Musa Paradisiacae*) on Oxidative Stress in Alloxan Induced Diabetic Rabbits. *Pakistan Journal of Nutrition*, **2010**, *9*, 1052–1057.

30. FAOSTAT. http://www.fao.org/faostat/en/#data/QC/visualize; Accessed on June 12, **2019**.

31. Ferreira, S. M. R.; de Mello, A. P.; dos Anjos, M. D. C. R. Utilization of Sorghum, Rice, Corn Flours with Potato Starch for the Preparation of Gluten-Free Pasta. *Food chemistry*. **2016**, *191*, 147–151.

32. Flores Silva, P. C.; Berrios, J. D. J.; Pan, J.; Osorio Díaz, P.; Bello Pérez, L. A. Gluten-Free Spaghetti Made with Chickpea, Unripe Plantain and Maize Flours: Functional and Chemical Properties and Starch Digestibility. *International Journal of Food Science and Technology*, **2014**, *49*, 1985–1991.

33. Gambus, H.; Gambuś, F.; Pastuszka, D. L. Quality of Gluten-Free Supplemented Cakes and Biscuits. *International Journal of Food Sciences and Nutrition*, **2009**, *60*, 31–50.

34. Giuberti, G.; Gallo, A.; Cerioli, C.; Fortunati, P.; Masoero, F. Cooking Quality and Starch Digestibility of Gluten-Free Pasta Using New Bean Flour. *Food Chemistry*, **2015**, *175*, 43–49.

35. Gupta, A.; Singh, R.; Prasad, R.; Tripathi, J.; Verma, S. Nutritional Composition and Polyphenol Content of Food Products Enriched with Millets and Drumstick Leaves. *International Journal of Food Fermentation and Technology*, **2017**, *2017*, 337–342.

36. Haas, J.; Bellows, L.; Li, J. Gluten-Free Diet Guide. *Food and Nutrition Series/ Health.* Fact Sheet 9.375; Colorado State University & U.S. Department of Agriculture; **2014**; www.ext.colostate.edu; Accessed on May 24, 2019.

37. https://www.amazon.in/Zero-G-Gluten-Free-Brown-Rice-Flour/dp/B00EZL7XAK/ref=sr_1_1_sspa?crid=1K3WDMQFSOEBT&keywords=gluten+free+products&qid=1559318717&s=gateway&sprefix=gluten+%2Caps%2C357&sr=8-1-spons&psc=1; Zero-G Gluten-free Brown Rice Flour by Q.E.D. Wellness; **2016**; Accessed on August 2, 2019.

38. https://www.amazon.in/Yogabar-Multigrain-Variety-Energy-Bars/dp/B00STGUXWM/ref=sr_1_3?crid=1K3WDMQFSOEBT&keywords=gluten+free+products&qid=1559318717&s=gateway&sprefix=gluten+%2Caps%2C357&sr=8-3; Yogabar Multigrain Energy Bars by Sprout-life Foods Private Limited; Accessed on August 2, 2019.

39. https://www.amazon.in/Peacock-Rice-Spaghetti-200g/dp/B00I7PWRF6/ref=sr_1_2_ sspa?crid=1SSZE1KV9BAFL&keywords=peacock+brown+rice+spaghetti&qid=15 60485026&s=gateway&sprefix=brown+rice+spaghetti+%2Caps%2C797&sr=8-2-spons&psc=1; Peacock Rice Spaghetti by Peacock; Accessed on August 2, 2019.

40. https://www.amazon.in/Green-Snack-Co-Cheese-Onion/dp/B016XMACDS/ref=sr_1_ 24?crid=1Q1KDNO66W2KE&keywords=gluten+free+products+snacks&qid=15593 71595&s=gateway&sprefix=gluten+free+products%2Caps%2C753&sr=8-24; Cheese onion by Green Snack Co.; Accessed on August 2, 2019.

41. https://www.amazon.in/Wheafree-Gluten-Free-Cake-grams/dp/B01MQKHGRO/ref= sr_1_7?crid=1K3WDMQFSOEBT&keywords=gluten+free+products&qid=1559318 717&s=gateway&sprefix=gluten+%2Caps%2C357&sr=8-7; Wheat Free Gluten-Free Cake; Accessed on August 2, 2019.

42. https://www.amazon.in/Pancake-Waffle-count-Simple-Almond/dp/B0176XTVTY/ref= sr_1_9?keywords=gluten+free+pancake+mix&qid=1560492636&s=gateway&sr=8-9; Pancake Waffle Count Simple Almond; Accessed on August 2, 2019.

43. https://www.amazon.in/Sattvic-Foods-Organic-Quinoa-Flour/dp/B018KPYOVW/ ref=sr_1_3?keywords=Sattvic-Foods-Organic-Quinoa-Flour&qid=1580395277&sr=8-3; Sattvic Foods Quinoa Flour; Accessed on August 2, 2019.

44. https://www.amazon.in/Naturevibe-Botanicals-Amaranth-Non-GMO-Maintain/dp/ B07M6HZMDJ/ref=sr_1_34?keywords=gluten+free+flour&qid=1559321557&s=gate way&sr=8-34; Amaranth Non-GMO; Accessed on August 2, 2019.

45. https://www.amazon.in/Dr-Gluten-Gluten-Free-Gulab-Jamun/dp/B07J19TNRY/ref=sr _1_52?keywords=gluten+free+flour&qid=1559321557&s=gateway&sr=8-52; Gluten-Gluten-Free-Gulab-Jamun; Accessed on August 2, 2019.

46. https://www.amazon.in/Ketofy-Biscuit-Yummy-Nutritious-Biscuits/dp/ B07QBC8NCG/ref=sr_1_2_sspa?keywords=Ketofy+Biscuits&qid=158039553 3&sr=8-2-spons&psc=1&spLa=ZW5jcnlwdGVkUXVhbGlmaWVyPUExMkdCM DVEUkU5U1NTJmVuY3J5cHRlZElkPUEwNjYyMzc4SjdZRlM2SEZZNUFNJm-VuY3J5cHRlZFkSWQ9QTA3NDYxMjYyRDlSM0lFSUE0REhVJndpZGdldE5h-bWU9c3BfYXRmJmFjdGlvbj1jbGlja1JlZGlyZWN0JmRvTm90TG9nQ2xpY2s9d-HJ1ZQ==; Yummy and Nutritious Keto Biscuits; Accessed on August 2, 2019.

47. https://www.amazon.in/NutraHi-Spinach-Gluten-Free-Pasta/dp/B07F2M7P22/ref=sr _1_34?crid=1Q1KDNO66W2KE&keywords=gluten+free+products+snacks&qid=1 559371595&s=gateway&sprefix=gluten+free+products%2Caps%2C753&sr=8-34; Spinach Gluten-free Pasta by NutraHi; Accessed on August 2, 2019.

48. https://www.amazon.in/Varyas-Gluten-Free-Coconut-Cookies/dp/B01D7H6A06/ref= sxbs_sxwds-stvp?keywords=gluten+free+flour&pd_rd_i=B01D7H6A06&pd_ rd_r=b0a1f933-14b4-49b9-8722-1e5ea70e343b&pd_rd_w=CeI2y&pd_rd_ wg=TH8A7&pf_rd_p=eb1a6561-d0c5-4c2b-a80c 5a575bb007be&pf_rd_r=5ZH2BB XXQA4DZBMGXT5V&qid=1559321557&s=gateway; Varya's Gluten-free Coconut Cookies; Accessed on August 2, 2019.

49. https://www.amazon.in/Green-Canteen-Chips-Gluten-Preservatives/dp/B07GZJ8TS8/ ref=sr_1_48?crid=1Q1KDNO66W2KE&keywords=gluten+free+products+snack s&qid=1559371595&s=gateway&sprefix=gluten+free+products%2Caps%2C75 3&sr=8-48; Green Canteen Chips Gluten Preservatives; Accessed on August 2, 2019.

50. https://www.amazon.in/Jioo-Organics-Tapioca-Starch Cassava/dp/B077HSXVZ3/ref=s r_1_101?keywords=gluten+free+flour&qid=1559362512&s=gateway&sr=8–101; Jioo Organics Tapioca Starch Cassava; Accessed on August 2, 2019.

51. https://www.amazon.in/NutriBuck-Buckwheat-Groats-Kuttu-Giri-Gluten-Free/dp/ B01EWVO4UY/ref=sr_1_114?keywords=gluten+free+flour&qid=1559362512&s=gat eway&sr=8–114; NutriBuck Buckwheat Groats Kuttu Giri Gluten-Free; Accessed on August 2, 2019.

52. Kamboj, R.; Nanda, V. Proximate Composition, Nutritional Profile and Health Benefits of Legumes—A Review. *Legume Research: An International Journal*, **2018**, *41* (3), 1–8

53. Kavitha, B.; Hemalatha, G.; Kanchana, S. Physicochemical, Functional, Pasting Properties and Nutritional Composition of Selected Black Gram (*Phaseolus mungo* L.) Varieties. *Indian Journal of Science and Technology*, **2013**, *6*, 5386–5394.

54. Kaur, R.; Ahluwalia, P.; Sachdev, P. A.; Kaur, A. Development of Gluten-Free Cereal Bar for Gluten Intolerant Population by Using Quinoa as Major Ingredient. *Journal of Food Science and Technology*, **2018**, *55*, 3584–3591

55. Khandelwal, P.; Upendra, R. S.; Kavana, U.; Sahithya, S. Preparation of Blended Low Alcoholic Beverages From Under-Utilized Millets with Zero Waste Processing Methods. *International Journal of Fermented Foods*, **2012**, *2012*, 77–86.

56. Krishnamoorthy, S.; Kunjithapatham, S.; Manickam, L. Traditional Indian Breakfast *(Idli* and *Dosa)* with Enhanced Nutritional Content Using Millets. *Journal of Nutrition and Dietetics,* **2013**, *2013*, 241–246.

57. Kumar, A.; Kaur, A.; Tomer, V.; Rasane, P.; Gupta, K. Development of Nutri-cereals and Milk-Based Beverage: Process Optimization and Validation of Improved Nutritional Properties. *Journal of Food Process and Engineering*, **2019**, *2019*, 1–9.

58. Kumar, A.; Tomer, V.; Kaur, A.; Kumar, V.; Gupta, K. Millets: A Solution to Agrarian and Nutritional Challenges. *Agriculture & Food Security*, **2018**, *7*, 31–34.

59. Kumar, A; Tomer, V.; Kaur, A.; Joshi, V. K. Synbiotics: A Culinary Art to Creative Health Foods. *International Journal of Food Fermentation and* Technology, **2015**, 5 (1), 1–14.

60. Kumar, P.; Yadava, R. K.; Gollen, B.; Kumar, S.; Verma, R. K.; Yadav, S. Nutritional Contents and Medicinal Properties of Wheat: A Review. *Life Sciences and Medicine Research*, **2011**, *22*, 1–10.

61. Lohekar, A. S.; Arya, A. B. Development of Value Added Instant *Dhokla* Mix. *International Journal of Food Nutrition and Science*, **2014**, *3* (4), 78–83.

62. Mazlien. Whiskey review: Koval single barrel millet whiskey; **2015**; https://thewhis-keywash.com/whiskey-styles/american-whiskey/whiskey-review-koval-single-barrel-millet-whiskey/; Accessed on December 31, 2019.

63. Mbithi-Mwikya, S.; Ooghe, W.; Van Camp, J. Amino Acid Profile After Sprouting, Autoclaving and Lactic Acid Fermentation of Finger Millet (*Elusine coracana*) and Kidney Beans (*Phaseolus vulgaris L.). Journal of Agricultural and Food Chemistry*, **2000**, *48*, 3081–3085.

64. Menon, R.; Gonzalez, T.; Ferruzzi, M.; Jackson, E.; Winderl, D.; Watson, J. Oats: From Farm to Fork. *Advances in Food and Nutrition Research*, **2016**, *77*, 1–55.

65. Milde, L. B.; Ramallo, L. A.; Puppo, M. C. Gluten-Free Bread Based on Tapioca Starch: Texture and Sensory Studies. *Food and Bioprocess Technology*, **2012**, *5*, 888–896.

66. Mishra, M. K.; Dubey, R. K.; Rao, S. K. Nutritional Composition of Field Pea (*Pisum sativum*). *Legume Research: An International Journal*, **2010**, *33*, 146–147.

67. Mordour Intelligence. Millet market segmented by Geography-growth, trends and forecast (2019–2024). **2019**; https://www.mordorintelligence.com/industry-reports/millets-market; Accessed on June 19, 2019.

68. Mudryu, A.N.; Yu, N.; Aukema, H. M. Nutritional and Health Benefits of Pulses. *Applied Physiology Nutrition Metabolism.* **2014**, *39*, 1197–1204.

69. Narayanan, J.; Sanjeevi, V.; Rohini, U.; Trueman, P.; Viswanathan, V. Postprandial Glycemic Response of Foxtail Millet Dosa in Comparison to A Rice Dosa in Patients with Type 2 Diabetes. *Indian Journal of Medical Research*, **2016**, *2016*, 712–717.

70. Nirmala, M., Rao, M. S., & Muralikrishna, G. Carbohydrates and Their Degrading Enzymes from Native and Malted Finger Millet (Ragi, *Eleusine coracana*, Indaf-15). *Food Chemistry*, **2000**, *69*, 175–180.

71. Ortiz, C.; Valenzuela, R.; Lucero Alvarez, Y. Celiac Disease, Non-Celiac Gluten Sensitivity and Wheat Allergy: Comparison of 3 Different Diseases Triggered by the Same Food. *Revista Chilena De Pediatría.* **2017**, *88*, 417–423.

72. Ostermann-Porcel, M. V.; Quiroga-Panelo, N.; Rinaldoni, A. N.; Campderrós, M. E. Incorporation of Okara Into Gluten-Free Cookies with High Quality and Nutritional Value. *Journal of Food Quality*, **2017**, *2017*, Article ID 4071585, pp. 8; online; https://doi.org/10.1155/2017/4071585.

73. Padalino, L.; Conte, A.; Del Nobile, M. Overview on the General Approaches to Improve Gluten-Free Pasta and Bread. *Foods*, **2016**, *5*, 87–90.

74. Pathak, Srivastava, S.; Grover, S. P. Development of Food Products Based on Millets, Legumes and Fenugreek Seeds and Their Suitability in the Diabetic Diet. *International Journal of Food Science and Nutrition*, **2000**, *51* (1), 409–414.

75. Preichardt, L. D.; Vendruscolo, C. T.; Gularte, M. A.; Moreira, A. D. S. The Role of Xanthan Gum in the Quality of Gluten-free Cakes: Improved Bakery Products for Coeliac Patients. *International Journal of Food Science and Technology*, **2011**, *46*, 2591–2597.

76. Rai, S.; Kaur, A.; Singh, B. Quality Characteristics of Gluten-free Cookies Prepared from Different Flour Combinations. *Journal of Food Science and Technology*, **2014**, *51*, 785–789.

77. Roopa, S. S.; Dwivedi, H.; Rana, G. K. Development and Physical, Nutritional and Sensory Evaluation of Instant Mix (*Dosa*). *Technofame.* **2017**, *6* (1), 109–113.

78. Saleh, A. S.; Zhang, Q.; Chen, J.; Shen, Q. Millet Grains: Nutritional Quality, Processing, and Potential Health Benefits. *Comprehensive Reviews in Food Science and Food Safety*, **2013**, *12*, 281–295

79. Sapone, A.; Bai, J. C.; Ciacci, C. Spectrum of Gluten-Related Disorders: Consensus on New Nomenclature and Classification. *BMC Medicine*, **2012**, *10*, 13–20.

80. Schalk, K.; Lexhaller, B.; Koehler, P.; Scherf, K. A. Isolation and Characterization of Gluten Protein Types from Wheat, Rye, Barley and Oats For Use As Reference Materials. *PLoS One*, **2017**, *12*, Article ID: 0172819; online; DOI:10.1371/journal.pone.0172819;

81. Sharma, S.; Saxena, D. C.; Riar, C. S. Nutritional, Sensory and *In-Vitro* Antioxidant Characteristics of Gluten-Free Cookies Prepared from Flour Blends of Minor Millets. *Journal of Cereal Science*, **2016**, *72*, 153–161.

82. Singh, P.; Raghuvanshi, R. S. Finger Millet for Food and Nutritional Security. *African Journal of Food Science*, **2012**, *6*, 77–84.

83. Sood, S.; Khulbe, R. K.; Gupta, A. K.; Agrawal, P. K. Barnyard Millet: Potential Food and Feed Crop of Future. *Plant Breeding*, **2015**, *134*, 135–147.

84. Statista. Worldwide Production of Grain in 2018/19, by Type (in million metric tons). https://www.statista.com/statistics/263977/world-grain-production-by-type/; 2019; Accessed on June 10, **2019**.

85. Suzuki, K.; Asano, S.; Iijima, K.; Kitamoto, K. Sake and Beer Spoilage Lactic Acid Bacteria: A Review. *Journal of the Institute of Brewing*, **2008**, *114*, 209–223.

86. Swinkels, J. J. M. Composition and Properties of Commercial Native Starches. *Starch Stärke*, **1985**, *37*, 1–5.

87. Sytar, O.; Biel, W.; Smetanska, I.; Brestic, M. *Bioactive Compounds and their Biofunctional Properties of Different Buckwheat Germplasms for Food Processing*. In: *Buckwheat Germplasm in the World*; New York: Academic Press; **2018**; pp. 191–204.

88. Torresi, S.; Frangipane, M. T.; Anelli, G. Biotechnologies in Sparkling Wine Production. Interesting Approaches for Quality Improvement: A Review. *Food Chemistry*, **2011**, *129*, 1232–1241.

89. Turkut, G.M.; Cakmak, H.; Kumcuoglu, S.; Tavman, S. Effect of Quinoa Flour on Gluten-Free Bread Batter Rheology and Bread Quality. *Journal of Cereal Science*. **2016**, *69*, 174–181.

90. Vanithasri, J.; Kanchana, S. Studies on The Quality Evaluation of Idli Prepared from Barnyard Millet (*Echinochloa frumentacaea*). *Asian Journal of Home Science*, **2013**, 373–378.

91. Vriezinga, S. L.; Schweizer, J. J.; Koning, F.; Mearin, M. L. Coeliac Disease and Gluten-Related Disorders in Childhood. *Nature Reviews Gastroenterology & Hepatology*, **2015**, *12*, 527.

92. World-atlas. The Leading Millet Producing Countries in the World. https://www.worldatlas.com/articles/the-leading-millet-producing-countries-in-the-world.html; **2019**; Accessed on June 19, 2019.

93. Wunderlich, S.; Back, W. *Overview of Manufacturing Beer: Ingredients, Processes, and Quality Criteria*. In: *Beer in Health and Disease Prevention*; New York: Academic Press; **2009**; pp. 3–16.

94. Zandonadi, R. P.; Botelho, R. B. A.; Gandolfi, L. Green Banana Pasta: An Alternative for Gluten-Free Diets. *Journal of the Academy of Nutrition and Dietetics*, **2012**, *112*, 1068–1072.

95. Zilic, S. Wheat Gluten: Composition and Health Effects. In: *Gluten*; Walter D B (Ed.); New York, USA: *Nova Science Publishers*, Inc.; **2013**; pp. 71–86.

CHAPTER 13

CEREALS FOR PREVENTION OF DISEASE CONDITIONS FOR BETTER HEALTH

SIMRAN KAUR ARORA

ABSTRACT

Cereals provide proteins, oils, and carbohydrates (including dietary fiber) and various vitamins and minerals; include several bioactive compounds with potential disease protective mechanisms. Whole grains can protect us from obesity, type 2 diabetes, cardiovascular diseases, cancers, constipation, and related disorders. This chapter reviews and explores the potential of achieving health benefits by adopting a balance in the intake of whole-grain cereals.

13.1 INTRODUCTION

Cereals are consumed as staple food in the form of gruels, *chapattis, nan, parathas*, bread, biscuits, cakes, *idlis, dosas*, burger, pasta, pizzas, steamed rice, *biryanis*, ready-to-eat breakfast cereals, porridges, and so many others. Compared to milk, meat, fruits, and vegetables, cereals are low in price and are affordable by a large proportion of the population. The consumption of cereal grains leads to several health benefits to prevent certain disease conditions.

According to Global Health Data Exchange, there is a paradigm shift for chronic diseases in India; and currently the lifestyle diseases are on the top of the list [29]. Currently with over 1.25 billion people in India, the demand for nutrition care is emerging as part of total preventive Healthcare.

This chapter explores the potential of achieving Healthcare through consumption of whole-grain cereals for preventing risks of various disease conditions.

13.2   CEREALS: STRUCTURE AND HEALTH COMPONENTS

Cereals (rice, wheat, sorghum, oats, maize (corn), rye, millets, teff, and barley) are also referred to as grains [22, 67]. The carbohydrates, protein, and fats in cereals can supply energy to the consumers, while vitamin E, vitamin B, calcium, magnesium, zinc, and selenium play a vital role in combatting micronutrient deficiencies. Pseudocereals (such as amaranth, quinoa, chia, and buckwheat) are considered as cereals [77]; and these have better nutritional value.

All cereal grains consist of (1) the multilayered outer portion known as bran, which is rich in dietary fibers (DF); (2) the germ portion, which is rich in micronutrients and lipids; and (3) the endosperm, which is rich in starch and protein. They are healthy choices if consumed as whole grain due to the presence of bran, which is rich in antioxidants, phytochemicals, minerals, and vitamins along with DF.

According to the American Association of Cereal Chemists International, high level of bioactive compounds is present in the aleurone layer of the bran [1]. The bioactive compounds like vitamins, minerals (like zinc, selenium, and magnesium), phytochemicals, and complex carbohydrates (lignans, β-glucans, inulin, resistant starch) have been related with various health functions and potential disease protective mechanisms [47]. Lignans have been related to lowering the risk of cancer and heart diseases. Phytic acid is found to reduce the glycemic index (GI) and may help to control diabetes. It also assists to safeguard the colon against the instigation of cancer cells. The tocotrienols, saponins, and oryzanol can lower the blood cholesterol levels, whereas the phenolic compounds have been reported to possess antioxidant effects [32, 66] due to their free radical scavenging and metal-chelation activities. The γ-oryzanol in rice, alkyl resorcinol in rye, β-glucans in oats and barley and avenanthramide, avenacosides and saponins in oats [21] play vital role in maintaining human health. Recently, anthocyanins present in the colored cereal grains (black, purple, blue, pink, red, and brown) have been claimed to provide antioxidation, anticancer, glycemic and body-weight regulation, neuroprotection, retinal protection, hypolipidemia, hepatoprotection, and antiageing activities [87].

Refined cereals consist mainly of the endosperm. A high proportion of DF, vitamins, minerals, and phytochemicals located in the outermost layers and/ or skin are lost during the processing, milling, and refining [21]; for example, a large drop in phenolic compounds have been observed after the milling of maize [10]. Ktenioudaki et al. [43] reviewed milling losses of various phenolic compounds, flavonoids, carotenoids, tocopherols, and sterols during the processing of different cereals [43]. In one study, very high losses were reported in the content of DF, magnesium, zinc, selenium, and vitamin E (i.e., 58%, 83%, 79%, 92%, and 79%, respectively) after refining of whole-grain flour into white flour [74]. Although milling reduces the mineral content, yet their bioavailability is improved due to reduction in antinutrient contents [53]. The products prepared from such flours are considered good for the people suffering from micronutrient deficiencies.

Also, refined grains containing low insoluble fibers are recommended for people suffering from Crohn disease and ulcerative colitis commonly known as inflammatory bowel diseases (IBD), because there is an increased requirement for energy and protein in some patients [24]. These disorders cause chronic inflammation in the gastrointestinal tract. In the IBD, the immune system works inappropriately, causing inflammation along with symptoms like diarrhea—abdominal pain, cramping, rectal bleeding, loss in body weight, and general fatigue [12]. Normally, intake of DF is avoided in patients having intestinal structuring, but DF like germinated barley, *Plantago ovata* seeds and husk (also known as Psyllium) may be useful in the maintenance of remission in some patients with ulcerative colitis [24]. However, DF is ineffective in the maintenance of gastrointestinal microflora in Crohn disease.

13.3 HEALTH BENEFITS OF CEREALS TO REDUCE RISK OF DISEASES

Extensive research works have been conducted to establish the health benefits from whole grains. Whole grains can protect us from the occurrence of diabetes (type-2) and obesity [57]. Johnsen et al. [39] reported that whole grains and whole-grain products are linked with lowering of cause-specific and all-cause mortality in Scandinavian HELGA cohort. Meta-analysis by Aune et al. [5] provides evidence that whole-grain intake is associated with a reduced risk of cardiovascular diseases, occurrence of different kinds of cancers, and mortality from respiratory diseases, infectious diseases, diabetes, and all noncardiovascular and noncancer causes. Another study by Edwards et al. [17] demonstrated that the whole-grain consumption also supports the cognitive control. They found that the individuals with higher whole-grain intake exhibited superior selective attention in terms of behavior variability and underlying neuroelectric activity. This section discusses health benefits from the consumption of whole grain.

13.3.1   RELIEF IN CONSTIPATION AND RELATED DISORDERS

DFs are present in whole-grain cereals and it is carbohydrate polymer with a degree of polymerization not less than 3 so that these do not get hydrolyzed by the endoenzymes in the small intestine of humans [70]. It includes undigested polysaccharides, lignins, fructo-oligosaccharides, and associated plant substances. The lack of DF (such as β-glucans and inulin) in the daily diet is associated with the occurrence of constipation, hemorrhoids, heart disease, gall bladder disease, appendicitis, diverticular disease, hiatus hernia, gastroesophageal reflux disease, abnormally dilated veins, and deep vein thrombosis. Health effects of DF also depend on the degree of fermentation by microorganisms in the gut.

The DF (both soluble and insoluble) have β-1–4 covalent bonds, which cannot be cleaved by the digestive enzymes; however, some fibers get fermented by the microorganisms in the colon. Many of the soluble DF can hold water and are fermentable in the colon. Therefore a substantial quantity of bacterial mass accumulates to form soft and bulky feces because of intake of soluble fiber. This affects the transit time of feces providing relief in constipation and eases or prevents hemorrhoids [35]. Also, the undegraded or unfermented fibers add to the volume of feces. The least fermentable fibers (such as wheat products, usually bran, and methylcellulose and carboxymethylcellulose, corn bran, psyllium husk/ispaghula, or some hydrocolloids like gum acacia) show high water-holding capacity and thereby work as efficient laxatives and bring relief from constipation [13, 70]. In one such study, it was found that the consumption of whole grains relieved constipation and showed improved gut microbiota in comparison to intake of refined grains [79]. Due to ease in the defecation process, the feces get removed from the digestive system in a faster and easy way leading to reduction in exposure period of gut walls to harmful and toxic/carcinogenic compounds.

13.3.2   PREBIOTIC EFFECT

Dietary fiber and whole grains that can be metabolized by the gut microbiota have shown to selectively support the growth of probiotic microorganisms (the beneficial bacteria, e.g., lactobacilli and bifidobacteria) in the gut [57]. Oligosaccharides, β-glucan, resistant starch, and arabinoxylan can selectively stimulate the growth of probiotic microorganisms in the colon and are known as prebiotics. In the large intestine, soluble DF gets fermented by the intestinal microflora, whereas insoluble DF shows very little fermentation.

The primary effect of some soluble DF like inulin (an oligofructose) on gut ecology is to stimulate bifidobacterial growth [61].

It has been reported that *Bifidobacteria* may increase by 1 log compared to the baseline [47] by adding 15 g inulin/day to the diet for 15 days. Oats also have similar effects on the gut microbiology, due to the presence of β-glucan [63]. A 6-week research study on the intake of whole-grain rye and wheat in overweight adults also showed positive improvement in the markers representing gut health [81]. Vitaglione et al. [80] found that the whole-grain wheat consumption for 4–8 weeks resulted in a quadruple increase in serum dihydro ferulic acid and double increase in fecal ferulic acid in comparison to refined wheat consumption in a randomized controlled trial on adults with unhealthy lifestyle. After whole-grain wheat consumption, increased fecal ferulic acid was correlated with increased Bacteroidetes and Firmicutes and reduced Clostridium suggesting the role of whole grains as a rich source of DF and other bioactive compounds, which might affect the gut microbiology and health of the consumer.

13.3.3 DIABETES MANAGEMENT

Diabetes mellitus is depicted by either deficient production of insulin (type 1 diabetes) or combined resistance to insulin-secretory response or insulin action (type 2 diabetes), which brings changes in carbohydrate, protein, and lipid metabolism in the body resulting in hyperglycemia. It is regarded as the most recurrent endocrine disorder [65]. The age-standardized disability-adjusted life-years rate for the corresponding time period for diabetes has increased by 39.6% [72]. Increased intake of whole grains can help manage healthy blood glucose and insulin levels. It has been found that a higher whole-grain intake is associated with a lower risk of type 2 diabetes among middle-aged persons [44]. A viscosity created by soluble fiber slows transit time of semidigested food (chime) in the digestive tract, which causes reduction in nutrients absorption rates and thus lowers the blood concentration of nutrients along with the delay in stomach emptying. Hence, spreading out the absorption of sugar over longer time period. Soluble DF also attaches bile and cholesterol in the digestive region, preventing their recirculation and reabsorption by the human body [37].

On the basis of in-vivo digestibility, different carbohydrate rich foods may be classified with respect to their effect on postmeal glycemia [25]: (1) low GI foods (<55); (2) medium GI foods (55–69); and high GI foods (>70). Several prospective observational studies have shown that the consumption of a low-GI diet can safeguard against the generation of colon cancer, obesity, and breast cancer, whereas the consumption of diet over many years with a high glycemic load (=GI × dietary carbohydrate content) is independently linked with a high risk of developing heart disease, type 2 diabetes, and certain cancers. Low-GI foods are beneficial for enhancing the glycemic control in diabetic subjects and for the prevention and control of metabolic risk factors, such as coronary heart diseases and obesity [8]. In a study on both obese and nonobese people, lower serum glucose levels and decreased insulin production have been observed after consumption of low-GI products [30].

Among new product development scientists/managers, there is a considerable enlarged interest in reducing the GI of highly digestible starchy products by evaluating and promoting the use and consumption of various millets. The constant intake of low-GI millets promotes only a small elevation in blood glucose levels after a meal [28]. The efficiency of glucose receptors and insulin in the body is enhanced by the sufficient levels of magnesium content in millets, which helps in controlling diabetes [59]. The presence of soluble DF in millets increases the viscosity of the bowel contents affecting insulin resistance and postprandial (after-the-meal) blood sugar.

Pearl millet grown in several African and Asian countries is known to have a low GI and hence it can be used a replacement food for control of weight and to lower the risk of *diabetes mellitus* [50].

On the basis of in vitro starch digestibility, starch has been categorized into resistant starch [RS, rapidly digestible starch (RDS)] and slowly digestible starch (SDS) [14]. SDS and RDS are likely to be completely digested in the human small intestine, but at variable rates [19]. Because SDS results in a comfortable blood glucose level with time [58] and controls hunger [46], it is therefore usually linked with managing diabetes. RS is that portion of starch which remains indigestible in the small intestine of healthy human beings; but may get partially or fully fermented in large intestine generating short-chain fatty acids, like butyric acid to fight against cancer [26]. As stated by the physicochemical parameters and the source of enzyme resistance, RS has been classified as [15]:

1. RS1 (resistant starch 1 is naturally occurring starch physically entrapped in an inaccessible matrix to enzymes, found in partly milled grains, seeds, and legumes);
2. RS2 (resistant starch 2 is a native granular indigestible starch found in raw banana and raw potato);

3. RS3 (resistant starch 3 is retrograded starch formed by the process of retrogradation that takes place naturally during the process of cooking and cooling of starch, for example, in cooked and cooled *chapatti*, in cooked and cooled rice);
4. RS4 (resistant starch 4 is chemically modified starch which are commercially manufactured); and
5. RS5 (resistant starch 5 is formed due to the complex formation of amylose with lipid).

Dona et al. [14] suggested that a change in the physical nature of the suspended starch samples (substrate) determines the kinetics of digestion. Multiple nutritional reports indicated that the consumption of RS can prevent colorectal cancer, promote hypoglycemic effects, lower plasma cholesterol and triglyceride concentrations, inhibit fat accumulation, and enhanced vitamin and mineral absorptions [58]. Latest dietary guidelines recommend the intake of slowly digestible carbohydrates so that RS should be at least 14% of the total starch consumed [18]. Researchers are putting efforts to increase the content of RS to realize its potential in Healthcare. In one such recent study, Dupuis et al. [16] reported that the RS content can range from 51.71 to 77.16 g/100 g in modified starches obtained by the application of vanillic acid on potato starch.

13.3.4 WEIGHT MANAGEMENT AND PREVENTING OBESITY

Obesity is significantly correlated with high consumption of carbohydrates and contributes to a risk factor for diabetes [72] and different organ cancers (e.g., esophagus, colorectum, breast, endometrium, and kidney) [36, 40]. Persons on whole-grain diet than on the refined flour products have healthier body weights and gain less weight over time. Higher consumption of dietary fibers has been correlated with reduced body weight and fat in women [75]. Women consuming minimum one serving of whole grain had a significantly less waist circumference and mean BMI than the women with no whole-grain consumption [30].

Whole grains are rich source of DF whose intake decreases the obesity [2, 9]. Soluble DF reduces the calories provided through a diet because of the feeling of fullness thereby promoting satiety for longer duration in comparison to intake of starch and simple sugars [11, 55]. Increasing satiety may result from several factors including the intrinsic properties of dietary fiber (high viscosity, bulking, and gel formation), which can reduce the rate of gastric emptying and macronutrient absorption, and also with the effects on certain gut hormones, for example, ghrelin and cholecystokinin, which transmits the signal for satiation [84]. Also, high-fiber foods are less energy dense compared with high-fat and high-protein food. Hence, high dietary fiber in foods can displace energy to reduce body weight. In some research works, the β-glucan has impacted weight loss by increasing satiety, has influenced the absorption efficiency in small intestine, and has also lowered cholesterol [42, 62]. Increased intake of whole grain was also found to lower waist circumference and BMI in the analysis of the data from 1999–2004 NHANES containing 13,000 adults (19–50 years old) [54].

Refined grain flours are concentrated source of calories with high GI values. Another study comparing whole and refined grain consumption showed that both subcutaneous and visceral adipose tissues were directly associated with refined grain intake while negatively associated with whole-grain intake [52]. In a randomized cross-over trials, Roager et al. [60] reported that whole-grain-rich diet reduces body weight and systemic low-grade inflammation without inducing major changes of the gut microbiome [60]. In a randomized controlled trial, Vitaglione et al. [80] related polyphenols (present in the bound form to the cereal dietary fiber) in reducing inflammation in subjects who were leading unhealthy dietary and lifestyle and were suffering from overweight or obesity [80].

13.3.5   PROTECTION AGAINST CARDIOVASCULAR DISEASES

Various studies have shown that whole grains have the potential to reduce the probability of occurrence of heart disease. Soluble fibers from barley bran, oat bran, and psyllium husk have been reported to have the tendency to decrease blood lipid levels [51]. The Federal Drug Administration (1997) has approved a health claim stating about the reduction in level of blood cholesterol with the daily intake of 3 g of -glucan soluble fiber [23]. Also, due to the presence of magnesium in cereals (particularly millets), there are reports of lowering of blood pressure and heart strokes particularly in patients suffering from atherosclerosis. Furthermore, by the action vasodilation potassium in cereals and millets assists in reducing blood pressure, thereby reducing heart-related problems, and the high fiber also plays a major role in attenuating cholesterol levels. Interestingly, microflora present in the human digestive system can convert the plant lignans present in millets into animal lignans which may protect against certain cancers and heart disease [59]. A study by Heidemann et al. [34] revealed that with the planned intake of balanced diet (based on whole grains, fruits, vegetables, poultry, and fish), the incidence of cardiovascular diseases and related deaths get reduced significantly. After analyzing the diet records of 14 years of 27,000 men (age between 40 and 75 years), Jensen et al. [38] concluded that the regular consumption of whole grains (40 g/day) could lower the chances of occurrence of heart diseases by 20%. Phenolic compounds present in whole grains can provide effective means to prevent oxidation of low-density lipoprotein-cholesterol to an atherogenic form and thus helps against heart diseases [38]. The ability of free-radical scavenging by simple phenols and their derivatives like flavonoids, phenolic acids, stilbenes, tannins, lignans, and lignin can prevent and/or treat many free radical-mediated degenerative diseases, such as atherosclerosis and ischemia, inflammation, infection, diabetes, cancer, radioactive damage, and even Parkinsonism. Also, some properties of phenolics like antiplatelet aggregation and vasodilation might exert a protective effect on the heart.

13.3.6   PROTECTIVE EFFECT OF CERTAIN CANCERS

About 20% of the cases of cancer are caused by being obese, along with the high risk of malignant tumors that develop into cancers, which is influenced by the level of physical activity, diet pattern, change in weight, and body fat distribution [56]. In obese people, insulin and insulin-like growth factor-I, steroids, and adipokines are considered as four main potential factors that can cause cancer. In a recent study, it was observed that higher BMI individuals with greatest dietary potential for hyperinsulinemia or chronic inflammation (characterized by an increased level of insulin in the blood than normal) had the greatest rise in multiple myeloma [45]. Also, obesity is linked with the secretion of rich levels of proinflammatory cytokines by immune cells and endogenous growth factors. Reports from the International Agency for Research into Cancer (2002) and the World Cancer Research Fund (2007) have shown a strong link between obesity and occurrence of different types of cancers like postmenopausal breast, endometrial, esophageal colorectal, adenocarcinoma, prostate, and renal [36, 85].

In a population of 1.2 billion Indians, more than 1 million new cases of cancer are reported every year [49]. The cancer mortality rate in India is very high (68% of the annual incidence). To slow down the increase in lifestyle-associated cancers and in order to prevent obesity among the Indian population, public health initiative aims at providing better nutrition and introducing designing, structuring, and maintenance of urban places to provide green areas/parks as exercise space. One way of improving nutrition is to consume whole grains that help in reducing the risk of cancers of the stomach and colon [33]. Consumption of millet grains may also be practiced in the regular diet as millets are known as rich in compounds like tannins, phenolic acids, and phytate [73], which have been found to reduce the risk cancer of colon and breast in animal studies [31]. Incidence of breast cancer among Indian women is also high [49]. Recent research has revealed that an increase consumption of DF is the best and practical mode to control the onset of breast cancer. It is believed that eating more than 30 g of DF/day, the possibility of breast cancer may be reduced by higher than 50%. DF is also present in millets (e.g., sorghum) along with the phenolic compounds that can also help in lowering the incidence of esophageal cancer [78].

13.4   HEALTH VERSUS TASTE ATTRIBUTES

Whole foods such as whole grains with bran, the peels or skins of the fruits may serve as a rich source of several compounds that may be associated with some health benefits [68]. To simplify the understanding of whole grain for consumers, Ross et al. [64] defined a whole-grain food is one for which the product is prepared with 30% of the ingredients is whole grain and content of whole-grain ingredients is more than refined ingredients of the grain; furthermore, when a product meets these requirements only then it can be labelled as "a whole-grain food" with the display of the whole-grain stamp along with the accurate proportion of whole-grain content. However, food products which contain whole grains face challenge of sensory acceptability in terms of taste, color, mouthfeel, and texture over those which are made up of refined flours. In the developed countries most of the consumers are affluent enough to buy highly processed expensive and refined foods. These are rich sources of concentrated energy and protein but lacks the bulking part of dietary fiber which on fortification from outside may sometimes impart coarse or chewy mouthfeel [4, 69] and/or dark color [69, 41, 48]. Seeing the growing burden of noncommunicable diseases on the economy of the country, there is a need that food nutritionists/scientists shall advise people in their connect to consume more of those products that consist the wholesomeness of whole grains and less of refined foods where the healthful components of whole grains get removed during processing [20]. Food companies, however, face challenge to develop new food products with whole grains without compromising on the taste and likeliness/preference of the consumers. Various innovative food products have been developed at laboratory scale mostly by using untrained panels that balance the taste and health attributes through DF fortification of flour, meat, and dairy products [86]. In one such study, β-glucan was incorporated into milk beverages, yogurt, and low-fat ice creams to increase viscosity and increase gel-forming capacity in the aqueous systems [7]. Similarly, to provide the health benefits of DF to consumers, two different DF blends comprising of different ratios of wheat fiber, oat fiber, microcrystalline cellulose, psyllium husk, and inulin were reported to be developed to fortify yoghurt and *kheer* (an Indian rice dessert) [3, 4]. In order to utilize pearl millet, Awolu [6] had developed composite flour with 85% of pearl millet flour with kidney beans and Tigernut flour in bread production. Martins et al. [50] had reviewed multiple types of nutrient sweet and savory products that can be prepared with grains of pearl millet as the local dishes eaten in India and Africa such as gluten-free *kibbe*, biscuits, extrudates, pastas, and nondairy probiotic drinks.

13.5 RECOMMENDATIONS

India is a growing economy and it struggles to provide good health-care systems to its huge population. Though effective personal medical care can help improve the health of people and reduce mortality, the growing population with increasing incidences of lifestyle-related diseases like diabetes, heart ailments, and cancer is offering challenges to its health-care system. In terms of health-care access and quality (HAQ), India has shown improvements on the HAQ Index (stands at 145th position among 195 countries) over the period of time but still there are high variations in availability of infrastructure for medical care, provision of public access to health-care facilities, and scale-up of medical technologies [27]. In such a scenario, it will be a big relief to the government if citizens can prevent the occurrence of some diseases by adopting small changes in their food preferences and lifestyle. Keeping in view the benefits of DF in preventing several noncommunicable diseases, National Institute of Nutrition (Hyderabad, India) recommends a daily allowance of 40 g DF/day for Indians [4].

Recently in 2018, WHO has emphasized on a healthy diet comprising of fruits, vegetables, and whole grains providing at least 25 g of DF every day [82]. In 2013–2020 roadmap under global action plan of WHO stress on taking action at all levels (local to global) was given in order to attain a 25% relative reduction in premature death from cardiovascular diseases, cancer, diabetes, or chronic respiratory diseases by 2025 [83]. Also, in order to ensure normal gastrointestinal function and to prevent chronic diseases, the Dietary Guidelines for Americans (2015–2020) encourage the consumption of whole grains even among children (1.5–4 ounce equivalents for 1–18 years old, respectively) so as to develop a habit of consuming DF rich foods in them since childhood [76].

While following the above recommendations, one should keep in mind that today the world is facing the burden of double malnutrition. People who are rich and affluent consume high calorie refined foods and face the health burdens of overweight, obese, diabetes, and cardiovascular ailments that are almost absent among the consumers in the other section who are deprived of even two meals a day with dependence being entirely on coarse grains but faces the burden of energy, protein, and micronutrients deficiency diseases like kwashiorkor, marasmus, anemia, night-blindness, scurvy, *beri-beri*, pellagra, and others. For example, Stein and Qaim [71] gave an estimate that about 9 million disability life-years are lost in India due to deficiency of iron, zinc, vitamin A, and iodine in the diet, which otherwise could have added 0.8%–2.5% of India's gross domestic product. The irony is that those who are malnourished and need both energy and micronutrients in large amounts in diet, for example, protein-energy malnourished (PEM) people in Africa or in Asia, are exceptionally consuming more of whole grains (rich in antinutritional factors, like phytates and oxalates) and has low access to energy-dense foods like chocolates and cakes/pastries/burgers/pizzas and fortified refined food products (normally are either free or are with reduced content of dietary fibers and phytates/oxalates etc.), which are expensive food items than coarse millets or cereals. Looking at the benefits of both whole-grain products as well as products made with refined flour, consumer should try to strike a balance between the intakes of the two. Whole grains should be consumed more as a part of healthy diet by the people who are overweight or obese, have history of diabetic patients/heart diseases in the family or suffer from constipation, and lead a sedentary mode of lifestyle. However, intake of refined flour products by them should be only up to limited extent. On the other hand, consumers who are working actively (moderately or heavily) and those who need more protein and energy in their diet, for example, construction workers, actively growing kids, adults, and children from poverty-driven areas (who are unable to get enough food to satisfy their hunger), may opt for some energy dense refined/processed flour products (which being comparatively free from antinutritional factors like phytates and oxalates, may extend better micronutrient bioavailability) along with some fruits and vegetables in addition to the food they were already consuming, that is, whole coarse grains. This may help them in adopting appropriate nutrition (the right balance of nutrients) to combat hunger and overcoming PEM and deficiency diseases.

Mallath et al. [49] also advocates adopting some transition strategy for India that can address the problem of malnutrition and decrease the rising obesity rate as in other growing countries like Mexico and Egypt.

13.6 SUMMARY

Consuming whole grains as a part of healthy diet may help in preventing type 2 diabetes, (colon/colorectal) cancer, cardiovascular diseases, and helps in reducing weight gain, obesity, respiratory diseases, and infectious diseases. The intake of whole cereals exerts protective effects by reducing the inflammation, increasing the viscosity of chime, enhancing the response of insulin, improving blood lipid profiles, and maintaining gut health. However, the role of food products made from refined flour in overcoming several micronutrient deficiencies and/or energy deficiency is also important. Thus the consumer should have a balance in the intake of whole-grain cereals and refined flour products in line with the individual's activity level and state of health.

KEYWORDS

- cancer
- cardiovascular disease
- cereals
- constipation
- diabetes
- dietary fiber
- micronutrient deficiency
- obesity

REFERENCES

1. AACCI. *Wholegrain Comments*. **2006**; http://www.aaccnet.org/definitions/pdfs/AACC IntlWholeGrainComments.pdf.

2. Alfieri, M. A. H.; Pomerleau, J.; Grace, D. M.; Anderson, L. Fiber Intake of Normal Weight, Moderately Obese and Severely Obese Subjects. *Obesity Research,* **1995**, *3* (6), 541–547.

3. Arora, S. K.; Patel, A. A. Development of Yoghurt Rich-in Dietary Fiber and its Physicochemical Characterization. *International Journal of Basic and Applied Agricultural Research,* **2015**, *13* (2), 148–155.

4. Arora, S. K.; Patel, A. A. Effect of Fiber Blends, Total Solids, Heat Treatment, Whey Protein Concentrate and Stage of Sugar Incorporation on Dietary Fiber-Fortified Kheer. *Journal of Food Science and Technology,* **2017**, *54* (11), 3512–3520.

5. Aune, D.; Keum, N.; Giovannucci, E.; Fadnes, L. T. Wholegrain Consumption and Risk of Cardiovascular Disease, Cancer, and All Cause and Cause Specific Mortality: Systematic Review and Dose-Response Meta-Analysis of Prospective Studies. *British Medical Journal,* **2016**, *353*, i2716; online; doi: 10.1136/bmj.i2716.

6. Awolu, O. O. Optimization of the Functional Characteristics, Pasting and Rheological Properties of Pearl Millet-Based Composite Flour. *Heliyon,* **2017**, *3* (2), E-article 00240; DOI:10.1016/j.heliyon.2017.e00240;

7. Bangar, S. Effects of Oat Beta Glucan on the Stability and Textural Properties of Beta Glucan Fortified Milk Beverage. Doctoral dissertation; University of Wisconsin—Stout; **2011**; p. 219.

8. Brand-Miller J. C.; Stockmann, K.; Atkinson, F.; Petocz, P.; Denyer, G. Glycemic Index, Postprandial Glycemia, and the Shape of the Curve in Healthy Subjects: Analysis of a Database of More Than 1,000 Foods. *American Journal of Clinical Nutrition,* **2009**, *89* (1), 97–105.

9. Burkitt, D. P.; Trowell, H. C. *Refined Carbohydrate Foods and Disease.* London: Academic Press; **1975**; p. 356.

10. Butts-Wilmsmeyer, C. J.; Mumm, R. H.; Rausch, K. D. Changes in Phenolic Acid Content in Maize During Food Product Processing. *Journal of Agricultural and Food Chemistry,* **2018**, *66*, 3378–3385.

11. Connolly, M. L.; Lovegrove, J. A.; Tuohy, K. M. *In Vitro* Evaluation of the Microbiota Modulation Abilities of Different Sized Whole Oat Grain Flakes. *Anaerobe,* **2010**, *16*, 483–488.

12. Crohn's & Colitis Foundation. Diet Nutrition. **2013**; https://www.crohnscolitisfoundation. org/sites/default/files/legacy/assets/pdfs/diet-nutrition-2013.pdf; Accessed on July 9, 2019.

13. Cummings, J. H. The Effect of Dietary Fiber on Fecal Weight and Composition. In: *CRC Handbook of Dietary Fiber in Human Nutrition;* Spiller, G.A. (Ed.); Boca Raton, FL: CRC Press; **2001**; pp. 183–252.

14. *Dona,* A. C.; Pages, G.; Gilbert, R. G.; Kuchel, P. W. Digestion of Starch: In Vivo and In Vitro Kinetic Models Used to Characterize Oligosaccharide or Glucose Release. *Carbohydrate Polymers,* **2010**, *80*, 599–617.

15. Dupuis, J. H.; Liu, Q.; Yada, R. Y. Methodologies for Increasing the Resistant Starch Content of Food Starches: A Review. *Comprehensive Reviews in Food Science and Food Safety,* **2014**, *13*, 1219–1234.

16. Dupuis, J. H.; Tsao, R.; Yada, R.Y.; Liu, Q. Physicochemical Properties and In Vitro Digestibility of Potato Starch after Inclusion with Vanillic Acid. *LWT—Food Science and Technology,* **2017**, *85*, 218–224.

17. Edwards, C.; Walk, A.; Baumgartner, N. Relationship between Wholegrain Consumption and Selective Attention: A Behavioral and Neuroelectric Approach. *Journal of the Academy of Nutrition and Dietetics,* **2017,** *117,* A93-A101.
18. EFSA (EFSA Panel on Dietetic Products Nutrition and Allergies). Scientific Opinion on the Substantiation of Health Claims Related to Resistant Starch and Reduction of Post-Prandial Glycemic Responses (ID 681); Digestive Health Benefits (ID 682); and Favors A Normal Colon Metabolism (ID 783) Pursuant to Article 13. *EFSA Journal,* **2011,** *9,* 2024–2030.
19. Englyst, H. N.; Kingman, S. M.; Cummings, J. H. Classification and Measurement of Nutritionally Important Starch Fractions. *European Journal of Clinical Nutrition,* **1992,** *46,* S33–S50.
20. Fardet, A. New Approaches to Studying the Potential Health Benefits of Cereals: From Reductionism to Holism. *Cereal Foods World,* **2014,** *59,* 224–229.
21. Fardet, A. New Hypotheses for the Health-Protective Mechanisms of Whole-Grain Cereals: What Is Beyond Fiber? *Nutrition Research Reviews,* **2010,** *23,* 65–134.
22. Food and Agriculture Organization of the United Nations (FAO). *Definition and Classification of Commodities: Cereals and Cereal Products.* Available online: http://www.fao.org/es/faodef/fdef01e.htm; Accessed on July 9, 2019.
23. Food and Drug Administration, Food Labeling: Health Claims: Oats and Coronary Heart Disease. Rules and Regulations, *Federal Register,* **1997,** *62,* 3584–3601.
24. Forbes, A. E. J; Hébuterne, X.; Kłęk, S. Espen Guideline: Clinical Nutrition in Inflammatory Bowel Disease. *Clinical Nutrition,* **2017,** *36,* 321-347.
25. Foster-Powell, K., Holt, H. A. S., Brand-Miller, J. C. International Table of Glycemic Index and Glycemic Load Values. *American Journal Clinical Nutrition,* **2002,** *76,* 5–56.
26. Fuentes-Zaragoza, E.; Sánchez-Zapata, E.; Sendra, E. Resistant Starch as Prebiotic: A Review. *Starch/ Staerke,* **2011,** *63,* 406–415.
27. Fullman, N.; Yearwood, J.; Abay, S. M. Measuring Performance on the Healthcare Access and Quality Index for 195 Countries and Territories and Selected Subnational Locations: A Systematic Analysis from the Global Burden of Disease Study 2016. *Lancet,* **2018,** *391,* 2236–2271.
28. Giuberti, G.; Gallo, A. Reducing the Glycemic Index and Increasing the Slowly Digestible Starch Content in Gluten-Free Cereal-Based Foods: A Review. *International Journal of Food Science and Technology,* **2018,** *53* (1), 50–60.
29. Global Health Data Exchange, GHDx. https://vizhub.healthdata.org/gbd-compare/; Accessed on July 6, 2019.
30. Good C. K.; Holschuh, N.; Albertson, A. M.; Eldridge, A. L. Wholegrain Consumption and Body Mass Index in Adult Women: An Analysis of NHANES 1999–2000 and the USDA Pyramid Servings Database. *Journal of American College Nutrition,* **2008,** *27,* 80–87.
31. Graf, E.; Eaton, J. W. Antioxidant Functions of Phytic Acid. *Free Radical Biology and Medicine.* **1990,** *8*(1), 61–79.
32. Gry, J.; Black, L.; Eriksen, F. D.; Pilegaard, K.; Plumb, J. EuroFIR-BASIS: A Combined Composition and Biological Activity Database for Bioactive Compounds in Plant-Based Foods. *Trends in Food Science and Technology,* **2007,** *18,* 434–444.
33. Haas, P. Effectiveness of Wholegrain Consumption in the Prevention of Colorectal Cancer: Meta-Analysis of Cohort Studies. *International Journal of Food Science and Nutrition,* **2009,** *21,* 1–13.

34. Heidemann, C.; Schulze, M. B.; Franco, O. H. Dietary Patterns and Risk of Mortality from Cardiovascular Disease, Cancer, and All-Causes in a Prospective Cohort of Women. *Circulation,* **2008,** *118,* 230–237.

35. Hillemeier, C. An Overview of the Effects of Dietary Fiber on Gastrointestinal Transit. *Pediatrics,* **1995,** *96,* 997–999.

36. IARC. *Weight Control and Physical Activity.* In: *IARC Handbook of Cancer Prevention;* Vainio, H. and Bianchini, F. (Eds.); Volume 6; Lyon, France: International Agency for Research on Cancer (IARC) Press; 2002; pp. 1–315.

37. Jenkins, D. J. A.; Kendall, C. W. C.; Augustin, L. S. A. Effect of Wheat Bran on Glycemic Control and Risk Factors for Cardiovascular Disease in Type 2 Diabetes. *Diabetes Care,* **2002,** *25,* 1522–1528.

38. Jensen, M. K.; Koh-Banerjee. P.; Hu, F. B. Intakes of Wholegrains, Bran, and Germ and the Risk of Coronary Heart Disease in Men. *American Journal of Clinical Nutrition,* **2004,** *80*(6), 1492–1499.

39. Johnsen, N. F.; Frederiksen, K.; Christensen, J.; Skeie, G. Whole-Grain Products and Whole-Grain Types are Associated with Lower All-Cause and Cause-Specific Mortality in the Scandinavian HELGA Cohort. *British Journal of Nutrition,* **2015,** *114,* 608–623.

40. Key, T. J.; Schatzkin, A.; Willett, W. C.; Allen, N. E. *Diet, Nutrition and the Prevention of Cancer. Public Health Nutrition,* **2004,** *7* (1A), 187–200.

41. Kim, B. K.; Chun, Y. G.; Cho, A. R.; Park, D. G. Reduction in Fat Uptake of Doughnut by Micro-Particulated Wheat Bran. *International Journal of Food Science and Nutrition,* **2012,** *63,* 987–995.

42. Kim, H.; Behall, K. M.; Vinyard, B.; Conway, J. M. Short-term Satiety and Glycemic Response after Consumption of Wholegrains with Various Amounts of β-Glucan. *Cereal Foods World,* **2006,** *51* (1), 29.

43. Ktenioudaki, A.; Alvarez-Jubete, L.; Gallagher, E. Review of the Process-Induced Changes in the Phytochemical Content of Cereal Grains: The Bread Making Process. *Critical Reviews in Food Science and Nutrition,* **2015,** *55,* 611–619.

44. Kyrø, C.; Tjønneland, A.; Overvad, K.; Olsen, A.; Landberg, R. Higher Whole-Grain Intake Is Associated with Lower Risk of Type 2 Diabetes among Middle-Aged Men and Women: The Danish Diet, Cancer, and Health Cohort. *Journal of Nutrition,* **2018,** *148,* 1434–1444.

45. Lee, D. H.; Fung, T. T.; Tabung, F. K. Dietary Pattern and Risk of Multiple Myeloma in Two Large Prospective US Cohort Studies. *JNCI Cancer Spectr*um, **2019,** *3*(2), pkz025; online; https://academic.oup.com/jncics/article/3/2/pkz025/5480692;

46. Lehmann, U.; Robin, F. Slowly Digestible Starch Its Structure and Health Implications: A Review. *Trends in Food Science and Technology,* **2007,** *18,* 346–355.

47. Liu, R. H. Wholegrain Phytochemicals and Health. *Journal of Cereal Science,* **2007,** *46,* 207–219.

48. Makhlouf, S., Jones, S.; Ye, S. H. Effect of Selected Dietary Fiber Sources and Addition Levels on Physical and Cooking Quality Attributes of Fiber-Enhanced Pasta. *Food Quality and Safety,* **2019,** *3* (2), 117–127.

49. Mallath, M. K.; Taylor, D. G.; Badwe, R. A. Cancer Burden and Health Systems in India, Part I: Growing Burden of Cancer in India: Epidemiology and Social Context. *Lancet Oncology,* **2014;** online; http://dx.doi.org/10.1016/S1470–2045(14)70115–9.

50. Martins, A. M. D.; Pessanha, K. L. F.; Pacheco, S. Potential Use of *Pearl Millet* (*Pennisetum glaucum* (L.) R. Br.) in Brazil: Food Security, Processing, Health Benefits and Nutritional Products. *Food Research International,* **2018,** *109,* 175–186,

51. Martos, M. V., Marcos, M. C. L., L´opez, J. F. Role of Fiber in Cardiovascular Diseases: A Review. *Comprehensive Reviews in Food Science and Food Safety*, **2010**, *9*, 240–258.

52. McKeown, N. M., Troy, L. M., Jacques, P. F. Whole and Refined-Grain Intakes are Differentially Associated with Abdominal Visceral and Subcutaneous Adiposity in Healthy Adults: The Framingham Heart Study. *American Journal of Clinical Nutrition*, **2010**, *92*, 1165–1171.

53. Oghbaei, M. and Prakash, J. Effect of Fractional Milling of Wheat on Nutritional Quality of Milled Fractions. *Trends in Carbohydrate Research*, **2013**, *5*, 53–58.

54. O'Neil, C. E., Zanovec, M., Cho, S. S., and Nicklas, T. A. Wholegrain and Fiber Consumption are Associated with Lower Body Weight Measures in US adults: National Health and Nutrition Examination Survey 1999–2004. *Nutrition Research*, **2010**, *30*, 815–822.

55. Pereira, M. A., and Ludwig, D. S. Dietary Fiber and Body-Weight Regulation: Observations and Mechanisms. *Pediatric Clinics of North America*, **2001**, *48*, 969–980.

56. Pergola, G. D.; Silvestris, F. Obesity as a Major Risk Factor for Cancer. *Journal of Obesity*, **2013**, *1*, 2013.

57. Quigley, E. M. Prebiotics and Probiotics: Their Role in the Management of Gastrointestinal Disorders in Adults. *Nutrition in Clinical Practice*, **2011**, *27*, 195–200.

58. Raigond, P.; Ezekiel, R.; Raigond, B. Resistant Starch in Food: A Review. *Journal of the Science of Food and Agriculture*, **2015**, *95*, 1968–1978.

59. Rao, B. D.; Bhaskarachary, K.; Christina, G. D. A. Nutritional and Health Benefits of Millets. ICAR, Indian Institute of Millets Research, Rajendra Nagar, Hyderabad; **2017**; p. 112.

60. Roager, H. M.; Vogt, J. K.; Kristensen, M. Wholegrain-Rich Diet Reduces Body Weight and Systemic Low-Grade Inflammation without Inducing Major Changes of the Gut Microbiome: A Randomized Cross-Over Trial. *Gut*, **2017**, *68*, 83–93.

61. Roberfroid, M. Prebiotics: The Concept Revisited. *The Journal of Nutrition*, **2007**, *137*(3), 830S–837S.

62. Rondanelli, M.; Opizzi, A.; Monteferrario, F. Beta-Glucan or Rice Bran-Enriched Foods: A Comparative Crossover Clinical Trial on Lipidic Pattern in Mildly Hypercholesterolemic Men. *European Journal of Clinical Nutrition*. **2011**, *65*, 864–871.

63. Rose, D. J. Impact of Wholegrains on the Gut Microbiota: The Next Frontier for Oats? *British Journal of Nutrition*, **2014**, *112*, 44–49.

64. Ross, A.B.; van der Kamp, J.W.; King, R.; Lê, K.A. Perspective: A Definition for Whole-Grain Food Products: Recommendations from the Health grain Forum. *Advance Nutrition*, **2017**, *8*, 525–531.

65. Saleh, A. S. M.; Zhang, Q.; Chen, J. Millet Grains: Nutritional Quality, Processing, and Potential Health Benefits. *Comprehensive Reviews in Food Science and Food Safety*, **2013**, *12*, 281–295.

66. Sarwar, M. Evaluating Wheat Varieties and Genotypes for Tolerance to Feeding Damage Caused by *Tribolium castaneum* (Herbst). *Pakistan Journal of Seed Technology* **2009**, *2* (13&14), 94–100.

67. Serna-Saldivar, S.O. Cereal Grains. In: *Cereal Grains: Properties, Processing, and Nutritional Attributes*; London, UK: CRC Press; **2016**; pp. 1–40.

68. Shahidi, F. Nutraceuticals and Functional Foods: Whole Versus Processed Foods. *Trends in Food Science & Technology*, **2009**, *20*, 376–387.

69. Sohaimy, S. A. E.; Shehata, M. G.; Mehany, T.; Zeitoun, M. A. Nutritional, Physico-chemical, and Sensorial Evaluation of Flat Bread Supplemented with Quinoa Flour. *International Journal of Food Science and Nutrition*, **2019**, E-article: 4686727; doi: 10.1155/2019/4686727.

70. Spiller, G.A. *Handbook of Dietary Fiber in Human Nutrition*. Boca Raton, FL: CRC Press; **2001**; p. 736.

71. Stein, A. J.; Qaim, M. The Human and Economic Cost of Hidden Hunger. *Food Nutrition Bulletin*, **2007**, *28*, 125–134.

72. Tandon, N.; Anjana, R. M.; Mohan, V. The Increasing Burden of Diabetes and Variations among the States of India: The Global Burden of Disease Study 1990–2016. *Lancet Global Health*, **2018**, *6* (12), 1352–1362.

73. Thompson, L. U. Potential Health Benefits and Problems Associated with Antinutrients in Foods. *Food Research International Journal*, **1993**, *26*, 131–149.

74. Truswell, A. S. Cereal Grains and Coronary Heart Disease. *European Journal of Clinical Nutrition*. **2002**, *56*, 1–14.

75. Tucker, L. A.; Thomas, K. S. Increasing Total Fiber Intake Reduces Risk of Weight and Fat Gains in Women. *Journal of Nutrition*, **2009**, *139*, 576–581.

76. U.S. Department of Health and Human Services: U.S. Department of Agriculture (USDA). 2015–2020 *Dietary Guidelines for Americans*. 8th ed.; Health and Human Services Department and Agriculture Department, Washington, DC, USA; **2015**; http://health.gov/dietaryguidelines/2015/guidelines/; Accessed on June 19, 2019.

77. Van der Kamp, J. W.; Poutanen, K. The Healthgrain Definition of 'Wholegrain'. *Food & Nutrition Research*, **2014**, *58* (1), 22100.

78. Van Rensburg, S. J. Epidemiological and Dietary Evidence for a Specific Nutritional Predisposition to Esophageal Cancer. *Journal of the National Cancer Institute*, **1981**, *67*, 243–251.

79. Vanegas, S. M.; Meydani, M.; Barnett, J. B. Substituting Wholegrains For Refined Grains in a 6-Wk Randomized Trial Has a Modest Effect on Gut Microbiota and Immune and Inflammatory Markers of Healthy Adults. *American Journal of Clinical Nutrition*, **2017**, *105*, 635–650.

80. Vitaglione, P.; Mennella, I.; Ferracane, R. Whole-Grain Wheat Consumption Reduces Inflammation in a Randomized Controlled Trial on Overweight and Obese Subjects with Unhealthy Dietary and Lifestyle Behaviors: Role of Polyphenols Bound to Cereal Dietary Fiber. *American Journal of Clinical Nutrition*, **2015**, *101*, 251–261.

81. Vuholm, S.; Nielsen, D. S.; Iversen, K. N. Whole-Grain Rye and Wheat Affect Some Markers of Gut Health Without Altering the Fecal Microbiota in Healthy Overweight Adults: A 6-Week Randomized Trial. *Journal of Nutrition*, **2017**, *147*, 2067–2075.

82. WHO. A Healthy Diet Sustainably Produced: Information Sheet; **2018**; p. 8; https://apps.who.int/iris/bitstream/handle/10665/278948/WHO-NMH-NHD-18.12-eng.pdf?ua=1; Accessed on January 25, 2019.

83. WHO. *Global Action Plan for the Prevention and Control of NCDS 2013–2020*; **2013**; https://www.who.int/nmh/publications/ncd-action-plan/en; Accessed on July 2, 2019.

84. Woods, S. Signals that Influence Food Intake and Body Weight. *Physiology & Behavior*, **2005**, *86*, 709–716.

85. Wiseman, M. The Second World Cancer Research Fund/American Institute for Cancer Research Expert Report. Food, Nutrition, Physical Activity, and the Prevention of

Cancer: A Global Perspective: Nutrition Society and BAPEN Medical Symposium on 'Nutrition Support in Cancer Therapy'. *Proceedings of the Nutrition Society*, **2008**, *67* (3), 253–256.

86. Yang, Y. Y.; Ma, S.; Wang, X.; Zheng, X. Modification and Application of Dietary Fiber in Foods. *Journal of Chemistry*, **2017**, *2017*, Article I.D. 9340427, 8 pages; https://doi.org/10.1155/2017/9340427;

87. Zhu, F. Anthocyanins in *Cereals:* Composition and *Health* Effects. *Food Research International*, **2018**, *109*, 232–249.

CHAPTER 14

CEREAL-BASED LOW GLYCEMIC INDEX FOODS FOR HEALTHCARE

GURSHARAN KAUR and RITU PRIYA

ABSTRACT

This chapter highlights the benefits of low glycemic diets especially rich in wholegrains on Healthcare. Wholegrain cereals have less potential to raise the blood glucose levels compared to refined cereals. Wholegrains not only have low glycemic index but also contain various bioactive components, micronutrients, resistant starch, soluble and insoluble fibers, which play crucial role in promoting the health of an individual. Various studies have shown that the health benefits of regular consumption of a low glycemic diet rich in wholegrains play a crucial role in dietary treatment of type 2 diabetes mellitus, polycystic ovarian syndrome, heart diseases, and others.

14.1 INTRODUCTION

In our daily diet, the total consumption of carbohydrates ranges from 40% to 80%, based on energy content (FAO/WHO 1998; [8]). As per current recommendations, 55% of the total energy required by the body should come from carbohydrates. In both developed and underdeveloped countries, cereals are main sources of carbohydrates, which represent more than 50% of energy. Fresh produce, for example, root vegetables, fruits, milk, and free sugars are other imperative sources of carbohydrates [30].

The prevalence of diabetes could be attributed to an alteration in nourishment habits (junk food, preprepared foods), nutrition transition, food digestibility, sedentary lifestyle, obesity, glucose intolerance, hypertension, dyslipidemia and hindered fibrinolytic ability, and others [6, 17, 30]. It is important to understand the effects of refining and high-temperature processing on levels of blood glucose and hormone insulin, particularly in the diabetic patients.

Consumption of carbohydrate elevates the level of blood sugar or blood glucose (glycemia). Therefore, glycemic response is the effect of dietary intake on the levels of blood glucose. Numerous methods, such as glycemic index (GI), glycemic load (GL), and glycemic glucose equivalents (GGE), have been developed to find the effects of food on glycemic response.

This chapter focuses on health benefits of low GI foods (such as wholegrain cereals); the foods with low GI and GL that are vital drivers in the management of type 2 diabetes through alliteration in regular dietary patterns.

14.2   GLYCEMIC INDEX

GI measures the extent by which a simple or complex carbohydrate elevates the level of glucose in the blood after entering the bloodstream. The real GI estimation relates the standard blood glucose-increasing impact of consuming 50 g of a standard food. GI is measured after plotting a graph from the data collected after feeding the subjects with selected food for every 2 h and the increase or decrease in the blood sugar; and insulin levels are estimated as the area under the curve (AUC).

The AUC for glucose is randomly allocated the value of 100 for every single subject. In other words, GI is defined as the ratio of AUC with the intake of test food sample to AUC with an intake of standard glucose [8]. The GI is a technique of stating the effect of food or snack on the blood sugar level. After the intake of meal, the assimilable carbohydrates are directly absorbed through the blood capillaries in the small intestine and owing to this direct absorption of glycemic carbohydrates the blood sugar concentration is elevated. The insulin hormone facilitates the return of the increased blood sugar level to or below fasting levels within a short duration of time. The blood glucose response differs with the type of carbohydrates.

14.3   GLYCEMIC LOAD

GL provides information on the amount of available carbohydrate with regard to its serving size in contrast to GI, which compares the fixed amount of accessible carbohydrate in a test sample of food to a standard glucose. In addition to the glycemic quality of the carbohydrate, the basic concept of GL is to integrate the amount of glycemic carbohydrates in the serving. Hence, it can be utilized as a sign of hormonal (insulin) demand because it provides actual information about the increase in blood sugar level in a definite period of time after the dietary intake. GL offers the facts on the amount of glycemic carbohydrates in grams present in the food and their effect on raising the levels of blood sugar [34]: 1 unit of glycemic load (GL) = 1 g of glucose.

The regular diet of people consists of 60 to 180 GL units for each day. The GI of the food and its portion size are two important factors, on which GL is dependent. The GI of the food can be reduced to obtain a low GI food by eradicating maximum carbohydrates from the daily diet [25].

14.4 GLYCEMIC GLUCOSE EQUIVALENT

The GGE is an alternative method to measure the glycemic effect of food. GGE combines GI, composition and quantity of food, so that it reflects the effect of a specific portion of food intake on the increase in blood glucose level. Precisely, it provides the hypothetical quantity of glucose required to increase the levels of GGE to that increased by the specified weight of consumed food. Main advantage of GGE is that it provides the reliable information on the impact of consuming a definite quantity of food [28].

14.5 CLASSIFICATION OF CARBOHYDRATES BASED ON THEIR EFFECT ON BLOOD GLUCOSE

Traditionally, carbohydrates are classified as monosaccharides, disaccharides, oligosaccharides, and polysaccharides based on their chemical structure. This classification provides feeble information on the behavior of carbohydrates in the human gut, such as their ability to be assimilated in the bloodstream and their impact on the level blood sugar. It is very imperative to completely understand the effects of this variable physiological property of carbohydrates on health. Dietary carbohydrates can be categorized on the basis of their postprandial effect on the blood sugar, such as glycemic carbohydrates or nonglycemic carbohydrates.

14.5.1 GLYCEMIC CARBOHYDRATES

Glycemic carbohydrates (such as glucose, galactose, fructose, sucrose, maltose, lactose, and starch) are those that directly enter the bloodstream after digestion absorption in the intestine to be transported as glucose to the body. Bioavailability of major glycemic carbohydrates depends on various factors. For example, disaccharides (such as sucrose, lactose, etc.) must be hydrolyzed with the help of enzymes into the monosaccharides (glucose) to get absorbed in the small intestine because cells utilize carbohydrates only in the form of glucose to produce energy.

Lower postprandial rise in blood sugar is caused by fructose and galactose because they are first metabolized by the liver into glucose and then enter the bloodstream. Starch is the main storage of polysaccharide and glycemic carbohydrate of cereal grains and is a polymer of glucose with alpha-1–4 linkages in monomer chain and alpha-1–6 glycosidic linkages at branch points in monomer chains that are hydrolyzed by carbohydrate splitting enzyme in the human gut [31].

The amount of carbohydrates in the form of starch is negligible in the bran, germ, and aleurone layer of cereal grains, whereas its content in the endosperm is maximum. Physical properties (e.g., size, shape, and crystallinity of starch granules) differ with the variety of cereal. In most of the cereal grains, approximately 1 g/100 g of carbohydrates are present in the form of monosaccharides, thus these simple sugars have negligible effect on the glycemic response.

Amylose and amylopectin are two components of starch, which are present in the ratio of 1:3. Both of these components are made up of glucose molecules: in case of amylose, the polymer chains are smaller and less branched with 100 to 10,000 glucose units; whereas in amylopectin, the polymer chain is long and highly branched with 10,000 to 100,000 glucose units. Glucose is rapidly released by amylopectin, because its highly branched structure provides free ends for the enzyme amylase to act. The other two factors for the fast action of amylase on amylopectin are the size and branches of its molecule, which reduces the gelatinization temperature and provides the highly porous structure [31].

14.5.2 NONGLYCEMIC CARBOHYDRATES

Nonglycemic carbohydrates are not digested in the alimentary canal; and without getting absorbed in the small intestine, they reach to the colon where the gut microflora ferment them. As a result of microbial action, nonglycemic carbohydrates yield short-chain fatty acids (SCFAs) [38]. Examples of nonglycemic carbohydrates are certain types of oligosaccharides, sugar alcohols, resistant starch (RS), and nonstarch polysaccharides (dietary fiber).

14.5.2.1 *DIGESTIBLE AND RESISTANT STARCHES*

The ratio of amylopectin and amylose varies with varieties and cultivars of a crop. Higher amylose content slows down the delivery of glucose to the bloodstream and digestion of starch may be incomplete [4]. Due to incomplete digestion, undigested fragments of starch molecules pass unabsorbed from the small intestine to the colon and here they are converted into "RS" (type of dietary fiber) [9, 21].

Starch crystallinity also affects the rate of digestion. because crystalline bonding inhibits the permeation of enzymes amylases, which in turn hinder the digestion rate. Retrogradation is a process in which some of the water which is present between the polymer chains of starch is eradicated by frequent heating and cooling of starchy food, which results in the formation of a crystalline network in the starch. Retrograded starch also becomes a type of RS, because it also resists digestion.

14.5.3 **DIETARY FIBER**

Almost all carbohydrates found in the bran and germ, and the endosperm cell walls are termed as nonglycemic carbohydrates. As these pass through the small intestine without digestion and absorption, they exert positive physiological effects due to their fermentation in the colon by microorganism [10, 19]. Cellulose is the chief dietary fiber present in plant-based foods. Cellulose is composed of monomer units of glucose, which are joined by β-linkages, but human enzymes are unable to hydrolyze it because they lack cellulase enzyme. Therefore, it is a key constituent of bran of cereals and efficiently improves the overall gut health by aiding gut motility, increasing laxation and stool weight because of its slight fermentability.

Pentosans are the polymers that create a "mortar-like mixture" in between the cell-walls. At present, these are characterized as arabinoxylans, β-glucan, pectins, fructans, and a variety of oligosaccharides and almost all are fermentable and capable to form SCFAs [1]. Few among them are viscous polymers that form a viscous layer in the small intestine like β-glucan present in barley and oats; possibly they will reduce the rate of absorption by entrapping numerous dietary constituents. Owing to these mechanisms, dietary fiber helps to maintain healthy serum cholesterol and blood glucose levels [22]. There are many benefits of dietary fiber [16]:

- It reduces the postprandial glycemia, the blood pressure, total and low-density lipoprotein (LDL) cholesterol levels, transit time, and weight.
- It helps in increasing satiety, SCFA production by colonic fermentation, and laxation.

14.6 FOOD CLASSIFICATION ON THE BASIS OF GLYCEMIC INDEX

Foods can also be categorized based on their GI, such as [12] high GI (>70), medium GI (56–69), and low GI (<55), few examples are mentioned in Table 14.1. In addition to the GI, the quantity of carbohydrates should also be considered. Hence, GL of diet is based on the carbohydrate content measured in grams (g), multiplied by the food's GI, and divided by 100. And the GL are categorized as high GL (>20), medium GL (11–19), and low GL (<10).

TABLE 14.1 Classification of Foods Based on Their Glycemic Indices

Glycemic Index	Fruits	Vegetables	Cereals	Misc.
Low	Apples, pears, oranges, bananas, cherries, grapefruit, plums, prunes, tomatoes.	Beans, peas, cabbage, lettuce, carrots, broccoli, spinach, mushrooms, onions.	Bread: whole wheat, multigrain, rye, oatmeal. Wholegrain crackers, cooked pasta, oatmeal, brown rice.	Dairy products: low-fat yoghurt, whole/low-fat/ skim milk eggs, meats.
Medium	Mangoes, pineapples, apricots, sultanas, melon.	Potatoes (boiled, new, tinned), beetroot.	Wholegrain bread, oat bran, muesli, hamburger buns, white rice. Popcorn, bars, honey.	-
High	Watermelon	Mashed potatoes, parsnips.	White bread, bagels, Coco pops, rice cakes, cornflakes, puffed wheat, steamed white rice. Chips/french-fries, jellybeans	-

14.7 FACTORS AFFECTING GLYCEMIC INDEX

GI of a food depends on various factors, such as the variety of starch, the gelatinization of the starch, the type of sugar, the type of fiber, the presence of fat or protein in a mixed meal, and the processing method of a food.

14.7.1 SUGAR

The sugar has a profound effect on the GI (Table 14.2). For instance, sucrose (table sugar) has a lower GI than the glucose because sucrose molecule is made up of glucose and fructose; and this fructose is first metabolized by liver into glucose then enters the bloodstream and thus results in a very slight rise in blood sugar concentration. For example, GI of glucose is 100, the GI of sucrose is 68; maltose is 105; saccharose is 75; fructose is 30. With the addition of sugar into the food, the GI of the meal can be lowered, but, unexpectedly, this fact does not stand true in various research works [42].

14.7.2 STARCH

14.7.2.1 CHEMICAL COMPOSITION OF STARCH

Species and variety of starch affect the contents of amylose and amylopectin. Foods with high amylopectin content have high GI in contrast to foods that are rich in amylose or they are modified to raise its content have lower GI; because amylose molecules are more resistant to digestion than amylopectin molecules (Table 14.2). For example, long-grain basmati rice has more amylose, due to which its GI values are lower compared to regular or short-grain rice. Similarly, consumption of waxy potatoes results in low blood sugar and insulin concentration and have GIs in the moderate range, whereas amylose rich varieties of potatoes have GI in the higher range. Therefore, amylose to amylopectin ratio affects the GI of food: with the increase in the ratio, the GI of the food is decreased. For example, the GI is 87 for instant rice compared to 58 for basmati rice [7].

14.7.2.2 SOURCE OF STARCH

The properties of starch granule differ with the variety of a food. Size of starch granule, structural design, and permeability of lipid-protein membrane depend on the source of starch. Lipid-protein membrane encloses the intrinsic raw starch architecture and comprises of amorphous and crystalline regions. The amorphous regions are easily available for enzyme attack compared to crystalline regions. Moreover, the integral granule of starch obstructs the rate of digestion by hindering the entrance of water molecule through the lipid-protein membrane and results in delaying the action of amylase. Consequently, rates of digestibility of starch depend on its source, owing to the architecture of the starch and availability of their constituents in the granule. Hence, the GI of native intact starches is lower than that of processed starches.

14.7.3 COOKING METHOD

The GI of foods also depends on the amount of water and cooking time and temperature, because the starch granules to different extents are inflated after the cooking, temperature, and quantity of water. Cooking in the presence of water leads to swelling of starch granules, which are gelatinized; such gelatinized starches are more easily broken down by amylases compared to ungelatinized starches (Table 14.2). Therefore, food with gelatinized starch increases the level of blood sugar and has higher GI value compared to the foods containing less or ungelatinized starch granules, such as al dente spaghetti, brown rice, and oatmeal.

TABLE 14.2 Main Food Factors Affecting the GI of Foods and Meals

Factors	GI	Factors	GI
Acid addition	Decreases	Fructose-to-glucose ratio	Decreases
Addition of table sugar	Decreases	Gelatinization of starch	Increases
Amylopectin	Increases	Grinding	Increases
Amylose	Decreases	Heat treatment	Decreases
Dietary fiber	Decreases	Retrogradation of starch	Decreases
Fermentation	Decreases	Ripening	Increases

Notes: *GI*, glycemic index.

Moreover, the GI also depends on the processing and storage method, for example, the GI of potato is varied when it is processed and stored by different methods. The GI of a boiled potato is more if eaten hot and GI of same boiled potato gets lowered with cooling due to the retrogradation of the starch molecules (which occur on cooling) and lead to the formation of crystalline regions; this crystallized starch is less accessible to the enzyme compared to the ungelatinized starch. Hence, it has a lower GI.

It has been observed in many studies that instantiation elevates the levels of GI in foods, such as ready-to-eat cereals, rice, or puddings. Moreover, popping and puffing treatment of grains makes the starch more easily available in popcorn or puffed rice, to amylase by opening the structure and therefore the GI of such snacks increases. Consequently, the GI of wholegrain cereals depends on various factors, such as the processing and storage methods along with the variety and constituents of grains.

14.7.4   FIBER

Fiber protects the native cereal grain from a harsh environment and similarly in the digestive tract it protects the starch from the prompt action of digestive enzymes and thus hinders the digestion process. Most of the dietary fibers are not digestible by humans, for example, cellulose is major dietary fiber and is indigestible because it contains glucose molecule linked by β-linkages, which are not able to be broken by amylase. They reach to large intestine as such, where they get fermented by gut microflora and produce SCFAs. It was hypothesized by some scientists that viscous and soluble fibers (legumes and oats) form a viscous mass of food in the digestive tract due to which activity of enzyme reduces, thus affecting the insulin response; and as a result, GI gets lowered [22, 42]. Oats and barley contain β-glucans (soluble viscous polymers), which help in reducing the rate of absorption by entrapping numerous dietary constituents and therefore lowers the GI of foods (Table 14.2).

14.7.5   PROCESSING METHODS

Decreasing the particle size by processes like crushing, rolling, or milling of cereals increases their capacity to absorb more water easily; thus, they easily get attacked by digestive enzymes. Generally, during processing or refining, most of the fiber is removed from the grains and thus carbohydrate is no longer able to resist the digestive enzymes and as a result they rapidly get metabolized into glucose. Owing to this reason, coarse particles of stone grounded flour have lower GI values than finely grounded flours with high GI [7]. Moreover, the starch is intact in wholegrains and is not available readily to enzyme and thus become a RS1 in the digestive tract and lowers the GI of food. In the processed food, the structure of food affects the GI, for example, bread has porous food structures and starch is easily available to enzymes (amylases); hence bread has high GI as it delivers glucose to bloodstream quite quickly. In contrast, dense food structures, such as pasta hinders enzyme penetration and delivers glucose slowly (i.e., slowly available glucose), and thus such products have lower GI.

14.7.6 FATS AND PROTEIN

High amount of fats and protein in food can prolong the food retention in the stomach, and in turn, hamper the rate of digestion and absorption of carbohydrates. If two types of meals contain the same amount and type of carbohydrate but different amount of fat, it will have different GI values. For example, one with high-fat content will have lower GI while the other with low-fat content will have higher GI.

Food rich in fat-content obstructs the elevation of blood glucose by reducing the rate of digestion of carbohydrates in the intestine and results in lower GI in contrast to similar foods without fat. That is why boiled potatoes have higher GI values compared to French fries owing to frying, which raises the fat content of fries and lowers the GI. For example, the GI of potato chips is 57 compared to 75 for French fries and 85 for baked potato. However, that does not make potato chips a better choice than the more nutritious baked potato [37]. When protein and fats are eaten alone, they have slight impact on glycemic and insulin concentration. However, when combined with carbohydrate diet they have significant effect on blood sugar. Moreover, insulin secretion is stimulated by the presence of protein in the meal, which further decreases the blood glucose levels. Therefore, extent of digestion and absorption of carbohydrate and GI can be altered by the incorporation of fats and protein in the food.

14.7.7 ACIDITY

Acidity of a food also delays the emptying of stomach, which in turn reduces the rate at which the carbohydrates are digested, hence lowers the GI of a food (Table 14.2). Therefore, raising the acidity in a meal can reduce its GI and the blood sugar response. Hence, the incorporation of organic acid into the food will lower the GI of the food. Lower acidity in food can be achieved either by combining acidic food products, such as vinegar, lemon juice, and sourdough bread [37] or by fermenting food, which produces acid, that is, curd or sourdough process. All enzymes have maximum activity at their optimum pH; acids might hinder the enzyme activity by changing the pH and due to this rate of digestion of carbohydrates is affected to some extent and as a result the acid helps in reducing the GI of food.

14.8 MIXED MEALS

Incorporation of other foods or its components also has marked effect on the GI of a meal. By consuming dairy product or protein-rich product along with carbohydrates, the overall GI of the meal can be reduced. This reduction in GI is contributed by the fact that protein upsurges the stimulation of hormone insulin, which helps in clearing the glucose from blood. Moreover, fats also lower the GI of meals due to various reasons, such as by improving the food retention time in stomach in the first step, by decreasing the rate of penetration of enzyme amylase into the mixture of food in the intestine in the second step, and by having minimal impact on stimulation of insulin hormone in the last step.

14.9 MEASUREMENTS OF GLYCEMIC INDEX, GLYCEMIC LOAD, AND GLYCEMIC GLUCOSE EQUIVALENTS

14.9.1 MEASUREMENT OF GLYCEMIC INDEX

A standard method for measurement of GI has been established by the FAO/ WHO Joint Expert Consultation Carbohydrates in Human Nutrition [8]. Significant features of GI analysis comprise:

1. Select at least 10 healthy persons and estimate GI in the fasting state. Some steps should be taken to standardize the circumstances, such as

    *   Subjects are not allowed to eat and drink for at least 10 h prior to testing.
    *   Water consumption is allowed.
    *   Alcohol consumption is not permitted a day before testing.
    *   Volunteers should not exercise heavily on the test day.

2. Measurement of GI is taken every 2-h beginning at the consumption of test sample.
3. Sample should contain 50 g of digestible carbohydrates; if serving size is large then sample should contain at least 25 g of glycemic carbohydrate. However, sample with less 10 g of available carbohydrate is not appropriate for GI estimation.
4. GI of the sample is estimated on different days in every volunteer 2–3 times per day.

5. GI value to food should not be assigned by calculating GI value of individual constituents of food item; it should only be assigned after testing in vivo.

14.9.2   MEASUREMENT OF GLYCEMIC LOAD

GL is calculated from the GI value of a food as follows:

$$\text{Glycemic load} = \frac{[\text{Glycemic index} \times \text{digestible carbohydrate (g)}]}{100} \quad (14.1)$$

where

Digestible carbohydrate =Amount of total carbohydrate – Amount of total dietary fiber.

GL can be used to manage postprandial glycaemia by directing both the selection of carbohydrate-containing food and its serving size. Conversely, estimation of the serving size is practically a restraint, because portion size varies with the size of the packet and in case of foods, such as rice or pasta, where people choose their serving size.

14.9.3   MEASUREMENT OF GLYCEMIC GLUCOSE EQUIVALENT

GGE is determined as follows:

$$\frac{\text{GGE}}{g} = \left\{ \left[\frac{\text{IUAC of food}}{\text{IUAC of glucose}} \right] \times \left[\frac{\text{weight of glucose}}{\text{weight of food}} \right] \right\} \times 1\,g \quad (14.2)$$

where (IAUC of food/IAUC of glucose) = 1 of equi-glycemic.

14.10   LOW GLYCEMIC DIET

GI is a relative classification of carbohydrate in foods based on their effects on blood glucose levels. Foods with a low GI (GI $\leq$ 55) discharge glucose into the circulation at a slow sustainable rate and have recognized benefits for Healthcare. GI of some foods is mentioned in Table 14.3. Dietary advantages of low GI diet in many different settings include [11]:

1. Help in curbing hunger and provide satiety for long durations, eventually leads to loss in weight and better management of weight.

2. In case of diabetic patients and patients with insulin resistance, it helps in regulating blood glucose levels and enhances one's sensitivity toward insulin.
3. Help in reducing the levels of blood lipids in diabetic patients as well as in nondiabetics who have increased blood lipids.
4. Help in the determination of the precise fuel before and after exercise.

TABLE 14.3   Glycemic Index of Selected Foods

High GI Foods	GI	Medium GI Foods	GI	Low GI Foods	GI
		Glucose: 100			
Boiled potato	78	Boiled sweet potato	63	Vegetable soup	48
Corn flakes	81	Boiled pumpkin	64	Apple	36
Instant mashed potato	87	Soft drink	59	Soy milk	34
Instant oat porridge	79	Muesli	57	Lentils	32
Rice crackers	87	Popcorns	65	Milk	39
Rice porridge	78	French fries	53	Orange	43
Watermelon	76	Wheat roti	62	Strawberry jam	49
White bread	75	Brown rice (boiled)	68	Chickpeas	28
White rice	73	Millet porridge	67	Corn tortilla	46
Whole wheat bread	74	–	–	Barley	28

Notes: *GI*, glycemic index.

14.11   HEALTH BENEFITS OF CEREAL-BASED LOW GLYCEMIC DIETS

14.11.1   OBESITY AND WEIGHT LOSS

Obesity is usually related with high blood sugar, high blood pressure, an elevated level of cholesterol or fats in blood, and insulin resistance; and owing to these disorders, the risk for type 2 diabetes mellitus, cardiovascular disease, and early death of a person increases [14]. Conventionally, for losing weight a person should follow a diet pattern, which comprises of either low amount of fats or high amount of carbohydrates, with most of the calories more than 50% should be come from carbohydrates, not more than 30% from the lipids and remaining from proteins. Conversely, current research workers showed that similar results of reducing waist circumference and maintaining weight in case of overweight and obese patients can be attained by inducing dietary restriction on carbohydrates and fats [18].

Numerous factors may elucidate the impact of cereal-based low glycemic diet on the management of body weight. Low glycemic foods, such as wholegrains, may stimulate satiation for longer durations after the dietary intake due to their bulk, low-calorie concentration, and comparatively lower palatability. Wholegrains, like barley and oats, contain β-glucan (a soluble viscous fiber), which decreases the absorption of glucose and other nutrients in the bloodstream by delaying the rate of emptying of stomach.

Low GI diet based on wholegrains aids in reducing the body weight; and this weight loss indirectly reduces the risk of high blood pressure, risk of heart diseases, stroke, and diabetes. Consumption of low GI diet results in the initiation of a prompt weight loss partially due to restriction on food choices and partially due to low-calorie intake [13]. Moreover, low carb-diet containing high protein content also helps in this weight loss by elevating the level of β-hydroxybutyrate in bloodstream, which suppresses the craving and brings satiation. In addition to this, some of the early loss in weight of around 1–2 kg may be elucidated due to decrease of approximately 1 g/100 g of muscle mass and 5 g/100 g liver weight. Glycogen present in liver contains 3 g of water in 1 g of glycogen [41]; therefore, initial weight loss is because of declining glycogen stored in liver along with excretion of water through urination. The process of rapid initial weight loss may take approximately one to two weeks depending upon the extent of glycogen reduction; and after this period the weight loss decelerates [32].

Moreover, low GI diet produces SCFA in the large intestine, which affects the metabolism of sugar, fats, and cholesterol in numerous tissues. SCFA helps in preventing the obesity caused by diet owing to their capability to upsurge the breakdown of fatty acid in various muscle, which helps in reducing the fat storage in adipose tissue.

14.11.2   DYSLIPIDEMIA

In dyslipidemia, the level of cholesterol is elevated abnormally in the blood. High cholesterol levels mean increased levels of serum triglycerides, low levels of high-density lipoprotein (HDL-c) and prevalence of small dense low-density lipoprotein cholesterol (SD LDL-c). Mostly dyslipidemia is common in people with metabolic syndrome and diabetes, and also in those who follow lethargic lifestyle, such as unhealthy diet and lack of physical activity.

Increasing the soluble fiber through consumption of barley in a healthy diet can reduce cardiovascular risk factors. Diets rich in wholegrain foods tend to decrease LDL cholesterol, triglycerides, blood pressure, and increase HDL cholesterol. Relations have also been described between GI and both unfavorable lipid profiles and elevated inflammatory status [33]. Low-GI diets may reduce the insulin-stimulated activity of 5-hydroxy-3-methylglutaryl-CoA reductase, the rate-limiting enzyme involved in cholesterol synthesis. Dietary fiber is likely to reduce bile acid and cholesterol reabsorption from the ileum, which may further inhibit hepatic cholesterol synthesis.

14.11.3 DIABETES

Regular consumption of high GI diet will increase the blood sugar and secretion of hormone insulin and may impair the function of beta cells of the pancreas to secrete insulin hormone, which further leads to T2DM (type 2 *diabetes mellitus*) [43]. Various scientists have discovered that routine consumption of dietary fiber from cereal sources tends to have defensive mechanism in diminishing the dangers of T2DM, this type of protective mechanism was not seen in fiber from fruits and vegetables. A team of German research workers monitored around 15,000 adults for a period of more than 7 years and unveiled the fact that those people on a diet rich in dietary fiber from whole cereals tend to have approximately 30% less chances to develop T2DM in contrast to people consuming minimum amount of cereal fiber [35]. It was believed by some scholars that factors involve in reducing the risk of developing T2DM, depressing blood sugar and hormone insulin concentration are the constituents of cereal grains, such as Mg, dietary fiber, tocopherol, phytates, lectins, and polyphenols. A diet comprising of low GI and high cereal fiber foods tend to have 45%–50% less chances to develop diabetes compared to high GI and low wholegrain fiber. Besides, overweight and obese patients, who consume refined cereals or high GI foods, have 10 times more chances to develop T2DM compared to lean subjects on low GI diet [5].

14.11.4 CORONARY HEART DISEASES

Cardiovascular diseases (CVDs) are related to GI in numerous ways. Consumption of high GI foods might elevate the blood glucose and insulin concentration that further multiply the danger for heart diseases through syndrome-X. Syndrome-X, also known as metabolic syndrome or insulin resistance syndrome, is a group of disorder related to metabolism, which comprises of elevated concentration of insulin hormone in blood, low concentration of good cholesterol (HDL), increased levels of bad cholesterol (LDL), and high blood pressure. Moreover, diet on high GI foods was thought to encourage insulin and glucose concentration in blood, which in turn raises the blood pressure, abnormally high levels of blood lipids and cause atherosclerosis, all these conditions intensify the danger of cardiovascular disease [25]. The low GI diets can help in lowering the oxidative damage, thereby reducing the heart disease risk.

Research works recommend that the risk of coronary heart disease is considerably plummeted by dietary intake of wholegrain cereals. It has been seen that the maximum risk reduction measured by enriching diet with wholegrains is around 25%–30%. Many research workers showed that by cautiously following a healthy low GI diet pattern comprises of whole cereals, legumes, fresh produce, and lean meats may diminish the danger of heart disease and reduce death rate due to CVD [15]. Investigations have shown that people who consistently ingest wholegrains have approximately 28% lower chances of heart failure, whereas those who intake cereals 2–6 servings a week have 22% less chances to die due to heart failure [20].

High sugar levels in blood (referred to as hyperglycemia) is considered as a constant threat for cardiovascular diseases and even deaths. Blood sugar and lipids are affected by secretion of insulin, which triggers the glycolysis and β-oxidation, thus aid in regulating the production of pyruvic acid and synthesis of ATP. Hyperglycemia impairs the balance between free radicals and antioxidant in the body leading to oxidative stress, which leads to hypertension, atherosclerosis, and reduced endothelium-dependent blood flow and insulin resistance might get worsened.

Due to the extension of lean soft tissues and fat-storing adipose tissues, the total consumption of oxygen in the body is raised and this results in increased cardiac output. The total blood volume increases with increasing body weight in people suffering from obesity, which in turn elevates the resting cardiac output of a person, as it elevates the load on the left ventricle. Moreover, to meet the demand of increased cardiac output, the stroke volume should be increased. However, this elevated stroke volume is only obtained by increasing the rate of diastolic filling of left ventricle, which further leads to unusual hypertrophy due to the thickening of left ventricular walls with dilation. In addition, when the left ventricle becomes incapable to take any more blood overload, systolic dysfunction occurs. When the radius of left ventricular cavity dilates, its capacity to contract decreases, thus resulting in diastolic dysfunction. Hence, heart failure occurs due to both systolic and diastolic dysfunctions.

14.11.5   HYPERTENSION

The risk of hypertension in women of age 45 or above can be reduced by incorporating the wholegrain grain in their diet, because high cereal grain diet plays a crucial role in preventing hypertension, which further aids in preventing cardiovascular complications. Many research works showed that baseline systolic blood pressure ≤160 mmHg and diastolic blood pressure 80–95 mmHg can be considerably decreased by the consumption of dietary approach to stop hypertension diet, which comprises of healthy foods and wholegrains [2].

14.11.6   POLYCYSTIC OVARIAN SYNDROME

In polycystic ovarian syndrome (PCOS), the female ovaries become enlarged and have minor cysts due to the hormonal imbalance. It has been recommended to patients with PCOS to have a diet that contains more unsaturated fats, rich in dietary fiber, primarily consisting of low GI foods, such as milk, beans, soy-products, fruits, porridge, and others. Studies showed that obese and overweight women are more prone to PCOS, thus they are suggested to lose weight by changing their lifestyle through physical activity and healthy dietary intake. Mostly the PCOS patients are recommended to follow a diet that has been designed for T2DM patients. It has been seen that diet rich in dietary fiber with little or no refined cereals tends to have a positive impact on the hormone profile of females suffering from PCOS [23].

14.11.7   NONALCOHOLIC FATTY LIVER DISEASE

In nonalcoholic fatty liver disease (NAFLD), excess fat is accumulated in the liver of people who do not consume alcohol and it is a nonserious type of fatty liver. In contrast to NAFLD, nonalcoholic steatohepatitis (NASH) is a type of NAFLD that is dangerous leading to inflammation and damage in liver cells. Various research works showed that NAFLD and NASH can be treated by changing diet pattern, such as consumption of low GI diet because it helps in maintaining the levels of glucose, insulin, and fatty acids. The diet is designed according to the individual requirements, as there is no precise approach for all NAFLD and NASH patients [44]. Further research is required to elucidate the precise impact of variety of diets and their constituents on the health of such patients

14.11.8 CANCERS

Whole cereal contains certain starches and dietary fiber, which get fermented in large intestine and further aids in decreasing transit time and improve overall gut health. Moreover, they contain components with high antioxidative property, and this property aids in preventing the various types of cancers as antioxidants provide protection from damage caused by oxidation. In addition to this, whole cereals also contain certain other bioactive components, which influence the levels of hormones and likely to reduce the danger of cancers caused by hormonal disorders [36]. The fermentation of various carbohydrate polymers in the colon leads to the formation of SCFA compounds, such as acetic acid, propionic acid, and butyric acid, which aid in promoting overall health. In the large bowel, the production of SCFA not only reduces the pH that might be helpful in decreasing the dangers of cancer of large intestine, but it also encourages the growth and differentiation of healthy cells of colon, also aids in fixing the damaged DNA of colon cells and persuades the death of nonrepairable aberrant cells, which grow into tumors, which slowly spread throughout the body of patients. Hence, consumption of dietary fiber is associated with the security against colorectal cancer [3].

14.11.9 GALL BLADDER DISEASE

Outcomes of few research works indicate that the ailment of gall bladder might be associated with the GI. Research showed that a group of males and females who regularly consume high GI foods considerably has higher risk to develop gallstones compared to those who ate low GI diet [39,40]. Nevertheless, further epidemiological studies are required to discover the relationship between occurrence of gall bladder disease and GI.

14.12 SUMMARY

Unhealthy lifestyle and diet are major culprits in elevating syndrome X, diabetes, heart diseases, and obesity among young population. Low glycemic diet plays a substantial role in preventing and treating such chronic diseases because some carbohydrates rich foods induce less postprandial glycaemia than others. The dietary management of various heath related issues can be rectified by the accurate usage of GI. It has been shown positive effect of consuming low GI diet in managing diabetes but the awareness relating to the impact of low GI foods on health among people is still needed to be created. Moreover, studies do not find any side effects of following the low GI diet, rather results showed the beneficial effects of low GI diet on human health.

KEYWORDS

- blood glucose
- cereals
- glycemic index
- human health
- wholegrains

REFERENCES

1. Andersson, A. A.; Andersson, R.; Piironen, V. Contents of Dietary Fiber Components and their Relation to Associated Bioactive Components in Wholegrain Wheat Samples from the Health Grain Diversity Screen. *Food Chem*istry, **2013**, *136 (3–4)*, 1243–1248.
2. Appel, L. J.; Moore, T J.; Obarzanek, E. Clinical Trial of the Effects of Dietary Patterns on Blood Pressure: DASH Collaborative Research Group. *New England Journal of Medicine,* **1997**, *336*, 1117–1124.
3. Aune, D.; Chan, D. S.; Lau, R.; Vieira, R. Carbohydrates, Glycemic Index, Glycemic Load, and Colorectal Cancer Risk: Systematic Review and Meta-Analysis of Cohort Studies. *Cancer Causes Control*, **2012**, *23* (4), 521–535.
4. Behall, K. M.; Scholfield, D. J.; Canary, J. Effect of Starch Structure on Glucose and Insulin Responses in Adults. *American Journal of Clinical Nutrition,* **1988**, *47*, 428–432.
5. Bhupathiraju, S. N.; Tobias, D. K. Glycemic Index, Glycemic Load, and Risk of Type 2 Diabetes: Results from 3 Large US Cohorts and an Updated Meta-Analysis. *American Journal of Clinical Nutrition*, **2014**, *100* (1), 218–232.
6. Brand, J. C.; Nicholas, P. L.; Thorburn, A. W.; Truswell, A. S. Food Processing and Glycemic Index. *American Journal of Clinical Nutrition*, **1985**, *42*, 1192–1196.
7. Brand-Miller, Petocz, P.; Colagiuri, S. Meta-Analysis of Low Glycemic Index Diets in the Management of Diabetes. *Diabetes Care*, **2003**, *26* (12), 3363–3364.
8. Brouns, F.; Bjorck, I.; Frayn, K. N.; Gibbs, A. L. Glycemic Index Methodology. *Nutrition Research Review*, **2005**, *18* (1), 145–171.
9. Brown, I. Complex Carbohydrates and Resistant Starch. *Nutrition Review*, **1996**, *54* (11), 115–119.
10. *Codex Alimentarius Commission. Guidelines on Nutrition Labelling. Standard CAC/GL 2–1985*; **2013**; http://www.codexalimentarius.org/standards/list-of-standards. Accessed on July 10, 2019.

11. Djousse, L.; Gaziano, J. M. Breakfast Cereals and Risk of Heart Failure in the Physicians' Health Study I. *Archives of Internal Medicine*, **2007**, *167* (19), 2080–2085.

12. Englyst, K. N.; Liu, S.; Englyst, H. N. Nutritional Characterization and Measurement of Dietary Carbohydrates. *European Journal of Clinical Nutrition*, **2007**, *61* (1), 19–39.

13. Fiona, S.; Atkinson, R. D.; Foster-Powell, K. International Tables of Glycemic Index and Glycemic Load Values: 2008. *Diabetes Care*, **2008**, *31* (12), 2281–2283.

14. Flegal, K. M.; Kit, B. K.; Orpana, H.; Graubard, B. I. Association of All-cause Mortality with Overweight and Obesity Using Standard Body Mass Index Categories: A Systematic Review and Meta-Analysis. *Journal of American Medical Association*, **2013**, *309* (1), 71–82.

15. Heidemann, C.; Schulze, M. B.; Franco, O. H. Dietary Patterns and Risk of Mortality from Cardiovascular Disease, Cancer, and All Causes in a Prospective Cohort of Women. *Circulation*, **2008**, *118* (3), 230–237.

16. Howlett, J. F.; Betteridge, V. A. The Definition of Dietary Fiber: Discussions at the Ninth Vahouny Fiber Symposium: Building Scientific Agreement. *Food and Nutrition Research*, **2010**, *54*, 5750–5752.

17. Hu, F.B. Globalization of Diabetes: Role of Diet, Lifestyle and Genes. *Diabetes Care*, **2011**, *34*, 1249–1257.

18. Hu, T.; Mills, K. T.; Yao, L.; Demanelis, K. Effects of Low-Carbohydrate Diets versus Low-Fat Diets on Metabolic Risk Factors: Meta-Analysis of Randomized Controlled Clinical Trials. *American Journal of Epidemiology*, **2012**, *176* (7), 44–54.

19. Jenkins, D. J.; Kendall, C. W. Effect of a Low-glycemic Index or a High-Cereal Fiber Diet on Type-2 Diabetes. *Journal of American Medical Association*, **2008**, *300* (23), 2742–2753.

20. Jensen, M. K.; Banerjee, P. K.; Hu, F. B.; Franz, M. Intake of Wholegrains, Bran and Germ and the Risk of Coronary Heart Disease in Men. *American Journal of Clinical Nutrition*, **2004**, *80*, 1492–1499.

21. Jeon, J. S.; Ryoo, N.; Hahn, T. R.; Walia, H.; Nakamura, Y. Starch Biosynthesis in Cereal Endosperm. *Plant Physiology and Biochemistry*, **2010**, *48*, 383–392.

22. Jones, J. M. Dietary Fiber Future Directions: Integrating New Definitions and Findings to Inform Nutrition Research and Communication. *Advances in Nutrition*. **2013**, *4*, 8–15.

23. Liepa, G. U.; Sengupta, A.; Karsies, D. Polycystic Ovary Syndrome (PCOS) and Other Androgen Excess-Related Conditions: Can Changes in Dietary Intake Make a Difference? *Nutrition in Clinical Practice*, **2008**, *23* (1), 63–67.

24. Liljeberg, E. H. Resistant Starch Content in a Selection of Starchy Foods on the Swedish Market. *European Journal of Clinical Nutrition*, **2002**, *56*, 500–505.

25. Liu, S.; Willett, W.; Stampfer, M.; Hu, F. Prospective Study of Dietary Glycemic Load, Carbohydrate Intake and Risk of Coronary Heart Disease in US Women. *American Journal of Clinical Nutrition*, **2000**, *71*, 1455–1461.

26. Maaran, S.; Hoover, R.; Donner, E.; Liu, Q. Composition, Structure, Morphology and Physicochemical Properties of Lablab Bean, Navy Bean, Rice Bean, Tepary Bean and Velvet Bean Starches. *Food Chemistry*, **2014**, *152*, 491–499.

27. Monro, J. A.; Shaw, M. Glycemic Impact, Glycemic Glucose Equivalents, Glycemic Index, and Glycemic Load: Definitions, Distinctions and Implications. *American Journal of Clinical Nutrition*, **2008**, *87* (1), 237–243.

28. Munro, J. Expressing the Glycemic Potency of Foods. *Proceedings of Nutrition Society*, **2005**, *64* (1), 115–122.

29. Murphy, M. M.; Douglass, J. S.; Birkett, A. Resistant Starch Intakes in the United States. *Journal of American Dietetic Association,* **2008**, *108,* 67–78.

30. Ostman, E. *Fermentation as a Means of Optimizing the Glycemic Index – Food Mechanisms and Metabolic Merits with Emphasis on Lactic Acid in Cereal Products.* Lund Institute of Technology, Lund University, Lund, France; **2003**; online; p. 66; https://portal.research.lu.se/portal/files/5676031/1370585.pdf; Accessed on December 31, 2019.

31. Pena, R. J.; Hans, J.; Braun, J. M. *The Wheat and Nutrition Series: Compilation of Studies on Wheat and Health.* CIMMYT, Mexico, D.F.; **2017**; p. 212.

32. Sacks, F. M.; Obarzanek, E.; Windhauser, M. M. Rationale and Design of the Dietary Approaches to Stop Hypertension trial (DASH): A Multicenter Controlled-Feeding Study of Dietary Patterns to Lower Blood Pressure. *Annals of Epidemiology,* **1995**, *5* (2), 108–118.

33. Salmeron, J.; Ascherio, A.; Rimm, E. B. Dietary Fiber, Glycemic Load and Risk of NIDDM in Men. *Diabetes Care,* **1997**, *20* (4), 545–550.

34. Salmeron, J.; Manson, J. E.; Stampfer, M. J. Dietary Fiber, Glycemic Load and Risk of Non-Insulin-Dependent Diabetes Mellitus in Women. *Journal of American Medical Association,* **1997**, *277*, 472–477.

35. Schulze, M.B.; Schulz, M.; Heidemann, C. Fiber and Magnesium Intake and Incidence of Type 2 Diabetes: A Prospective Study and Meta-analysis. *Archives of Internal Medicine,* **2007**, *167* (9), 956–965.

36. Slavin, J.; Jacobs, D.; Marquart, L. Wholegrain Consumption and Chronic Disease: Protective Mechanisms. *Nutrition and Cancer,* **1997**, *27*, 14–21.

37. Sunyer Pi, F. X. Glycemic Index and Disease. *American Journal of Clinical Nutrition,* **2002**, *76* (1), 290–298.

38. Topping, D. L.; Clifton, P. M. Short-Chain Fatty Acids and Human Colonic Function: Roles of Resistant Starch and Non-Starch Polysaccharides. *Physiological Reviews,* **2001**, *81* (3), 1031–1064.

39. Tsai, C. J.; Leitzmann, M. F.; Willett, W. C.; Giovannucci, E. L. Dietary Carbohydrates and Glycemic Load and the Incidence of Symptomatic Gall Stone Disease in Men. *Gut,* **2005**, *54* (6), 823–828.

40. Tsai, C. J.; Leitzmann, M. F.; Willett, W. C.; Giovannucci, E. L. Glycemic Load, Glycemic Index, and Carbohydrate Intake in Relation to Risk of Cholecystectomy in Women. *Gastroenterology,* **2005**, *129* (1), 105–112.

41. Wang, L.; Gaziano, J. M.; Liu, S. Whole- and Refined-Grain Intakes and the Risk of Hypertension in Women. *American Journal of Clinical Nutrition,* **2007**, *86* (2), 472–479.

42. Willett, W. C. Eat, Drink, and Be Healthy: the Harvard Medical School Guide to Healthy Eating. *American Journal of Epidemiology,* **2001**, *154* (12), 1160–1167; online; https://doi.org/10.1093/aje/154.12.1160-a;

43. Willett, W.; Manson, J.; Liu, S. Glycemic Index, Glycemic Load, and Risk of Type 2 Diabetes. *American Journal of Clinical Nutrition,* **2002**, *76* (1), 274–280.

44. Zivkovic, A. M.; German, J. B.; Sanyal, A. J. Comparative Review of Diets for the Metabolic Syndrome: Implications for Nonalcoholic Fatty Liver Disease. *American Journal of Clinical Nutrition,* **2007**, *86* (2), 285–300.

INDEX

Printed and bound by CPI Group (UK) Ltd, Croydon, CR0 4YY

23/10/2024

01777701-0006